Decision Making and Policy Analysis in Transportation Planning

Decision Making and Policy Analysis in Transportation Planning

Editors

Mahyar Amirgholy
Jidong J. Yang

MDPI

Basel • Beijing • Wuhan • Barcelona • Belgrade • Novi Sad • Cluj • Manchester

Editors

Mahyar Amirgholy
Civil and Environmental Engineering
Kennesaw State University
Marietta
USA

Jidong J. Yang
School of Environmental, Civil, Agricultural & Mechanical Engineering
University of Georgia
Athens
USA

Editorial Office
MDPI AG
Grosspeteranlage 5
4052 Basel, Switzerland

This is a reprint of articles from the Special Issue published online in the open access journal *Systems* (ISSN 2079-8954) (available at: https://www.mdpi.com/journal/systems/special_issues/NZ262HH717).

For citation purposes, cite each article independently as indicated on the article page online and as indicated below:

Lastname, A.A.; Lastname, B.B. Article Title. *Journal Name* **Year**, *Volume Number*, Page Range.

ISBN 978-3-7258-1980-5 (Hbk)
ISBN 978-3-7258-1979-9 (PDF)
doi.org/10.3390/books978-3-7258-1979-9

Contents

About the Editors . . . . **vii**

Preface . . . . **ix**

Zhichao Cao, Avishai (Avi) Ceder, Zihan Wang, Silin Zhang and Yaoyao Wang
Management Policy in Urban Rail Transit System: Trade-Off between Social Distancing and Service Efficiency Using Simulation in the Post-Epidemic Era
Reprinted from: *Systems* **2024**, *12*, 151, doi:10.3390/systems12050151 . . . . **1**

Yi Zhu, Xiaofei Ye, Xingchen Yan, Tao Wang, Jun Chen and Pengjun Zheng
Exploring the Impact of Charging Behavior on Transportation System in the Era of SAEVs: Balancing Current Request with Charging Station Availability
Reprinted from: *Systems* **2024**, *12*, 61, doi:10.3390/systems12020061 . . . . **26**

Yunxiang Yang and Jidong J. Yang
Strategic Sensor Placement in Expansive Highway Networks: A Novel Framework for Maximizing Information Gain
Reprinted from: *Systems* **2023**, *11*, 577, doi:10.3390/systems11120577 . . . . **46**

Chun Yang, Chao Gu and Wei Wei
Does Robotaxi Offer a Positive Travel Experience? A Study of the Key Factors That Influence Consumers' Use of the Robotaxi
Reprinted from: *Systems* **2023**, *11*, 559, doi:10.3390/systems11120559 . . . . **59**

Mohammad Maleki and Janille Smith-Colin
Estimating Benefits of Microtransit for Social Determinants of Health: A Social Return on Investment System Dynamics Model
Reprinted from: *Systems* **2023**, *11*, 538, doi:10.3390/systems11110538 . . . . **72**

Florian Noto and Hamid Mostofi
Acceptance Analysis of Electric Heavy Trucks and Battery Swapping Stations in the German Market
Reprinted from: *Systems* **2023**, *11*, 441, doi:10.3390/systems11090441 . . . . **101**

Debao Dai, Yu Fang, Shihao Wang and Min Zhao
Prediction of China Automobile Market Evolution Based on Univariate and Multivariate Perspectives
Reprinted from: *Systems* **2023**, *11*, 431, doi:10.3390/systems11080431 . . . . **116**

Hung-Chia Yang, Ling Jin, Alina Lazar, Annika Todd-Blick, Alex Sim, Kesheng Wu, et al.
Gender Gaps in Mode Usage, Vehicle Ownership, and Spatial Mobility When Entering Parenthood: A Life Course Perspective
Reprinted from: *Systems* **2023**, *11*, 314, doi:10.3390/systems11060314 . . . . **141**

Jie Ma, Yilei Zeng and Dawei Chen
Ramp Spacing Evaluation of Expressway Based on Entropy-Weighted TOPSIS Estimation Method
Reprinted from: *Systems* **2023**, *11*, 139, doi:10.3390/systems11030139 . . . . **160**

Jiafeng Gu
High-Speed Rails and City Innovation System: Empirical Evidence from China
Reprinted from: *Systems* **2023**, *11*, 24, doi:10.3390/systems11010024 . . . . **176**

About the Editors

Mahyar Amirgholy

Dr. Mahyar Amirgholy is an Assistant Professor at the Southern Polytechnic College of Engineering and Engineering Technology, Kennesaw State University. Before joining KSU, Mahyar was a Postdoctoral Scholar in the Civil and Environmental Engineering Department at Cornell University. He received his Ph.D. in Civil and Environmental Engineering with a concentration in 'Transportation Systems' from the University of Massachusetts Amherst. Mahyar is a recipient of the Council of University Transportation Centers (CUTC) Milton Pikarsky Award for Outstanding Doctoral Dissertation in Science and Technology. Mahyar's research is primarily focused on the modeling and optimization of large-scale transportation systems.

Jidong J. Yang

Dr. Jidong J. Yang is an Associate Professor at the College of Engineering, University of Georgia. His multidisciplinary research integrates civil and environmental engineering, data science, computer vision, machine learning, and system optimization. Dr. Yang focuses on advancing smart mobility systems, sustainable infrastructure, and the application of artificial intelligence in engineering. His work is particularly known for leveraging big data and developing novel machine learning algorithms for various engineering applications, including traffic safety, autonomous vehicle systems, and infrastructure resilience.

Preface

The ongoing evolution in transportation planning necessitates a multifaceted approach to policy design and decision-making, integrating technological advancements, and adapting to shifting societal needs. This Special Issue of *Systems* on "Decision Making and Policy Analysis in Transportation Planning" features groundbreaking research that explores key elements of transportation planning, policy analysis, and system management. Each paper contributes new insights and solutions to contemporary challenges in the transportation sector.

"Management Policy in Urban Rail Transit System: Trade-Off between Social Distancing and Service Efficiency Using Simulation in the Post-Epidemic Era" by Cao et al. examines the balance between maintaining social distancing and ensuring service efficiency in urban rail systems. In "Prediction of China Automobile Market Evolution Based on Univariate and Multivariate Perspectives", Dai et al. analyze the evolving dynamics of China's automobile market. Gu's study, "High-Speed Rails and City Innovation System: Empirical Evidence from China", investigates the impact of high-speed rail non-urban innovation systems in China. Noto and Mostofi, in their paper "Acceptance Analysis of Electric Heavy Trucks and Battery Swapping Stations in the German Market", explore the economic and environmental factors influencing the adoption of electric heavy trucks and the associated battery swapping infrastructure in Germany. "Ramp Spacing Evaluation of Expressway Based on Entropy-Weighted TOPSIS Estimation Method" by Ma et al. develops a methodology for evaluating expressway ramps. In "Estimating Benefits of Microtransit for Social Determinants of Health: A Social Return on Investment System Dynamics Model" Maleki and Smith–Colin introduce a system dynamics model to assess the social return on investment of microtransit systems. Yunxiang Yang and Jidong J. Yang, in "Strategic Sensor Placement in Expansive Highway Networks: A Novel Framework for Maximizing Information Gain", present a novel approach to the Traffic Sensor Location Problem (TSLP). Yang et al. delve into gender differences in transportation choices in their paper, "Gender Gaps in Mode Usage, Vehicle Ownership, and Spatial Mobility When Entering Parenthood: A Life Course Perspective". Zhu et al. tackle the challenges of integrating autonomous electric vehicles into urban transportation systems in their paper, "Exploring the Impact of Charging Behavior on Transportation Systems in the Era of SAEVs".

This Special Issue offers a comprehensive overview of the latest research in transportation planning and policy analysis, presenting innovative solutions and frameworks that are essential for addressing contemporary challenges in the transportation sector. We hope that the insights presented in these papers will inspire further research and inform policy development, ultimately contributing to the creation of more sustainable, efficient, and equitable transportation systems.

Mahyar Amirgholy and Jidong J. Yang
Editors

systems

MDPI

Article

Management Policy in Urban Rail Transit System: Trade-Off between Social Distancing and Service Efficiency Using Simulation in the Post-Epidemic Era

Zhichao Cao [1,2,3,4], Avishai (Avi) Ceder [5], Zihan Wang [1], Silin Zhang [1,*] and Yaoyao Wang [1]

1 School of Transportation and Civil Engineering, Nantong University, Nantong 226019, China; caozhichao@bjtu.edu.cn (Z.C.); wangzihan199611@163.com (Z.W.); 1826031018@stmail.ntu.edu.cn (Y.W.)
2 Guangxi Key Laboratory of International Join for China-ASEAN Comprehensive Transportation, Nanning University, Nanning 530004, China
3 Jiaxing Key Laboratory of Smart Transportations, Jiaxing 314001, China
4 Technology and Equipment of Rail Transit Operation and Maintenance Key Laboratory of Sichuan Province, Chengdu 610041, China
5 Transportation Research Institute, Technion–Israel Institute of Technology, Haifa 32000, Israel; ceder@technion.ac.il
* Correspondence: 12114213@bjtu.edu.cn; Tel.: +86-188-10870559

Abstract: The past COVID-19 pandemic introduced the world to the necessity of dealing with the trade-off between minimizing probability of contagion, and providing people with services they need. This trade-off stipulates that a large person-to-person distance will reduce contagion probability, but will render service inefficient, and vice versa. This work focuses on the urban rail transit (URT) hub, as an example of a busy passenger area, from which we can derive an optimal preparedness policy to use during the pandemic time of any coronaviruses. We use simulation methodology, based on the classical social force model, to represent behaviors and characteristics of pedestrians. Passenger flow movement process is a mechanism we explore to figure out how the epidemic management policy and pedestrian psychological-related behaviors interact with the URT system. The systems' complexity regarding contagion-prevention distances are tested over a few scenarios: before/after the outbreak, and for different person-to-person distances demonstrating different crowd levels. A case study of Xinjiekou Station, Nanjing URT, China, enables assessment of passenger management policy with person-to-person distances of 0.5 m, 1.0 m and 2.0 m. Multi-scenario performance illustrates the trade-off in dynamic between the efficiency of pedestrians' walking behaviors and the distancing needs for preventing coronaviruses transmission. The results show that queuing length with social distancing of 1.0 m and 2.0 m is increased by 4.17% and 21.22%. The average delays in boarding are 14.1 s and 22.5 s for 1.0 m and 2.0 m, which leads to 15.29% and 22.39% increases, respectively, in comparison with ordinary social distancing of about 0.5 m.

Keywords: management policy; pedestrian simulation; contagion-prevention distancing; service efficiency

Citation: Cao, Z.; Ceder, A.; Wang, Z.; Zhang, S.; Wang, Y. Management Policy in Urban Rail Transit System: Trade-Off between Social Distancing and Service Efficiency Using Simulation in the Post-Epidemic Era. *Systems* **2024**, *12*, 151. https://doi.org/10.3390/systems12050151

Academic Editors: Mahyar Amirgholy and Jidong J. Yang

Received: 22 March 2024
Revised: 21 April 2024
Accepted: 24 April 2024
Published: 27 April 2024

1. Introduction

A finding published in the Lancet notes that "the risk for infection is highly dependent on distance to the individual infected and the type of face mask and eye protection worn" [1]. In many instances, either congestion or large passenger flow occurs at urban rail transit (URT) station platforms. However, there is a research gap regarding pedestrian social distancing necessary to prevent person-to-person transmission of COVID-19 or, in general, any coronavirus. Based on 504 questionnaires surveyed during the COVID-19 time, we identified multiple, observational simulation performances to investigate optimum distance for avoiding person-to-person virus transmission.

With the gradual resumption of the economy and transportation in China, passengers have been seeking efficiency-oriented travel, in addition to contagion prevention. A reasonable management policy at URT, to a certain extent, not only facilitates passengers' travel efficiency, but also reduces the risks of COVID-19 transmission. Note that URT was suspended in many cities in China during the most threatening period of COVID-19, while some URT has now returned to normalcy. However, some related management policies are hysteretic (and questionable), and need verification. The study of determining reasonable social distancing is then developed as our objective, where necessary, to balance both efficiency and safety to produce a tradeoff effect.

Employing some necessary contagion prevention management measures/policies, some service areas have resumed the previous normalization in a sustainable manner. One challenge the transport management encounters is to observe contagion prevention social distancing without losing the dominance of travel efficiency. However, seeking the tradeoff between COVID-19 pandemic prevention and improving orderliness for boarding/alighting/walking is multifaceted. Pedestrian simulation has been proposed in the literature, but few studies have investigated the tradeoff mechanism. A date-based simulation experiment is a useful approach, from the operator's perspective, for determining contagion prevention social distancing as requested in public, and enables efficient travel under the management policy.

This study focuses on pedestrian distancing by using investigation-based simulation that explores and investigates contagion prevention social distancing among individual passengers, and specifically examines their effectiveness. In view of COVID-19 transmission characteristics, one limitation of existing studies—and policies derived from them—is that they only recommend subjective measures, such as wearing masks, reducing the number of passengers serviced and continuously sterilizing the air. Indeed, there is little research on quantitative guidance for passenger flow control in URT that is framed against the backdrop of the COVID-19 pandemic. Therefore, a multicriteria decision-making framework that can take various system performance measures and users' path choices into account is warranted.

Taking the complexity of station layout and the unpredictability of pedestrian routes into consideration, a large-scale, city-wide and online questionnaire survey was conducted over three months. It was based on 504 samples considered on the WeChat electronic platform. The modeling parameters set and simulation layout are built explicitly from real data as the input of the experiment.

In this study, the pedestrian-restricted movement issue (density/time) is regarded as the key criterion of the management policy. The psychological, contagion prevention, social distancing between people in THE pedestrian flow is tested in the simulation. The conventional habits of distance at URT, prior to 2019, is obviously unable to satisfy the requirements of epidemic prevention. For instance, large passenger flow disposal and commuter passenger flow guidance are not yet able to prevent the spread of COVID-19. Based on the classical social force model (SFM) of pedestrian psychology [2], we use AnyLogic software (version 8.6, The AnyLogic Company formerly XJ Technologies, Lisboa, Portugal) to build simulation scenarios. Taking Xinjiekou Station, Nanjing URT, China as a case for empirical study and a statistical base, we illustrate the movement paths based on SFM by observing pedestrian distancing of 0.5 m, 1.0 m and 2.0 m, respectively. These experimental activities are based on SFM-oriented simulation frameworks.

1.1. Literature Review

With respect to passenger flow control management, the reasonable policy that enables reductions in common and excessive congestion is usually performed/interpreted by a passenger demand assignment problem. To address this problem, Yuan et al. [3] established a mixed integer linear programming model to optimize the solution with the research target of passenger waiting time. Jiang et al. [4] developed a new method of enhanced learning to optimize passenger flow at stations over a given period of time, and reduce passenger detention rate, thereby ensuring passenger safety. Li et al. [5] used the model prediction control valve method, combined with URT line and passenger flow control, for their study. This allowed for analyzing the optimal control strategy of train regulation and passenger flow control. Based on pedestrian simulation, an evaluation grid was constructed to control the intensity of passenger flow [6]. Burdzik et al. [7] focused on the passenger flow interactions with a minimum spread risk of COVID-19 and other infectious diseases. Crowding satisfaction and delays are modeled and estimated by valid simulation.

Passenger dynamic observations at URT can be modelled on a simulation approach to uncover the character of the unpredictable mobility. For further insight into papers using this approach, the reader may refer to Yao et al. [8], Zhao et al. [9], Jin et al. [10], and Shang et al. [11]. Yao et al. [8] propose a dynamic passenger allocation model based on the route selection. The results facilitate the dynamic supply and demand relationship in the public transport network. Zhao et al. [9] examine the movement characteristics of pedestrians at stations where the URT is connected to a shopping mall. Jin et al. [10] study the dynamic changes in one-way flow and bi-directional flow rates in high-density situations. In considering the discrete passenger flow states, Shang et al. [11] execute a space–time hyper network flow distribution, analyze pedestrian path selection behaviors and improve passenger flow organization.

We consider the operational URT efficiency aspect using train scheduling optimization problem. Li et al. [12] propose a service frequency optimization tactic to encourage some passengers to move their journeys from peak to off-peak hours and ease congestion. Wang et al. [13] use sequential quadratic programming and genetic algorithms to solve train scheduling problems in terms of departure time, arrival time, and passenger arrival rate. Combined with factors from Beijing URT operation, Wang et al. [14] propose passenger demand control methods by solving the mixed integer linear programming problem. For the purpose of reducing the operation cost and waiting time of passengers, Yin et al. [15] propose the train scheduling problem on the bi-directional URT, and validate the Beijing subway, based on the heuristic algorithm of Lagrange relaxation combined with the programming model.

This study develops the simulation-based approach for public management policy in the case of specific URT stations. The aim of the present work is grounded in the variability of contagion prevention social distancing in walking patterns subject to COVID-19 pandemic circumstances. Hoogendoorn and Bovy [16] set up a pedestrian utility function to simulate the different walking conditions and uncertainty levels derived from route/area selections. Tong and Cheng [17] propose a pedestrian flow simulation model based on multi-agent technology, which takes the factors of distance and pedestrian psychology into account. Weng et al. [18] develop a model based on the maximum likelihood regression tree from a pedestrian perspective to predict delayed responses to URT accidents. Xiao et al. [19] propose a heuristic-based pedestrian flow simulation performance that is compatible with the normal movement of passengers.

The majority of pedestrian simulation models are derived from the classical social force model (SFM) [20]. In focusing on export selection performances, Zheng et al. [21] simulate pedestrian micro behaviors and analyze the difference between actual data and experimental data. Li et al. [22] study the impact of passenger obstacles on walking efficiency. According to the platform queue, the subway station capacity was calculated by means of the Markov chain [23]. Jiang et al. [24] describe the typical phases of dynamic pedestrian evacuation. In considering station service capacity, Xu et al. [25] set different demand scenarios and combine data closure analysis to assess station service capacity. Using data obtained from the Vienna URT station data, pedestrian paths were simulated based on continuous observation of perceived times [26]. Chen et al. [27] develop a multi-agent, learning-based pedestrian simulation approach, whose adaptation was proved for pedestrian scenes. Zhang et al. [28] establish a micro-simulation model of off-board behavior based on cellular automata, which evaluates the facilities and passenger organization managements at URT stations. Feng et al. [29] use the p-space method to describe the basic topological space of the URT network. Zheng et al. [30] simulate the progress of pedestrian evacuation in an emergency, by testing the emergency diffusion fields according to the panic indexes.

The known commercial software AnyLogic was selected for our pedestrian simulation. AnyLogic [31] is a valid tool for analyzing pedestrian movements. The use of this common tool allows other scholars/readers to repeat the experimental results. From the literature we learn that Zuo et al. [32] used AnyLogic to simulate emergency evacuation performance, subject to the restricted usable spaces in the high-density pedestrian evacuating process. Liu and Hong [33] employ AnyLogic to track pedestrians' distribution and actions of entering, boarding and alighting at a URT station. Zhang et al. [34] adapt AnyLogic to perform transfer of passengers at a Wuhan URT station. Overall, we decided to use this tool because it takes regional density, average passenger flow speed and the time distribution of entrances and exits as the criteria, where the feasible passenger flow guidance is verified through simulation experiments.

1.2. Motivation and Contribution

1.2.1. Background, Motivation and Framework

At the global spread era of COVID-19, and potential spread of other coronaviruses in the post-epidemic era, URT operation is affected and regulated by the specific passenger management policy. Passengers are advised to maintain a certain social distance to prevent contagions. In order to reflect epidemic prevention requirements, the pedestrian flow paths at the URT stations are simulated. Hereby pedestrian distancing is set at 0.5 m, 1.0 m and 2.0 m, respectively. A comparative analysis of pedestrians' boarding, alighting and exiting behaviors was then simulated, testing different scales of distancing.

In the field of public health security, the trade-off between congestion and operational efficiency is a key issue for controlling coronaviruses. This distancing management is meaningful for devising the optimal passenger flow control policy. The trauma from what happened after the arrival of the COVID-19 pandemic is still scorched in people's minds. Thus, our motivation is to provide guidance for preparedness policy that can be used during the pandemic time of any coronaviruses.

Unlike the existing literature, which focuses on the effectiveness of pedestrian evacuation, and only takes bottleneck congestion into account, our study seeks to validate a reasonable passenger flow control regulation and contagion prevention social distancing observation. We seek to analyze pedestrians' restricted movements at a URT station platform by exploring different social distances through simulation experiments. In order to determine optimal pedestrian distancing and ensure both efficiency and contagion prevention, we develop a simulation-based methodology, to verify and calibrate the model parameters precisely. The purpose is to help clarify elements of the passenger management policy for the COVID-19 pandemic.

Figure 1 illustrates the simulation-based methodology framework. The whole process is divided into four steps, as shown in Figure 1, namely travel survey, facilities' assignment, simulation preparation, and a scenario-based experiment. Details of the four steps appear in Sections 3–5.

1.2.2. Contribution

The first contribution of our study, in terms of COVID-19 transmission prevention on URT, concentrates on the tradeoff between service-based efficiency and contagion-prevention distancing control. Our study utilizes a simulation-based approach. A case verification is achieved based on the SFM by an empirical survey of parameter calibration. Thus, our study, under COVID-19 pandemic circumstances, which is explicitly concerned with better reproduction of passenger behaviors, allows for congestion in a URT.

Improved accuracy of simulation is our second contribution. More precise parameter calibration is fulfilled in the AnyLogic context. These parameters include user heterogeneity, age, gender, purpose, path choice and familiarity degree. To obtain this information, a questionnaire survey was conducted for pedestrian flow under the epidemic control policy. The survey helps in studying and comprehending pedestrians' exiting choices, queue lengths, queuing behavior, and total alighting times.

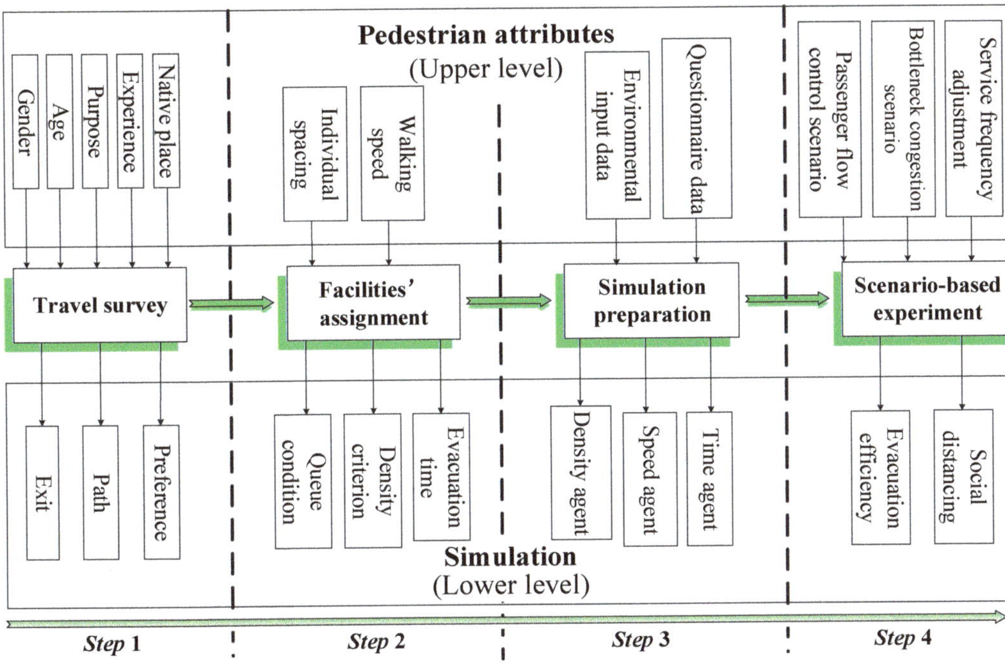

Figure 1. Framework of the proposed simulation-based methodology.

1.3. Outline

The paper structure is as follows: Section 2 presents the social force model, to calibrate the pedestrians' attributes. Section 3 introduces the simulation process under public management. Section 4 analyzes the simulation results by testing social distances in walking dynamics. Finally, a case study of Nanjing Xinjiekou URT station is validated in Section 5. Conclusions and future research prospects are presented in Section 6.

2. Methodology

In this section, the simulation model is described. The simulation context and SFM-based parameter calibration are presented.

2.1. Pedestrian Flow Model

Social force refers to the psychological mechanism of human response to stimulation, according to the psychological level. Pedestrians make the benefit-maximizing choice after a general assessment of different courses of action, so as to mentally produce the driving force of action and take practical actions. Pedestrians are familiar with daily walking, so the resulting reactions are unconscious but predictable. Helbing et al. [2] introduce the SFM. It states the microscopic forces between pedestrians and pedestrians, pedestrians and targets, pedestrians and obstacles, and is intended to make pedestrians move according to certain routes.

Derived from Newton's law and vector mechanics, *Pedestrian driving force* refers to pedestrians at a certain speed along a certain route to the destination. Friction exists in the process, due to physical contact with the high density of pedestrians or obstacles, which is the repulsive force from other pedestrians. According to Helbing et al. [2,35], the equation of motion of SFM can be described by

$$f_a(t) = m_a \frac{dv_a(t)}{dt} = f_a^0(t) + \sum f_{ai}(t) + \sum f_{aw}(t) + \sum f_{ar}(t) + \xi_n(t) \tag{1}$$

where $f_a(t)$ denotes the resultant force acting on pedestrian a. $f_a^0(t)$ is the pedestrian self-driving force (the direction of the force is consistent with the direction of the target). $f_{ai}(t)$ is pedestrian a subject to repulsion from pedestrian i. $f_{aw}(t)$ is set to pedestrian a subject to repulsion from obstacle w. $f_{ar}(t)$ implies pedestrian a attracted to the attraction element. $\xi_n(t)$ denotes a disturbance term.

Assume pedestrian mass in calculations $m_a = m = 1$, where all forces are equal to the acceleration at this time. In the case of SFM, the role of attracting elements or pedestrian groups is often ignored, and $f_{ar}(t)$ is omitted to make Equation (2):

$$f_a(t) = f_a^0(t) + \sum f_{ai}(t) + \sum f_{aw}(t) + \xi_n(t) \tag{2}$$

Define the path used by pedestrian a reaching his/her destination as a collection of $\left\{S_a^1, \ldots, S_a^k, \ldots, S_a^K\right\}$, where K represents the total number of pedestrians' paths. The components of the resultant force are analyzed separately, using Equations (1) and (2).

2.2. Interaction Effect

(1) Following Helbing and Molnar [20], self-driving force is expressed by:

$$f_a^0(t) = m_a \frac{v_a^0 e_a(t) - V_a(t)}{t_a} \tag{3}$$

where v_a^0 is the desired speed of pedestrian a. $e_a(t)$ denotes the unit vector of time t to target $S_a^K (e_a(t) = \frac{[S_a^K - S_a(t)]}{||S_a^K - S_a(t)||})$. $V_a(t)$ denotes the actual speed of pedestrian a at time interval t. t_a is time constant.

(2) The interaction force between pedestrians, described in Helbing and Molnar [20], is

$$f_{ai}(t) = m_a \frac{A_a}{B_a} \exp\left(-\frac{||S_{ai}(t)||}{B_a}\right) \frac{S_{ai}(t)}{||S_{ai}(t)||} \tag{4}$$

where $S_{ai}(t)$ is set to the vector difference between the pedestrian position at time interval t and i position ($S_{ai}(t) = S_a(t) - S_i(t)$). A_a is the interaction force's strength between pedestrians. B_a is the action distancing of the interaction force between pedestrians.

(3) The repulsion of pedestrians and obstacles, based on Helbing and Molnar [20], is

$$f_{aw}(t) = m_a \frac{A_w}{B_w} \exp\left(-\frac{||S_{aw}(t)||}{B_w}\right) \frac{S_{aw}(t)}{||S_{aw}(t)||} \tag{5}$$

where $S_{aw}(t)$ implies the vector difference between pedestrian a and obstacle w at the closest point to the pedestrian at time interval t ($S_{aw}(t) = S_a(t) - S_w^a(t)$). A_w is the force's strength between pedestrians and obstacles. B_w is the action distance between pedestrians and obstacles.

(4) Disturbance term

Disturbance term refers to the sum of all random fluctuations of pedestrians in the walking surroundings. Moreover, in the calculation formulas in this paper, it applies to pedestrians' reduced acceleration or deceleration due to random disturbance.

3. Simulation

3.1. Simulation Preparation

The AnyLogic simulation software is a tool that combines discrete system dynamics and multi-agents. More specifically, it has the advantage of a possible setting of simulation rules related to pedestrian distancing. It is widely used in supply chain, pedestrian traffic simulation, urban development, disease diffusion, and other fields. It adopts the latest design methodology of complex systems, introduces unified modeling language into the modeling simulation context, and can describe discrete and continuous behavior explicitly.

The pedestrian library module enables us to describe pedestrian flow, and bottleneck points in the restricted movement of passengers, and it can be found through the simulation for evaluating the station's traffic capacity. Regarding characteristics of the SFM, we put the corresponding modules from the pedestrian library into the model. Thereby, we depict passengers' restricted movements diagram by the walking paths. It should be noted that all passengers exiting the URT station obey the First-In-First-Out principle.

- Survey 1: Questionnaire survey, in which 504 samples of the travel-demand questionnaire were distributed randomly online for people who have lived in Nanjing for more than a few years.
- Survey 2: Field survey, from which the station infrastructure data measurement is collected.

3.2. Pedestrian Library Module

In the Pedestrian Library module, simulated station surroundings are built as a background diagram by using the survey data. Furthermore, we then create a simulation surrounding module by using spatial markers and functions to complete scene construction and the modeling of pedestrians' activities. The simulation modeling process of the Pedestrian Library is presented in detail in Figure 2.

We use the AnyLogic software's Pedestrian Library module to describe the pedestrian agents in the simulation. In addition, control functions in the Pedestrian Library, on the basis of *PedSource*, *PedGoTo*, *PedSink*, *PedService*, *PedWait*, are carried out to figure out the pedestrian decision-making process.

3.3. Restricted Movement

3.3.1. Pedestrian Agent

For the completion of scene construction and pedestrian agent modeling, it is necessary to comprehensively consider various factors for simulating restricted movements at the URT station. In other words, time, speed and density agents are observed so as to evaluate the restricted movement process throughout its experimental form.

(i) Time agent: Evaluate if there is any delay in the URT station exiting process. As a primary evaluation index, the time agent enables assessment of the speed of exiting flow, which is the intuitive indicator for passengers.
(ii) Speed agent: The walking speed of a pedestrian, in comparison with the agent's average speed.
(iii) Density agent: The density of the pedestrian flow is introduced to analyze the status of traffic. Certainly, the density at different facilities is variant, as an important indicator for analyzing congestion of the corridor or lack thereof.

Generally speaking, passengers at stations are dispersing spontaneously. This involves a natural social distancing because, by their very nature, people do not like to be too close to each other. We used the simulation parameter *density agent* to control the number of passengers allowed to enter the station and, as a result, to derive the space between people. Hence, if the space is quantitatively controlled by the simulation, the averagely restricted social distances can be satisfied.

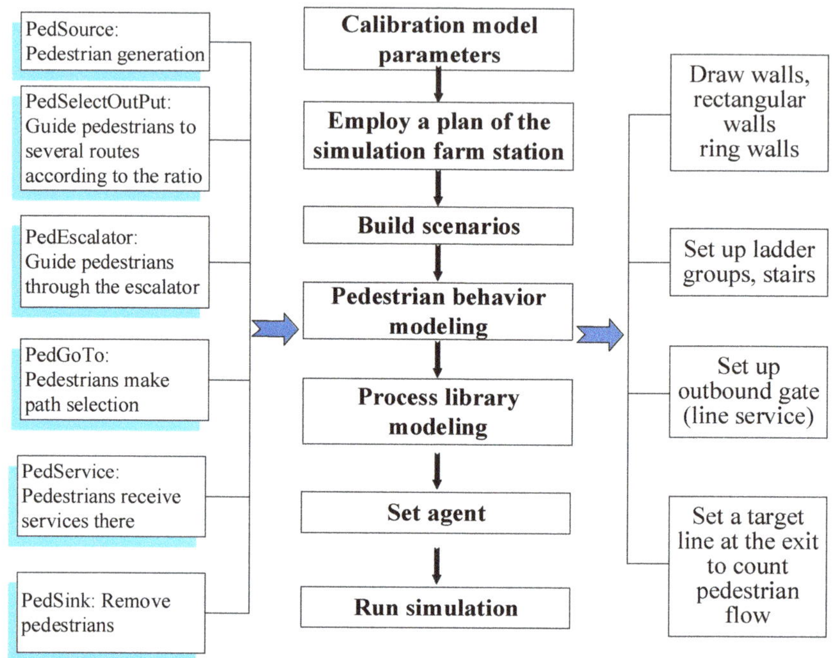

Figure 2. Pedestrian Library simulation modeling description.

3.3.2. Parameter Setting

A noticeable social distance, e.g., 2 m, is hard to impose. In our study we are referring to it as a maximal controlled average distance. A long distance such as this will only allow a certain number of passengers to enter the station. Thus, this will create longer passenger queues at the entrance.

(1) Individual space

According to the SFM nested into AnyLogic software, a hierarchical architecture defining the individual pedestrian space is needed. Particularly, repulsion is caused by the existence of individual spatial needs. In the process of individual space setting, three situations are set based on Zhang et al. [28], specifically, allowing for a pedestrian oval cellular space with a long axis of, 57.9 cm, and a short axis of 33 cm [36]. This corresponds with the elliptical area formula, such that an adult man needs an area of about 0.146 m^2, rather than the distensible rectangular area that is about 0.191 m^2. In the *PedSource* set, the default pedestrian diameter distribution is uniform, with a determination of 0.4 m or 0.5 m. The calculated area is 0.126 m^2 to 0.196 m^2, so the diameter's parameter is 0.4 m to 0.5 m.

Explicitly, the pedestrian distancing agent is set as triple-tuple. We set variable *sd* as social distance, and mark the case considered as '*k*', making it sd_k. In the case $k = 1, 2, 3$ for each group: (i) define $k = 1$ under normal circumstances, the default value of the diameter parameters is 0.4 m to 0.5 m; (ii) $k = 2$, in the case of an epidemic, the diameter of pedestrians that meets contagion-prevention social distancing is 0.9 m to 1.0 m. Furthermore, in (iii), we set $k = 3$ as, under the heavy influence of the pandemic, some pedestrians demonstrate higher demand for individual space than general pedestrian requirements. Taking the real needs and requests of pedestrians and contagion-prevention social distancing into consideration, the diameter for pedestrians is set at 1.5 m to 2.0 m. This setting is based on our assumption of uniform personal space or body diameters for simulating a mass number of moving pedestrians. In other words, we do not simulate individual user's behavior or characteristics.

(2) Walking speed

Pedestrian walking speed varies with factors such as passenger flow density, gender, and age. Referring to Chandra and Bhari [37], the surveyed average male pedestrian speed is 1.31 m/s, and the surveyed average female pedestrian is 1.16 m/s. According to the properties of the *Pedsource* module, we write the walking speed according to the investigation data.

4. Case Study

4.1. Input

4.1.1. Data

We use the Xinjiekou Station case to validate the effectiveness of the simulation. Xinjiekou Station is the transfer station of Nanjing URT between Line 1 and Line 2. By virtue of the station's location in the center of Nanjing, it leverages the advantages of geographical location to raise the possibility of large passenger flow. Note that Xinjiekou Station is the only subway station in Asia with 24 exits. The station platform is all island style, where Line 1 is located on the third floor, and Line 2 is located on the second floor. It should be noted that our study only considers exiting passenger paths. Line 1 is north–south, 311.872 m long and 24.2 m wide. On the other side, Line 2 is east–west, 440.3 m long and 21.6 m wide. By measuring the dimensions of the local elevator and stairs in the subway station, it is found that the elevator width is precisely between 1.30 m and 1.45 m. In addition, the stair width is between 2.34 m and 5 m. The dimensions of the elevator and stairs are based on a scale of one meter, which is approximately equal to 10 pixels. So far, the observation information stipulates the scale of the simulation environment.

In order to acquire the data of passengers in URT, we conducted an online questionnaire for a total of 504 questionnaires collected and validated. Allowing for unused/closed 17th, 18th, 19th and 24th exits of the subway station, they are excluded from the content of the questionnaire.

4.1.2. Scenario

According to the real structural diagram/data of Xinjiekou Station, the basic scene surroundings are constructed. Within the AnyLogic context, we use the *wall*, *rectangular wall* and *ring wall* tools in the Pedestrian Library to draw the peripheral wall of the URT station, and limit the pedestrian walking range. Then, we activate the *escalator group* and *ramp* to guide existing pedestrians to change floors. Let the *target line* label for generation and arrival of pedestrian flow, and *line service*, represent the number, location, and turnstile service. Given that the station is a three-layer platform, we precisely set up the layer layout and assign various service facilities, including security inspection equipment, ticket vending machine, and automatic fare gate, for simulation. Explicitly, the first and second layers are platform layers and the third layer is the URT hall. The pedestrian distribution ground layer is depicted in Figures 3 and 4.

4.1.3. Paths

In the real world, once passengers alight from trains to exit the station, they are supposed to make rational decisions based on expected destinations. Similarly, the walking process can be separated into several individual segments, such as path selection, stair elevator service, turnstile service and final route selection. Corresponding to the above segments, the pedestrian agent module is carried out by using *PedSource* and *PedGoTo*, *PedService*, respectively.

Based on the pedestrians' actual restricted movement path, we set the probability for selecting different ladder groups. We assigned pedestrians to the required exit from the corresponding turnstiles, according to the actual questionnaire survey results. The decision-making control modules are connected to show a complete flowchart of the passenger's exit path, as shown in detail in Figure 5.

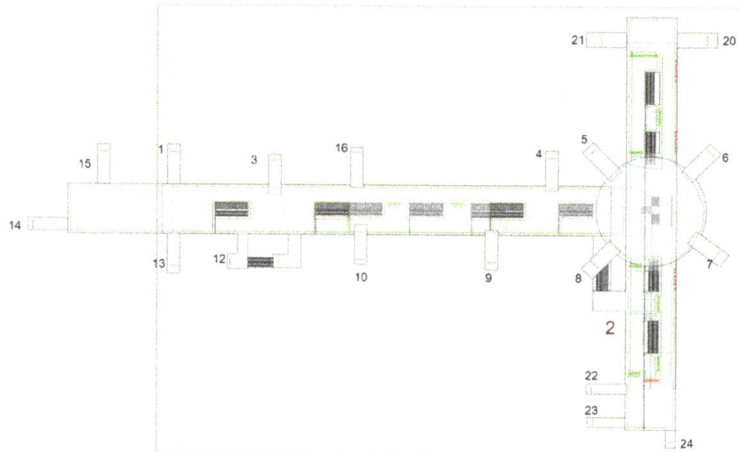

Figure 3. Xinjiekou Station simulation structure.

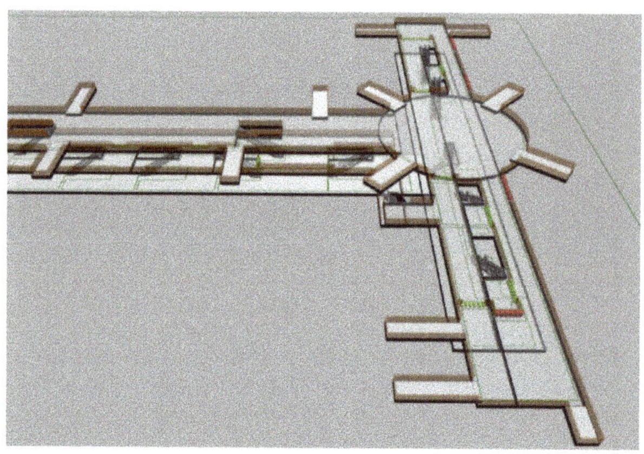

Figure 4. 3D view of modelling Xinjiekou Station.

4.1.4. Parameterization

(1) *Pedestrian parameters setting*

We used the survey data to calibrate the pedestrian agent. Table 1 illustrates the gender attributes of the passengers surveyed.

Table 1. Gender ratio of passengers at the Nanjing Xinjiekou Station.

Options	Subtotal	Proportion
Male	265	52.58%
Female	239	47.42%
Number of persons valid for filling in this question	504	100%

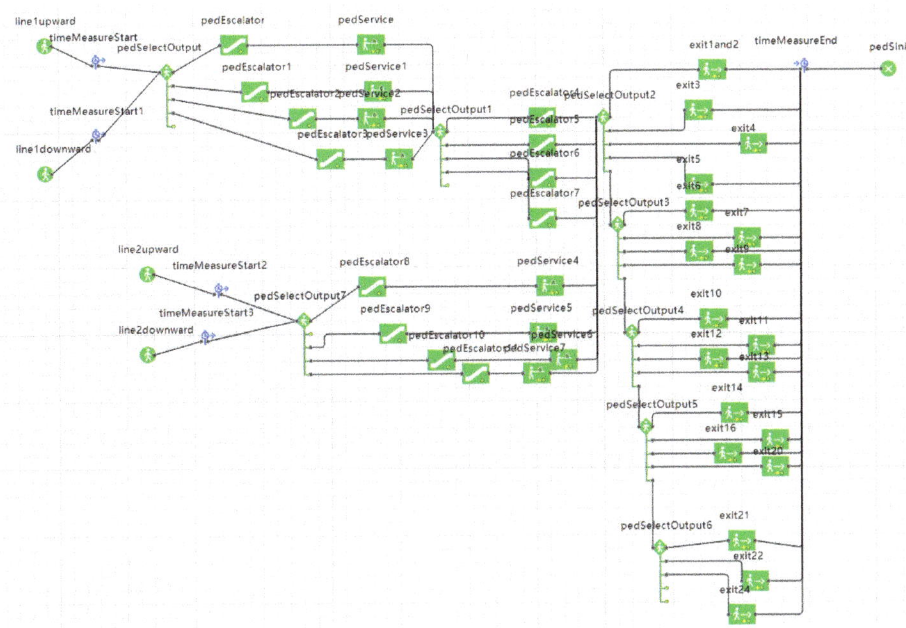

Figure 5. Flowchart of passengers' exiting process.

Passengers' choices for different exits are derived from the questionnaire statistics. Corresponding data are shown in Table 2. The 17th, 18th, 19th, and 24th exits are not taken into consideration.

Table 2. Passenger exit selection at Nanjing Xinjiekou Station.

Options	Subtotal	Proportion
Exit 1	23	4.56%
Exit 2	30	5.95%
Exit 3	25	4.96%
Exit 4	12	2.38%
Exit 5	33	6.55%
Exit 6	8	1.59%
Exit 7	28	5.56%
Exit 8	26	5.16%
Exit 9	13	2.58%
Exit 10	17	3.37%
Exit 11	24	4.76%
Exit 12	41	8.13%
Exit 13	22	4.37%
Exit 14	54	10.71%
Exit 15	28	5.56%
Exit 16	23	4.56%
Exit 20	28	5.56%

Table 2. *Cont.*

Options	Subtotal	Proportion
Exit 21	23	4.56%
Exit 22	30	5.95%
Exit 23	16	3.17%
Number of persons validated for filling in this questionnaire	504	100%

Based on the data above, taking the probability of passenger selection exit into account, we set activation rules for the pedestrian flow by parameter calibration. In addition, the basic overall questionnaire data are clearly provided in the tables in Appendix A.

(2) *Infrastructure data setting*

We have the *SetSpeed* module to set passengers' walking speed. Set ladder group data in the Escalator-group, such as ladder speed, width, length, etc. Pedestrian speed is based on observed average value, although in principle it is a random variable. Otherwise, the simulation is too complex to execute. The detailed parameters are shown in Table 3.

Table 3. Data parameter setup table.

Pedestrian Parameters	**Gender**	**Male**	**Female**
	Speed	**1.31 m/s**	**1.16 m/s**
Escalator group parameters	Width	1.4 m	
	Length	10 m	
	Speed	0.65 m/s	
Stair parameters	Width	3 m	
	Length	10 m	

4.2. Xinjiekou Station Simulation

As per the official data of the Nanjing Traffic Design Institute in 2018, Nanjing Subway Corporation presently uses Type-A train, marshalling six-car trains. Satisfying the URT safety standard, the upper bound of the loading threshold (capacity) is 2520 pax/train, where pax stands for passengers. At the time when the COVID-19 pandemic prevailed, the daily level of demand at Xinjiekou Station fell to hundreds of orders of magnitude. In effect, during a given day, the surveyed overall demand is 117,588 pax/day. It is assumed that the number of passengers desiring to exit is 1000 per train, based on the service frequency.

Regarding train agents, the arrival interval time of trains on Line 1 and Line 2 is 100 s, while the arrival interval time of the trains on the other side of Line 1 and Line 2 start from 200 s, meaning the arrival time interval is 300 s. As per the different pedestrian diameters (i.e., distancing), the multi-agent-based simulation is carried out respectively. In order to observe the impact of pedestrian distancing on congestion precisely, as large passenger flow exits, comparative analysis is conducted and the pedestrian behaviors at the same turnstile are intercepted for this purpose.

4.2.1. Different Distancing Queues for Pedestrians at the Same Turnstile

Figures 6–8 shows the pedestrian queuing situation on the right side of the 22nd exit for purposes of analysis and comparison. Given pedestrian distancing of 0.4–0.5 m, from the view of simulated 3D, the crowding degree of pedestrian flow is the largest, so the service rate of the turnstile is the lowest. Furthermore, when the distancing is 1.5–2.0 m, there is no need to wait in front of the turnstile, despite passenger movement efficiency being restricted. When distancing is 0.9–1.0 m, pedestrians are less likely to form a long queue. Subsequently, the service rate of the turnstile is higher than the other two distancing cases.

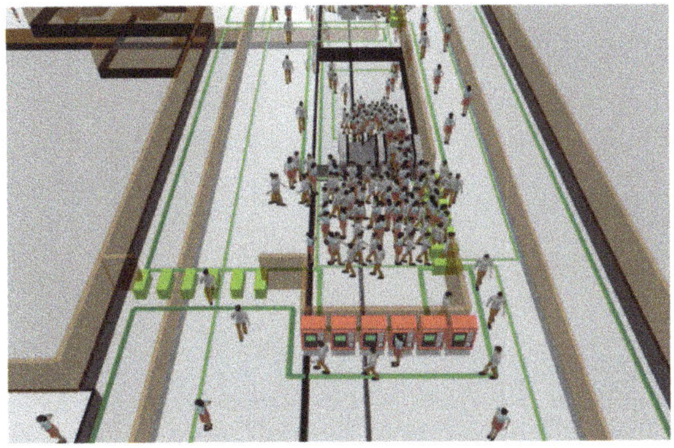

Figure 6. Queue condition of turnstile when distancing is 0.4–0.5 m.

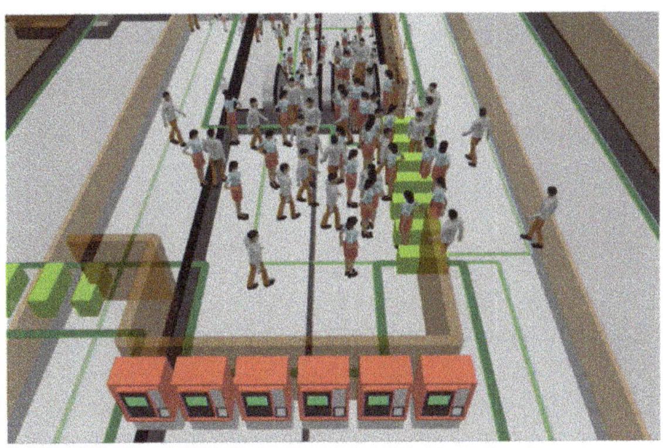

Figure 7. Queue condition of turnstile when distancing is 0.9–1.0 m.

Figure 8. Queue condition of turnstile when distancing is 1.5–2.0 m.

4.2.2. Simulation of Density Diagram with Different Distancing

(1) *Simulation of distancing observed between 0.4 m and 0.5 m*

Figure 9 explicitly shows the performance of pedestrian restricted movement when the running time of the simulation is at 200 s. Allowing for the impact of the COVID-19 pandemic on denseness, low-density aggregation appears in such areas as escalators, stairs, and existing turnstiles in the short queue. This distancing distinctly affects serving efficiency, in comparison to daily life prior to the pandemic. The physical distancing of less than 0.5 m cannot contribute to lowering the risk of infection [1], and the low bound of the distancing therefore hardly enables prevention, which gets even worse, especially when medical protection facilities are lacking. Hence, the infection risk remains with distancing of about 0.5 m, although rapid restricted movement is feasible.

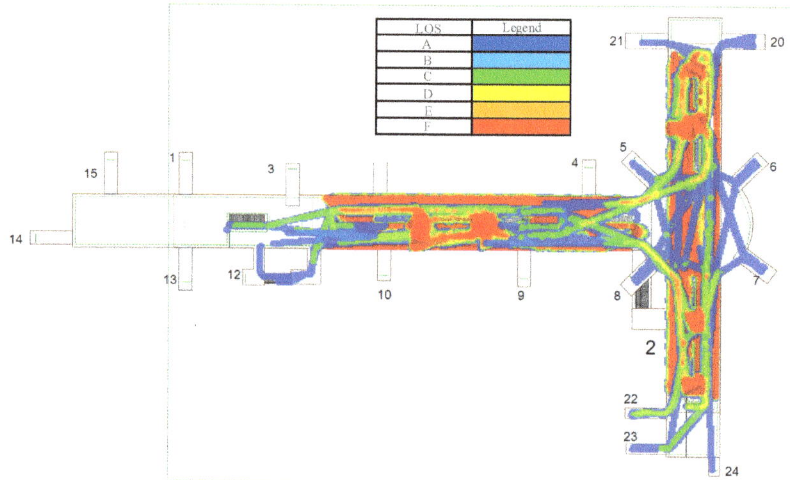

Figure 9. Simulation density diagram observing distancing between 0.4 m and 0.5 m at 200 s.

Next, considering the convenience of implementing a management policy, the observations of distancing are explored per approximate 0.5 m as a simulation step/unit. In Figures 9–11, six different colors denote multiple levels of passenger density: (i) serious congestion is depicted by red, (ii) normal dynamics are demonstrated by green, and (iii) sparse flow pattern are indicated by blue. Based on Fruin [38], Table 4 shows the densities and space associated with the six different colors of Figures 9–11.

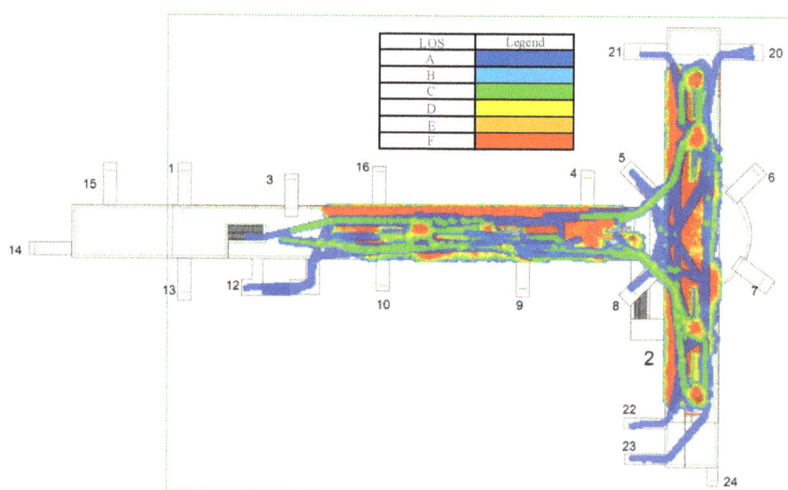

Figure 10. Simulation density diagram with observing distancing between 0.9 m and 1.0 m at 200 s.

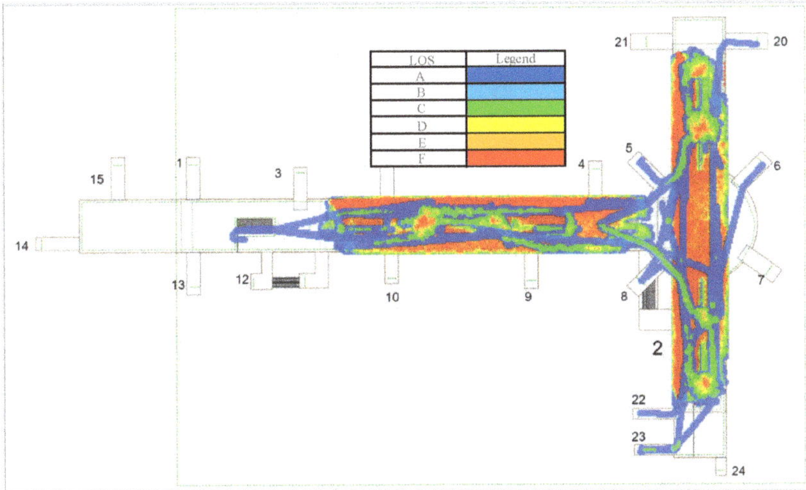

Figure 11. Simulation density diagram for observing distancing between 1.5 m and 2.0 m at 200 s.

Table 4. Color bar of pedestrians' density and space, as per the underlying criterion (referring to Fruin [38]).

LOS	Density (pax/m^2)	Space (m^2/pax)	Legend
A	$d \leq 0.309$	$s \geq 3.24$	
B	$0.309 < d \leq 0.431$	$3.24 > s \geq 2.32$	
C	$0.431 < d \leq 0.719$	$2.32 > s \geq 1.39$	
D	$0.719 < d \leq 1.075$	$1.39 > s \geq 0.93$	
E	$1.075 < d \leq 2.174$	$0.93 > s \geq 0.46$	
F	$d > 2.174$	$s < 0.46$	

Note: Adapted from J. Fruin, Pedestrian planning and design, published by Metropolitan Association of Urban Designers and Environmental Planners, New York, 1971.

(2) *Simulation observing distancing between 0.9 m and 1.0 m*

Figure 10 visibly captures the restricted movement situation of pedestrians under observational distancing between 0.9 m and 1.0 m. Allowing for the self-organized pedestrian crowd effect [35], optimized distancing (about 1.0 m) can reduce a clogging effect from the close distance (about 0.5 m) between pedestrians. Interestingly, the relationship between contagion-prevention social distancing and efficiency is observed to be significant. In other words, distancing dynamics still do not have a completely positive correlation with efficiency.

(3) *Simulation observing distancing between 1.5 m and 2.0 m*

Figure 11 shows the situation with distancing of 1.5–2.0 m. It can be clearly observed that the greater the distancing, the lower the degree of passenger congestion. However, excessive distancing leads to low platform utilization, hence distancing management between 1.5 m and 2.0 m is unfeasible for satisfying the walk efficiency of pedestrians.

4.3. Results

(1) Simulation observing queuing length

Figure 12 illustrates different queuing lengths at individual exits of Xinjiekou Station using different simulation distancing. The corresponding data are referenced in Appendix B. Passenger data are derived from Table 2. Evidently, the minimum average queuing length is reached under social distancing between 0.4 m and 0.5 m, including the relevant degree of deviation. In detail, Figure 12 shows that the social distancing queuing length between 0.9 m and 1.0 m increased by 4.17%, in comparison with the social distancing of about 0.5 m. We observe that 0.5 m is assumed as the ordinary distance without the COVID-19 effect. Hereby, based on queuing theory, the queuing length is defined as the queue of users waiting for the exit service, which discharges people at the rate of 2.6 sec/person [39] in an ordinary case without the virus's restrictions. In other words, the queuing length is the number of people in the queue using different social distancing. However, longer distancing, from 1.5 m to 2.0 m, leads to an increase in queuing length up to more than 3 pax (an average increase of 21.22% from the ordinary case). Indeed, this scenario demonstrates worsening service performance through an augmented queue length. This is particularly observed with a considerable delay in the 12th and 14th exits, with a 77.78% and 80.00% increase of complaints about inconvenience. In addition, while Figure 12 shows minor differences between the cases of 0.4–0.5 m and 0.9–1.0 m (green and blue), the latter results in more space, allowing for more movement flexibility. Generally speaking, the scenario of compromise distancing (about 1.0 m) indicates a somewhat superior tradeoff outcome for the contagion-prevention distance and queuing length (restricted movement efficiency) than the other two scenarios.

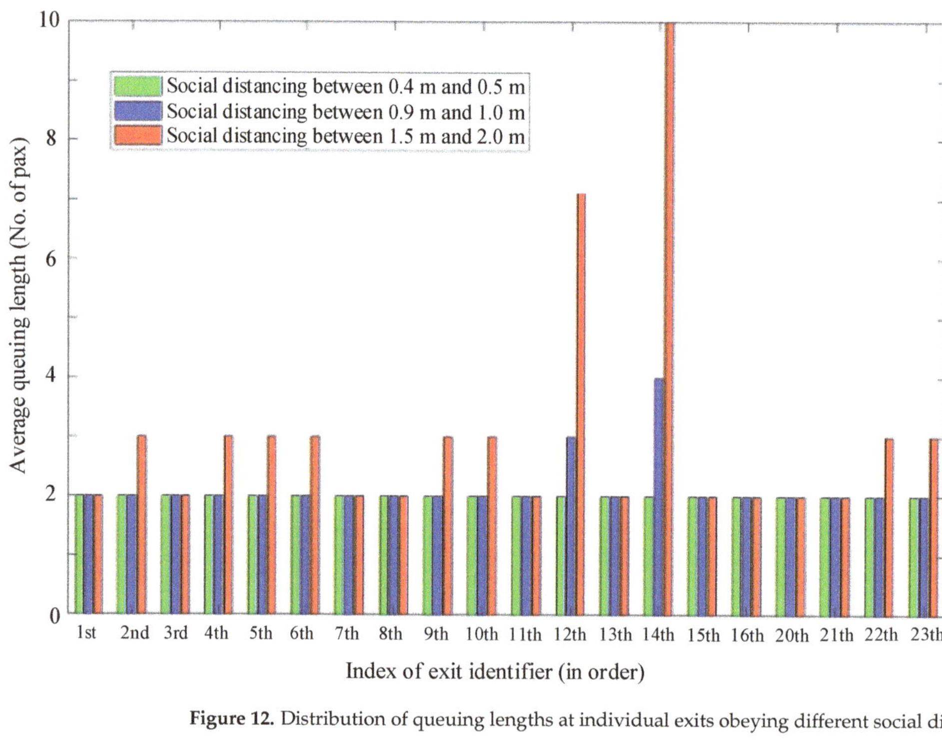

Figure 12. Distribution of queuing lengths at individual exits obeying different social distances.

(2) Simulation observing queuing behaviors and complete boarding times

According to the observation of *individual space* in Section 3.3.2 and restricted movement efficiency of Figure 9–11, we specify the similar queuing phenomena using different social distancing based on the simulation experiments. The experiments are inspired by Zhou et al. [40]. Snapshots of the pedestrians' queuing for alighting and boarding are shown in Figure 13. The simulation time points are discrete. Case (a) is from 10 s to 50 s, and shows every 10 s, i.e., the queuing situation after 10 s, after 20 s, and after 50 s. Cases (b) and (c) represent five snapshots, with the same number of waiting passengers to case (a), respectively. The entrance to the train is only in the middle, and passengers form a queue on the sides, each with a blue-colored circle, to allow disembarking passengers, with a green-colored circle, to leave the train. From the outset, pedestrians kept the regulated social distancing for queuing until the last passenger alighted. Finally, the two queues overlapped for boarding and still maintained the fixed social distancing. As a result, the greater the social distancing, the further the delay accumulated in the boarding process. This can be observed in Figure 13a–c.

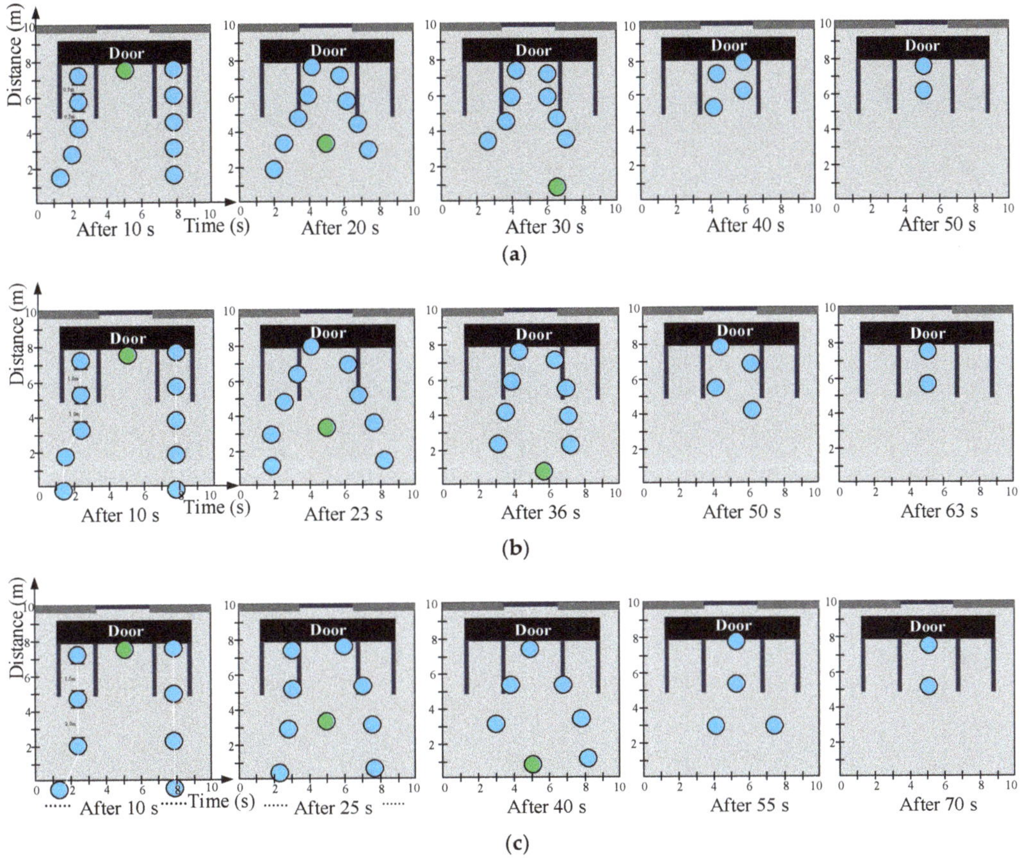

Figure 13. Snapshots of simulated passenger queuing and boarding process based on different degrees of social distancing. (**a**) Queuing situation for social distancing between 0.4 m and 0.5 m. (**b**) Queuing situation for social distancing between 0.9 m and 1.0 m. (**c**) Queuing situation for social distancing between 1.5 m and 2.0 m.

Figure 13 tests a simulation performance, in which ten passengers are waiting for boarding, and only one passenger is about to alight. Case (a) is the ordinary case without restrictions. In cases (b) and (c), we define the delay time as the extra needed time to complete boarding, in comparison with case (a). The average delay is the mean delay value of the five snapshots of Figure 13. The total increased delay and average increased delay per passenger of cases (b) and (c) are shown in Appendix C. Referring to [22], the average alighting and boarding times are 1.23 s and 1.18 s. Overall, the average increased delays per passenger for completing boarding are 14.1 s and 22.5 s for cases (b) and (c), respectively. Figure 14 illustrates the total increased delay time for the 10 pax, in comparison with case (a) for each of the five snapshots in Figure 13, with a total of 921 and 1005 s required for their waiting, or 92.1 s and 100.5 s average per passenger waiting time, for cases (b) and (c), respectively. Consequently, for case (b), with social distancing of 0.9–1.0 m, and case (c), with social distancing of 1.5–2.0 m, the percent of increased delay in the total waiting time, is 15.29% and 22.39% per passenger, respectively.

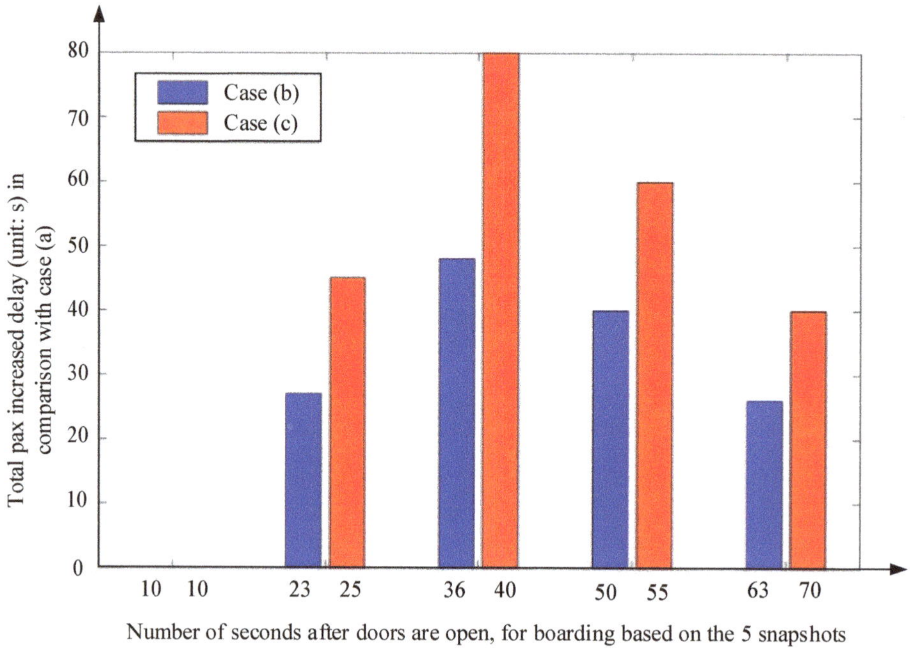

Figure 14. Comparison between increased pax delays of cases (b) and (c).

(3) Simulation summary

We observed pedestrian exiting behaviors. The peak value of the simulation is reached once the train arrives, and passengers begin to evacuate based on the multipath choice decisions. We find out that the ladder, turnstile, and corner are prone to congestion. While updating passengers' exiting process, congestion makes it possible to anticipate its gradual elimination, based on the iteration of the simulation. Indeed, this is consistent with the empirical survey of pedestrian exiting behaviors, indicating that the SFM-based simulation is able to precisely verify the dynamic performance of the in-station passengers.

This lends further validation to the simulation results from the different pedestrian distancing. The greater the pedestrian distancing, the lower the congestion level of the station turns out to be. Moreover, the insufficient utilization rate on the platform has a negative impact on the pedestrians' exiting and entering efficiency. In order to observe the queuing of pedestrian flow, we check the service efficiency at a given turnstile from the 3D perspective. Given the service efficiency of the fixed turnstile, we observe that the greater the pedestrian distancing, the fewer people can pass the turnstile. Accordingly, it is not easy to generate congestion in front of the turnstile.

5. Analysis

Pedestrian congestion varying in social distancing is primarily derived from different physical surroundings: (i) infrastructures, (ii) control policy and (iii) loading and unloading from trains, which regulate their behaviors and social interactions. To examine the management effectiveness at the distancing between 0.9 m and 1.0 m, we develop a three-phase parameter-based scenario for further analyzing environmental impacts.

5.1. Passenger Management Analysis

Given Xinjiekou Station is the largest transfer station in Asia and is located in the center of the city, the large passenger flow occurs unpredictably. The necessary passenger management is usually implemented. For example, operators can restrict the number of people entering the URT during peak hours. Furthermore, some operators provide real-time passenger control actions to guide the optimal passenger paths to reduce the bottleneck crowding.

To facilitate a direct policy of preventing person-to-person transmission of COVID-19, we investigate the effectiveness of lowering the number of passengers allowed into the URT. The number of passengers exiting from Xinjiekou Station is decreased from 1000 to 600 pax. As shown in Figure 15, we find that the efficiency of pedestrian-restricted movement is better than the *ordinary situation* without any control. A long queue should still occur at escalators, turnstiles, and other infrastructure mechanisms. The term *'ordinary situation'* describes normal pedestrian movements (freely in the post-epidemic era) without any distancing rules.

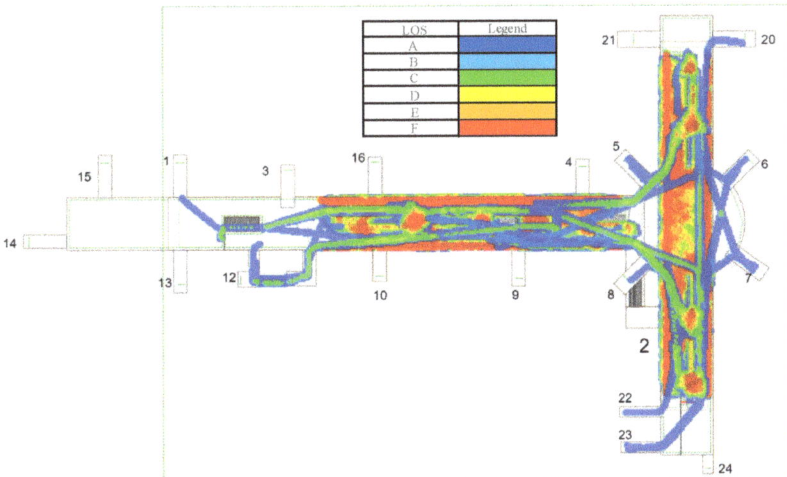

Figure 15. Restricted movement sketch of 600 people onto a platform, based on a consideration of 200 s.

5.2. Bottleneck Congestion Analysis

Our search identified the main infrastructure mechanism, including turnstiles, escalators, and stairs. Considering the infeasibility/difficulty of resetting the location and width of facilities in the real world, it is possible to switch the direction of the escalator as soon as passenger congestion rises. In this case, a typical feature of the bottleneck effect is a long queue in front of the escalator. By means of changing/adding one corridor of a restricted movement path (escalator), efficiency is improved. Complementarily, as the number of passengers queuing in front of the escalator gradually increases, the operators can further guide passengers to the stairs and other paths to evacuate.

5.3. Service Frequency Adjustment

The service frequency in the previous simulation is set to 100 s. This means that the passengers from trains belonging to both Line 1 and Line 2 arrive at the same time at 100 s of simulation. Here, we set the service frequency for another direction of both Line 1 and Line 2 to 300 s. Figure 16 illustrates the density graphic of the platform, in which the reduced frequency of service and the correlating arrival time on the other side of the train allows sufficient restricted movement time for the accumulated passenger flow, which considerably alleviates the degree of passenger congestion on the platform.

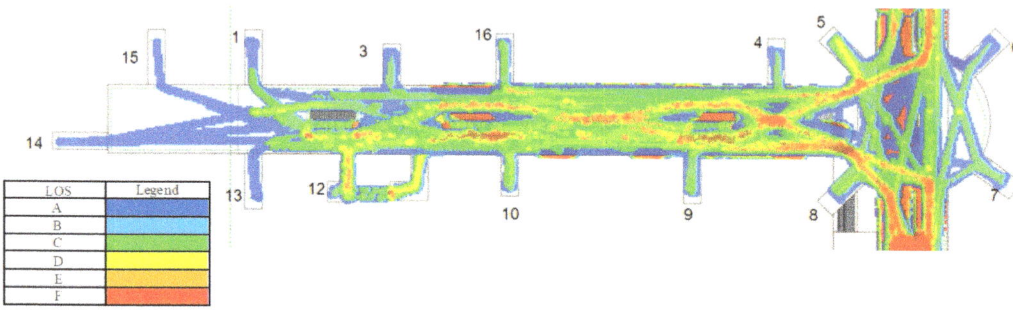

Figure 16. Density diagram of station platform after resetting arrival time.

6. Conclusions

It appears that some of the prior management policies at URT to protect public safety rest on the question of how to attain a tradeoff between restricted movement efficiency and social distancing. Note that the single strategy of limiting the number of passengers to enter URT is unable to lower the risk of transmission of COVID-19 without maintaining person-to-person distancing, especially taking certain dramatically congested, bottleneck facilities into account.

This paper focuses on an optimal public management problem between safety distancing and walk efficiency. It is expected to develop and substantiate comprehensive policies concerning the COVID-19 epidemic or other public health events. In the study, we investigated crowd behaviors at multiple service infrastructures, as per testing the different distancing.

We combine pedestrian psychology, pedestrian interaction force, self-driving force and other aspects to calibrate the parameters of SFM. They are utilized as the simulation parameter input and the simulation surroundings. Based on data collection from an online questionnaire survey, conducted in January–May in 2020, we simulate the experimental scenario of Nanjing Xinjiekou Station. By comparing different pedestrian distancing for public management policy in preventing COVID-19 spread, we investigated passengers' behaviors. In particular, we referred to the increased congestion phenomena at escalators and turnstiles. Furthermore, we developed optimal public management schemes as an auxiliary guide-based policy.

The queuing length with social distancing between 0.9 m and 1.0 m (case (b)), and between 1.5 m to 2.0 m (case (c)) are increased by 4.2% and 22.2%, respectively, in comparison with ordinary social distancing of about 0.5 m (case (a)). The average delays per passenger for completing boarding are 14.1 s and 22.5 s for cases (b) and (c), which leads to 15.29% and 22.39% increases, compared to case (a), for each passenger, respectively.

This study's methodology and results can, to some extent, be applied to other URT stations because of the similarities in facing a high-density environment. Further recommended will be to consider: (i) including human's perception, URT station design, topology, and more; and (ii) a comparison of different procedures used worldwide to decide on social distancing during a pandemic.

Author Contributions: Conceptualization, Z.C., A.C., Z.W., S.Z. and Y.W.; methodology, Z.C., A.C., Z.W., S.Z. and Y.W.; software, Z.W., Z.C. and A.C.; validation, Z.C., A.C., Z.W., S.Z. and Y.W.; formal analysis, Z.C., A.C., Z.W., S.Z. and Y.W.; investigation, Z.C., A.C., Z.W., S.Z. and Y.W.; resources, Z.W.; data curation, Z.W.; writing—original draft preparation, Z.C., A.C., Z.W., S.Z. and Y.W.; writing—review and editing, Z.C., A.C., Z.W., S.Z. and Y.W.; visualization, Z.C., A.C., Z.W., S.Z. and Y.W.;

supervision, Z.C., A.C., Z.W., S.Z. and Y.W.; project administration, Z.C. and S.Z.; funding acquisition, Z.C. and S.Z. All authors have read and agreed to the published version of the manuscript.

Funding: This study was supported by the National Natural Science Foundation of China (72101126), the Transportation Operation Subsidy Project of Guangxi Key Laboratory of International Join for China-ASEAN Comprehensive Transportation (No. 21-220-21), the National Natural Science Foundation of China (72101127), the Jiaxing Key Laboratory of Smart Transportations Open Project Funding (ZHJT202303), and the Technology and Equipment of Rail Transit Operation and Maintenance (2022YW004).

Data Availability Statement: Data are contained within the article.

Conflicts of Interest: The authors declare no conflicts of interest.

Appendix A. Survey-Extracted Statistical Results for Pedestrians Who Used the Exit Doors of Xinjiekou URT Station

Table A1. 1st Question. About gender (Single choice).

Gender	Replies	Percentage
Male	265	52.58%
Female	239	47.42%
Total valid answers	504	100.00%

Table A2. 2nd Question. About age (Single choice).

Age	Replies	Percentage
Under 20-years-old	86	17.06%
20–30	175	34.72%
30–40	150	29.76%
40–50	71	14.09%
50–60	22	4.37%
More than 60-years-old	0	0%
Total valid answers	504	100.00%

Table A3. 3rd Question. Which exit do you usually use to leave Xinjiekou? (Single choice).

Exit	Replies	Percentage
1st exit (Nandayang Mall, Nanjing, China)	23	4.56%
2nd exit (Nandayang Mall, Nanjing, China)	30	5.95%
3rd exit (Dongfang Mall, Nanjing, China)	25	4.96%
4th exit (Intersection Between Hanzhong Road and Hanzhong North Road, Nanjing, China)	12	2.38%
5th exit (Nanjing International Finance Center and China Merchants Bank, Nanjing, China)	33	6.55%
6th exit (West of Zhongshan Road and North of Hanzhong Road, Nanjing, China)	8	1.59%
7th exit (Deji plaza, ICBC, and Bank of China, Nanjing, China)	28	5.56%
8th exit (Nanjing Xinjiekou Department Store Co., Ltd., Nanjing, China)	26	5.16%

Table A3. *Cont.*

Exit	Replies	Percentage
9th exit (Nanjing Xinjiekou Department Store Co., Ltd., Nanjing, China)	13	2.58%
10th exit (Nanjing Xinjiekou Department Store Co., Ltd., Nanjing, China)	17	3.37%
11th exit (Zhenghong Street Underground Commercial Street, Nanjing, China)	24	4.76%
12th exit (Pedestrian Mall, and East of Zhongshan South Road, Nanjing, China)	41	8.13%
13th exit (Suning Tesco Xinjiekou Flagship Store and Dahua Theater, Nanjing, China)	22	4.37%
14th exit (Central Mall, Nanjing, China)	54	10.71%
15th exit (Dayang Mall, Nanjing, China)	28	5.56%
16th exit (Dongfang Mall, Nanjing, China)	23	4.56%
20th exit (Jinying Guoji Mall, Nanjing, China)	28	5.56%
21st exit (Underground Passage of Jinying Guoji Mall, Nanjing, China)	23	4.56%
22nd exit (Nanjing Xinjiekou Department Store Co., Ltd., Nanjing, China)	30	5.95%
23rd exit (Xinhua Bookstore, Nanjing, China)	16	3.17%
Total valid answers	504	100.00%

Table A4. 4th Question. What is your trip purpose? (Single choice).

Purpose	Replies	Percentage
Go to work	129	25.6%
Go to school	64	12.7%
Go shopping	232	46.03%
Commuting	79	15.67%
Total valid answers	504	100.00%

Table A5. 5th Question. Did you feel crowded at the exit? (Single choice).

Crowded	Replies	Percentage
Yes	472	93.65%
No	32	6.35%
Total valid answers	504	100.00%

Table A6. 6th Question. Are you a Nanjing native? (Single choice).

Native	Replies	Percentage
Yes	483	95.83%
No	21	4.17%
Total valid answers	504	100.00%

Appendix B. Queuing Lengths at Individual Exits under Different Simulated Social Distancing (SD), in Pax

Exit	SD between 0.4 m and 0.5 m (pax)	DD (%)	SD between 0.9 m and 1.0 m (pax)	DD (%)	SD between 1.5 m and 2.0 m (pax)	DD (%)
1st exit	2	0	2	0	2	0.00%
2nd exit	2	0	2	0	3	33.33%
3rd exit	2	0	2	0	2	0.00%
4th exit	2	0	2	0	3	33.33%
5th exit	2	0	2	0	3	33.33%
6th exit	2	0	2	0	3	33.33%
7th exit	2	0	2	0	2	0.00%
8th exit	2	0	2	0	2	0.00%
9th exit	2	0	2	0	3	33.33%
10th exit	2	0	2	0	3	33.33%
11th exit	2	0	2	0	2	0.00%
12th exit	2	0	3	33.33%	9	77.78%
13th exit	2	0	2	0	2	0.00%
14th exit	2	0	4	50.00%	10	80.00%
15th exit	2	0	2	0	2	0.00%
16th exit	2	0	2	0	2	0.00%
20th exit	2	0	2	0	2	0.00%
21st exit	2	0	2	0	2	0.00%
22nd exit	2	0	2	0	3	33.33%
23rd exit	2	0	2	0	3	33.33%
Average	2	0	$2.15 \approx 2$	4.17%	$3.15 \approx 3$	21.22%

Note: SD = simulation distancing; DD = degree of deviation.

Appendix C. Delay Times and Its Increased Percentage for the Three Cases Examined, in Comparison with the Base Case (a)

Subject	Boarding Time-Point, Corresponding to the Base Case (a)					Average Delay per Passenger (s)
Base Case (a)	10 s	20 s	30 s	40 s	50 s	--
Case (b)	10 s	23 s	36 s	50 s	63 s	--
Increased delay (s) per passenger	0	3	6	10	13	7.25
Total delay of pax (s)	0	27	48	40	26	14.10
Case (c)	10 s	25 s	40 s	55 s	70 s	--
Increased delay (s) per passenger	0	5	10	15	20	12.50
Total delay of pax (s)	0	45	80	60	40	22.50

Note: The increased delay relates to the comparison with case (a). For example, for the time-point of 30 s, for case (a) of Figure 13, one obtains for case (b) the time-point of 36 s, with a difference of 6 s in having 8 (out of 10) pax waiting to board. The total delay of pax is the increased delay per passenger multiplied by the number of passengers for each snapshot of Figure 13; for instance, in the 3rd snapshot for cases (b) and (c), this total delay is $8 \times 6 = 48$ s, and $8 \times 10 = 80$ s, respectively.

References

1. Chu, D.K.; Akl, E.A.; Duda, S.; Solo, K.; Yaacoub, S.; Schünemann, H.J.; El-Harakeh, A.; Bognanni, A.; Lotfi, T.; Loeb, M.; et al. Physical distancing, face masks, and eye protection to prevent person-to-person transmission of SARS-CoV-2 and COVID-19: A systematic review and meta-analysis. *Lancet* **2020**, *395*, 1973–1987. [CrossRef] [PubMed]
2. Helbing, D.; Keltsch, J.; Molnar, P. Modelling the evolution of human trail systems. *Nature* **1997**, *388*, 47–50. [CrossRef] [PubMed]
3. Yuan, F.; Sun, H.; Kang, L.; Wu, J. Passenger flow control strategies for urban rail transit networks. *Appl. Math. Model.* **2020**, *82*, 168–188. [CrossRef]
4. Jiang, Z.; Fan, W.; Liu, W.; Zhu, B.; Gu, J. Reinforcement learning approach for coordinated passenger inflow control of urban rail transit in peak hours. *Transp. Res. Part C Emerg. Technol.* **2018**, *88*, 1–16. [CrossRef]
5. Li, Y.; Wang, X.; Sun, S.; Ma, X.; Lu, G. Forecasting short-term subway passenger flow under special events scenarios using multiscale radial basis function networks. *Transp. Res. Part C Emerg. Technol.* **2017**, *77*, 306–328. [CrossRef]

6. Dubroca-Voisin, M.; Kabalan, B.; Leurent, F. On pedestrian traffic management in railway stations: Simulation needs and model assessment. *Transp. Res. Procedia* **2019**, *37*, 3–10. [CrossRef]
7. Burdzik, R.; Chema, W.; Celiński, I. A study on passenger flow model and simulation in aspect of COVID-19 spreading on public transport bus stops. *J. Public Transp.* **2023**, *25*, 100063. [CrossRef]
8. Yao, X.; Han, B.; Yu, D.; Ren, H. Simulation-Based Dynamic Passenger Flow Assignment Modelling for a Schedule-Based Transit Network. *Discret. Dyn. Nat. Soc.* **2017**, *2017*, 2890814. [CrossRef]
9. Zhao, Z.; Liang, D. Pedestrian Flow Characteristic of Metro Station along with the Mall. *Procedia Eng.* **2016**, *135*, 602–606. [CrossRef]
10. Jin, C.; Jiang, R.; Wong, S.C.; Xie, S.; Li, D.; Guo, N.; Wang, W. Observational characteristics of pedestrian flows under high-density conditions based on controlled experiments. *Transp. Res. Part C Emerg. Technol.* **2019**, *109*, 137–154. [CrossRef]
11. Shang, P.; Li, R.; Guo, J.; Xian, K.; Zhou, X. Integrating Lagrangian and Eulerian observations for passenger flow state estimation in an urban rail transit network: A space-time-state hyper network-based assignment approach. *Transp. Res. Part B Methodol.* **2019**, *121*, 135–167. [CrossRef]
12. Li, C.; Ma, J.; Tom HLuan Zhou, X.; Xiong, L. An incentive-based optimizing strategy of service frequency for an urban rail transit system. *Transp. Res. Part E Logist. Transp. Rev.* **2018**, *118*, 106–122. [CrossRef]
13. Wang, Y.; Tang, T.; Ning, B.; Van Den Boom, T.J.; De Schutter, B. Passenger-demands-oriented train scheduling for an urban rail transit network. *Transp. Res. Part C Emerg. Technol.* **2015**, *60*, 1–23. [CrossRef]
14. Wang, Y.; D'Ariano, A.; Yin, J.; Meng, L.; Tang, T.; Ning, B. Passenger demand oriented train scheduling and rolling stock circulation planning for an urban rail transit line. *Transp. Res. Part B Methodol.* **2018**, *118*, 193–227. [CrossRef]
15. Yin, J.; Yang, L.; Tang, T.; Gao, Z.; Ran, B. Dynamic passenger demand oriented metro train scheduling with energy-efficiency and waiting time minimization: Mixed-integer linear programming approaches. *Transp. Res. Part B Methodol.* **2017**, *97*, 182–213.
16. Hoogendoorn, S.P.; Bovy, P.H. Pedestrian route-choice and activity scheduling theory and models. *Transp. Res. Part B Methodol.* **2004**, *38*, 169–190. [CrossRef]
17. Tong, W.; Cheng, L. Simulation of Pedestrian Flow Based on Multi-Agent. *Procedia-Soc. Behav. Sci.* **2013**, *96*, 17–24. [CrossRef]
18. Weng, J.; Zheng, Y.; Qu, X.; Yan, X. Development of a maximum likelihood regression tree-based model for predicting subway incident delay. *Transp. Res. Part C Emerg. Technol.* **2015**, *57*, 30–41. [CrossRef]
19. Xiao, Y.; Gao, Z.; Qu, Y.; Li, X. A pedestrian flow model considering the impact of local density: Voronoi diagram based heuristics approach. *Transp. Res. Part C Emerg. Technol.* **2016**, *68*, 566–580. [CrossRef]
20. Helbing, D.; Molnar, P. Social force model for pedestrian dynamics. *Phys. Rev. E* **1995**, *51*, 4282.
21. Zheng, X.; Li, H.; Meng, L.; Xu, X.; Chen, X. Improved social force model based on exit selection for microscopic pedestrian simulation in subway station. *J. Cent. South Univ.* **2015**, *22*, 4490–4497. [CrossRef]
22. Li, Z.; Lo, S.M.; Ma, J.; Luo, X.W. A study on passengers' alighting and boarding process at metro platform by computer simulation. *Transp. Res. Part A Policy Pract.* **2020**, *132*, 840–854. [CrossRef]
23. Xu, X.; Liu, J.; Li, H.; Hu, J. Analysis of subway station capacity with the use of queueing theory. *Transp. Res. Part C Emerg. Technol.* **2014**, *38*, 28–43. [CrossRef]
24. Jiang, Y.; Chen, B.; Li, X.; Ding, Z. Dynamic navigation field in the social force model for pedestrian evacuation. *Appl. Math. Model.* **2020**, *80*, 815–826. [CrossRef]
25. Xu, X.Y.; Liu, J.; Li, H.Y.; Jiang, M. Capacity-oriented passenger flow control under uncertain demand: Algorithm development and real-world case study. *Transp. Res. Part E Logist. Transp. Rev.* **2016**, *87*, 130–148. [CrossRef]
26. Stubenschrott, M.; Kogler, C.; Matyus, T.; Seer, S. A dynamic pedestrian route choice model validated in a high density subway station. *Transp. Res. Procedia* **2014**, *2*, 376–384. [CrossRef]
27. Chen, X.; Li, H.; Miao, J.; Jiang, S.; Jiang, X. A multiagent-based model for pedestrian simulation in subway stations. *Simul. Model. Pract. Theory* **2017**, *71*, 134–148. [CrossRef]
28. Zhang, Q.; Han, B.; Li, D. Modeling and simulation of passenger alighting and boarding movement in Beijing metro stations. *Transp. Res. Part C Emerg. Technol.* **2008**, *16*, 635–649. [CrossRef]
29. Feng, S.; Xin, M.; Lv, T.; Hu, B. A novel evolving model of urban rail transit networks based on the local-world theory. *Physica A* **2019**, *535*, 122227. [CrossRef]
30. Zheng, L.; Peng, X.; Wang, L.; Sun, D. Simulation of pedestrian evacuation considering emergency spread and pedestrian panic. *Physica A* **2019**, *522*, 167–181. [CrossRef]
31. AnyLogic. Commercial Simulation Software Including AI for Transportation and Pedestrians. 2018. Available online: https://www.anylogic.com/resources/educational-videos/simulating-a-pedestrian-crossing/ (accessed on 5 March 2018).
32. Zuo, J.; Shi, J.; Li, C.; Mu, T.; Zeng, Y.; Dong, J. Simulation and optimization of pedestrian evacuation in high-density urban areas for effectiveness improvement. *Environ. Impact Assess. Rev.* **2021**, *87*, 106521. [CrossRef]
33. Liu, L.; Chen, H. Microscopic Simulation-Based Pedestrian Distribution Service Network in Urban Rail Station. *Transp. Res. Interdiscip. Perspect.* **2021**, *9*, 100313. [CrossRef]
34. Zhang, L.; Li, M.; Wang, Y. Research on Design Optimization of Subway Station Transfer Entrance Based on AnyLogic. *Procedia Comput. Sci.* **2022**, *208*, 310–318. [CrossRef]
35. Helbing, D.; Buzna, L.; Johansson, A.; Werner, T. Self-organized pedestrian crowd dynamics: Experiments, simulations, and design solutions. *Transp. Sci.* **2005**, *39*, 1–24. [CrossRef]

36. Zhang, Q.; Han, B. Simulation model of pedestrian interactive behavior. *Phys. A Stat. Mech. Its Appl.* **2011**, *390*, 636–646. [CrossRef]
37. Chandra, S.; Bharti, A.K. Speed distribution curves for pedestrians during walking and crossing. *Procedia-Soc. Behav. Sci.* **2013**, *104*, 660–667. [CrossRef]
38. Fruin, J.J. *Pedestrian Planning and Design*; Metropolitan Association of Urban Designers & Environmental Planners: New York, NY, USA, 1971.
39. Chang, D. Quantified Study of Microscopic Pedestrian Behavior Parameters in Subway. Ph.D. Thesis, Beijing Jiaotong University, Beijing, China, 2010.
40. Zhou, M.; Dong, H.; Wang, F.Y.; Zhao, Y.; Gao, S.; Ning, B. Field observations and modeling of waiting pedestrian at subway platform. *Inf. Sci.* **2019**, *504*, 136–160. [CrossRef]

systems

MDPI

Article

Exploring the Impact of Charging Behavior on Transportation System in the Era of SAEVs: Balancing Current Request with Charging Station Availability

Yi Zhu [1], Xiaofei Ye [2,*], Xingchen Yan [3], Tao Wang [4], Jun Chen [5] and Pengjun Zheng [2]

1 Faculty of Maritime and Transportation, Ningbo University, Fenghua Road 818#, Ningbo 315211, China; zhuyi19991125@163.com
2 Ningbo Port Trade Cooperation and Development Collaborative Innovation Center, Faculty of Maritime and Transportation, Ningbo University, Fenghua Road 818#, Ningbo 315211, China; zhengpengjun@nbu.edu.cn
3 College of Automobile and Traffic Engineering, Nanjing Forestry University, Nanjing 210037, China; xingchenyan.acad@gmail.com
4 School of Architecture and Transportation, Guilin University of Electronic Technology, Jinji Road 1#, Guilin 541004, China; wangtao@guet.edu.cn
5 School of Transportation, Southeast University, Nanjing 211189, China; chenjun@seu.edu.cn
* Correspondence: yexiaofei@nbu.edu.cn

Abstract: Shared autonomous electric vehicles (SAEVs) can offer safer, more efficient, and more environmentally friendly real-time mobility services with advanced autonomous driving technologies. In this study, a multi-agent-based simulation model considering SAEVs' vehicle range and charging behavior is proposed. Based on real-world datasets from the Luohu District in Shenzhen, China, various scenarios with different fleet sizes, charging rates, and vehicle ranges are established to evaluate the impact of these parameters on parking demand, charging demand, vehicle miles traveled (VMT), and response time in the era of SAEVs. The results show there would be much more charging demand than parking demand. Moreover, a larger fleet size and longer vehicle range would lead to more parking demand, more charging demand, and more VMT while increasing the charging rate can dramatically reduce the charging demand and VMT. Average response time can be reduced by increasing the fleet size or the charging rate, and a larger vehicle range leads to longer response time due to the longer time spent recharging. It is worth noting that the VMT generated from relocating from the previous request destination to the origin of the upcoming request accounts for nearly 90% of the total VMT, which should be addressed properly with appropriate scheduling. A charging policy considering current requests and the availability of charging stations was proposed and verified in terms of reducing the response time by 2.5% to 18.9%.

Keywords: shared autonomous electric vehicles; charging behavior; multi-agent-based simulation model; vehicle miles traveled; response time

Citation: Zhu, Y.; Ye, X.; Yan, X.; Wang, T.; Chen, J.; Zheng, P. Exploring the Impact of Charging Behavior on Transportation System in the Era of SAEVs: Balancing Current Request with Charging Station Availability. *Systems* **2024**, *12*, 61. https://doi.org/10.3390/systems12020061

Academic Editors: Mahyar Amirgholy and Jidong J. Yang

Received: 16 January 2024
Revised: 12 February 2024
Accepted: 13 February 2024
Published: 17 February 2024

1. Introduction

Shared autonomous vehicles (SAVs), as a new mode of transportation that combines the sharing economy and autonomous driving technology, have gained significant attention from both commercial and academic communities [1,2]. Many scholars believe that SAVs will revolutionize the future of urban transportation and provide much more convenient, safer, and economical mobility services [3,4]. Rapidly increasing research and development efforts are contributing to the expected presence of SAVs on the road, particularly in terms of parking, which they can free from the constraint of distance due to their self-driving ability [5]. A further way to enhance the advantages of SAVs is by replacing traditional fuel-powered vehicles with electric ones, so here comes SAEVs [6–9]. With the gradual decrease in the price of electric vehicles and the continuous development of charging infrastructure [1], the prospects for SAEVs are becoming increasingly promising. However, the limitations in vehicle range and the time required for recharging SAEVs can influence the planning and operation of the transportation system [10], making it crucial to consider these two new constraints when trying to model the transportation system in the era of SAEVs.

Currently, most charging stations and parking lots are built separately, but for SAEVs, parking and charging can occur simultaneously [11]. For SAEVs that provide real-time mobility services, the reasons for parking are (1) low travel demand at the current period which does not need such a great number of SAEVs driving on the roads and (2) their current battery level is satisfied which needs no recharging at charging stations. Moreover, most existing parking lots have not considered installing charging facilities or only partially modifying some spaces to fulfill the demand for recharging, mainly due to safety and economic concerns. Charging electric-powered vehicles is a time-consuming process that inevitably occupies charging station spaces for extended periods of time. Compared to SAEVs that are parked in parking lots with no charging needs, if vehicles could be parked at charging stations, it would couple both demands and allow for charging while parking. Therefore, it is essential to study the impact of charging behavior on parking demand.

Although there are lots of studies on SAVs and SAEVs about their impact on the transportation system [12–16], to the authors' best knowledge, there are few studies that incorporate the constraints of vehicle range and charging behavior into a simulation model. Likewise, research focusing on the impact of SAEVs on parking demand can hardly be found which indicates the omission of this vital issue. Since the relationship between charging-related indicators (e.g., vehicle range and charging rate) and traditional transportation metrics (e.g., parking demand, VMT, and awaiting time) is of great significance to the sustainable development of SAEVs service, and a suitable charging policy would certainly be an assistance to the SAEVs service performance, this study aims to fill these research gaps by:

1. Developing a comprehensive multi-agent simulation model that considers both vehicle range and charging behavior of SAEVs to reveal the relationship between SAEVs' fleet size, charging rate, vehicle range, parking demand, charging demand, VMT, and average response time in the era of SAEVs based on the output of the experiment using real-world datasets in Shenzhen, China;
2. Dividing the total VMT into different parts based on the different origins and destinations and analyzing the specific VMT parts that counted most to provide suggestions for reducing these parts;
3. Proposing a charging policy that considers the balance between current requests and the availability of charging stations, and reveals its effectiveness on these mentioned metrics.

The remainder of this paper is structured as follows. In the next Section, previous studies related to these topics were comprehensively summarized. Then, we give a brief description of the dataset we used and the specifications of the proposed multi-agent-based simulation model. Next, the results of different experiment scenarios are discussed, and finally, conclusions from the discussion are presented and suggestions for future work are given.

2. Previous Studies

Many works have been completed focusing on SAVs' fleet size to analyze their impact on urban transportation systems, especially on parking demand [17–20]. An agent-based simulation model was proposed to estimate the potential impact of SAV in terms of urban parking demand. They pointed out that 90% of parking demand would be eliminated at a relatively low market penetration rate of 2%, but there would be an increase in VMT because of empty cruising [3]. After that, SAVs' parking demand in the City of Atlanta was further examined using a discrete-event, agent-based simulation model and the results suggested that parking demand can be reduced by over 90% for households who would give up private vehicles and use SAVs. Researchers also pointed out that the shift of urban parking demand to adjacent areas may result in larger VMT, more congestion, and longer response time [4]. This model was further developed by focusing on the future trajectories of reduced parking demand [21]. It was pointed out that in the most optimal SAV adoption scenario, the parking demand will decrease by over 20% after 2030, especially in core urban areas. Taking parking preference date into consideration, a discrete event simulation model was proposed using the example of the University of the West of England, Frenchay campus as a case study to examine the impacts of SAVs. The results indicated that the parking demand decreased dramatically, leaving over 2500 m^2 of existing parking space unused [22]. Considering the fact that reducing parking demand may cost another price such as the increase in VMT and road congestion, reference [23] analyzed such side effects of SAVs from a transportation analysis zones (TAZs) perspective based on a dynamic traffic flow simulator named SOUND. The results show consistency with previous research that parking demand was reduced the most in residence-dominant zones in terms of quantity and office-dominant zones in terms of proportion, at the cost of a dramatic increase in empty VMT generated because of SAVs' relocation behavior which would therefore result in road congestion. Reference [24] explored the effects of SAVs based on a data-driven modeling approach using the dataset from Langfang, China. The simulation experiments contain two kinds of sharing schemes called "ride-sharing" and "car-sharing", respectively. The results indicated these two schemes would reduce parking demand by reducing car ownership, but VMT would increase regardless of sharing schemes while car-sharing alone increases much more VMT than ride-sharing. In terms of SAV service performance, reference [25] developed an agent-based simulation model for SAVs based in the city of Amsterdam, the Netherlands. Three different proactive relocation strategies named "Demand-Anticipation", "Supply-Anticipation", and "Demand-Supply-Balancing" were introduced and analyzed with regard to passenger waiting times and operational efficiency. The results show that "Demand-Anticipation" leads to the highest waiting time while the "Demand-Supply-Balancing" leads to the most favorable results in waiting time and operational efficiency. While these studies have simulated the impact of SAVs on urban parking demand, to further enlarge the benefit of SAVs by replacing traditional fuel with electricity, it is necessary to take the vehicle range and charging behavior of these service providers into consideration to reveal the relationship between charging related parameters (e.g., vehicle range and charging rate) and traditional metrics (e.g., parking demand, VMT and waiting time). Also, even though the mentioned research has confirmed that the SAVs would increase the total VMT due to empty travel, limited research has divided the total VMT into different parts to find out which part counts the most. This should not be overlooked since a clear understanding of this issue would be of great significance for operators to increase the overall performance of the SAV fleet by reducing empty travel.

Combining the upward trend of vehicle electrification and the promise of automation comes SAEVs [26,27]. A regional, discrete-time, agent-based model was proposed to explore the management of a fleet of SAEVs and the results indicated that SAEVs can serve nearly all requests with an average response time between 7 and 10 min [2]. A novel agent-based simulation framework was developed for electric vehicles by researchers considering the queuing issue in the fast charging stations [28]. By simulating an adaptive strategy based on implicit communication through booking systems in the charging station, the experiment results verified the effectiveness of the proposed approach in terms of route planning and reduction in total travel times within the whole system. Using MATSim, a multi-agent modeling platform, simulated performance characteristics of the SAEV fleet serving travel requests across the Austin, Texas 6-county region. It had been pointed out that reducing charging time and increasing fleet size can lower the response time but improving the vehicle range did not appear to do the same [6]. In the meantime, in order to reveal the impacts of SAEV in regard to socioeconomic heterogeneity, reference [29] modeled this new kind of mobility service with different pricing schemes with the help of MATSim. The results indicated that compared with reducing fares, reducing travel time for customers plays a much more important role in SAEV service usage. They also pointed out that women tend to use SAEVs for shorter trips in regard to gender. As for the environmental effects of this new travel mode, reference [30] extends a multimodal transport model to simulate an increase in the market share of EVs to reveal their impacts from the environmental perspective. Their work based on MATSim pointed out that even the reduction in emissions could be limited if only short trips were served by EV. The impact can be higher if the government is able to target users with longer trips, at the cost of an optimized deployment charging stations for the sustainable development of such mobility service. After that, by taking economic metrics into consideration, the previous model was modified and pointed out that SAEV with longer vehicle range and charging station with fast charging rate can not only provide the best service for travelers but also the most profitable choice for companies providing mobility services using SAEV [7]. An agent-based simulation of SAEVs was performed across the Rouen Normandie metropolitan area in France to explore the impacts of different charging types and vehicle battery capacities on service efficiency. The results indicated that a faster charging rate results in higher performance, which means shorter response time. They also reveal the importance of choosing the right battery capacity to avoid the overlaps between demand and charging peak times [8]. By coupling charging and repositioning events and verifying the rightness of this kind of synergy using an agent-based model, the response time was 39% lower after coupling these two events [9]. Focusing on the environmental impact of SAEV and the First-Mile-Last-Mile system, research results showed that improving vehicle range can provide a better service [1]. The charging dispatching problem is also important for the mobility service provided by SAEVs, reference [31] tries to address this issue through their proposed traffic simulation framework based on a simulator Simulation Urban Mobility (SUMO) to improve the efficiency of charging station usage and save time for SAEVs users. Various Deep Reinforcement Learning (DPL) algorithms were performed to verify the robustness of their developed framework. Reference [12] evaluated the impact of charging infrastructure on SAEVs' service performance using an open-source agent-based simulation platform MATSim. The charging station within the simulation was generated to minimize the distance between the demand point and the charging station. The simulation results indicated that the combination of faster charging and such charging station planning would perform better than other scenarios but the battery swapping station would be better. Combining agent-based simulation and hybrid algorithm, reference [32] first allocated the charging demand based on the simulation output and then, used the proposed algorithm to site and size charging stations to meet the charging demand. However, even though most of the aforementioned research has revealed the impact of SAEV on response time, the parking demand, which is another important parameter to consider when it comes to urban transportation management, is still omitted. Additionally, when it comes to the charging policy, existing studies paid little attention to the relationship between the requests and the availability of charging stations, which should not be overlooked if the service providers and policymakers hope to increase the fleet performance to meet the high travel demand in peak time.

3. Materials and Methodology

3.1. Data Description

The simulation experiments designed in this study are based on several real datasets, including the parking lot dataset and charging station dataset provided by the Shenzhen government (https://opendata.sz.gov.cn, accessed on 25 August 2023) and a dataset containing trajectory information of the taxis running within the city for one day (https://people.cs.rutgers.edu/~dz220/data.html, accessed on 25 August 2023). Due to the nature of these former two datasets, the parking lots and charging stations in this research are assumed to be separated. Therefore, the situation of parking lots with charging functions is not considered here. The research area was limited to a circle located at the center of Luohu District, with a two miles diameter not only to reveal the effect of the introduction of SAEVs on the city's critical area but also because of the fact that the selected area would generate much more travel requests and the research of this kind of area is necessary. The pre-processing process of these datasets and the processed data are elaborated as follows.

3.1.1. Travel Data

The following Table 1 shows an example of the unprocessed taxi trajectory dataset.

Table 1. Overview of the Original Taxi Trajectory Dataset.

Taxi ID	Time	Longitude	Latitude	Occupancy Status	Speed
34745	20:27:43	113.8068	22.62325	1	27
34745	20:24:07	113.8099	22.6274	0	0
...	...	...	...	...	...
28265	21:35:13	114.3215	22.7095	0	18
28265	9:08:02	114.3227	22.6817	0	0

Note: For Occupancy Statue, 1-with passengers and 0-without passengers. And the unit for Speed is mile/h.

The dataset includes a total of 1,155,654 pieces of records. According to Table 1, the dataset does not give the specific origin and destination of these trips, instead, the GPS location information and the corresponding instantaneous speed of the vehicle at certain points are given. In order to transform this dataset into a usable travel dataset, TransBigData, an open-sourced toolkit (https://transbigdata.readthedocs.io/en/latest/index.html, accessed on 25 August 2023) is used to exclude the anomalous data and to obtain the specific origin and destination of these trips. Fours steps including

1. Filtering to exclude data outside the research area;
2. Excluding data with instantaneous changes in Occupancy Status;
3. Rasterizing the GPS data and counting the amount of data in each raster;
4. Extracting the origin and destination points from the GPS data are taken before 13,380 pieces of OD (Original-Destination) data are shown in Table 2.

Table 2. Overview of the OD dataset after TransBigData.

Trip ID	Start Time	Start LON.	Star LAT.	End Time	End LON.	End LAT.
0	0:19:41	114.013016	22.664818	0:23:01	114.0214	22.663918
1	0:41:51	114.021767	22.6402	0:43:44	114.02607	22.640266
...	...	...	...	...	...	...
13378	23:03:45	114.118484	22.547867	23:20:09	114.133286	22.61775
13379	23:36:19	114.112968	22.549601	23:43:12	114.089485	22.538918

Note: Start and End refer to origin and destination, respectively, while LON. and LAT. refer to longitude and latitude, respectively.

3.1.2. Parking Lots Data

The original parking lot dataset contains 716 off-street parking lots in Luohu District, Shenzhen, with a total of 111,845 parking spaces. In the process of spatial visualization of these parking lots, it is found that not all the parking lots in the dataset are located within the Luohu District, therefore, the parking lots outside of Luohu District are excluded by using the tbd.clean_outofshape function in the TransBigData toolkit. After manually removing the parking lots with the help of a visualization map, a total of 308 parking lots with 3557 parking spaces within the study area were finally selected as shown in the following table (Table 3).

Table 3. Overview of the parking lot dataset after TransBigData and manual modification.

ID	Price (RMB) *	Longitude	Latitude	Capacity
1	5	114.1408886	22.5565399	8
2	5	114.1378183	22.5588927	6
...	...	...	...	...
307	5	114.112597	22.5791201	25
308	15	114.1068256	22.5834787	20

* Note: The exchange rate between RMB and USD on the day of data collection (2023.08.25) is 0.1385 which indicated that 1 RMB equal to 0.1385 USD.

The following figure (Figure 1) shows the spatial distribution of parking lots according to their parking capacity and parking price.

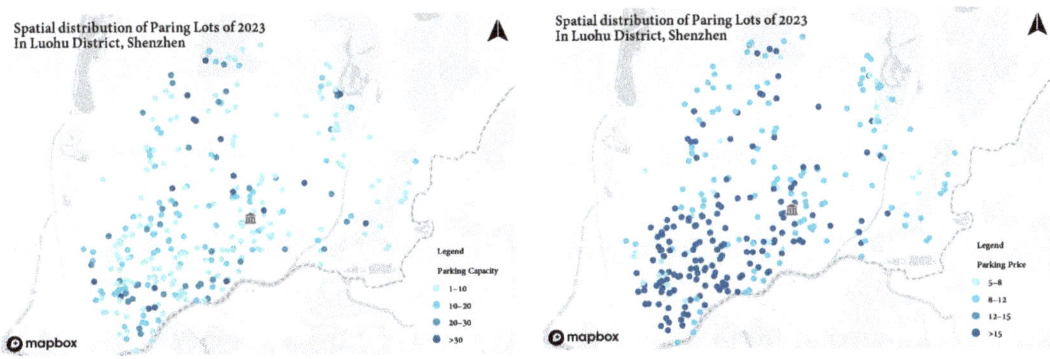

Figure 1. Parking capacity structure (**Left**) and parking price structure (**Right**) in Luohu District, Shenzhen.

3.1.3. Charging Stations Data

The initial charging station dataset includes 904 charging stations, providing 11,034 charging spaces. However, it only contains the address information of the stations, and there is no GPS location, which is not friendly to the subsequent experiments. Therefore, by using a Python script and the open-source Gaode API, these addresses are converted into points with GPS location information. After the tbd.clean_outofshape step, the number of charging stations within the study area was ultimately determined to be 267 with 2075 charging spaces, as shown in Table 4, and the spatial distribution of these stations according to their charging capacity is shown in Figure 2.

Table 4. Overview of the charging station dataset after TransBigData and manual modification.

ID	Property	Longitude	Latitude	Capacity
1	public	114.117447	22.545063	2
2	public	114.118273	22.543836	2
...	...	...	...	...
266	public	114.181697	22.559031	10
267	public	114.183742	22.612863	4

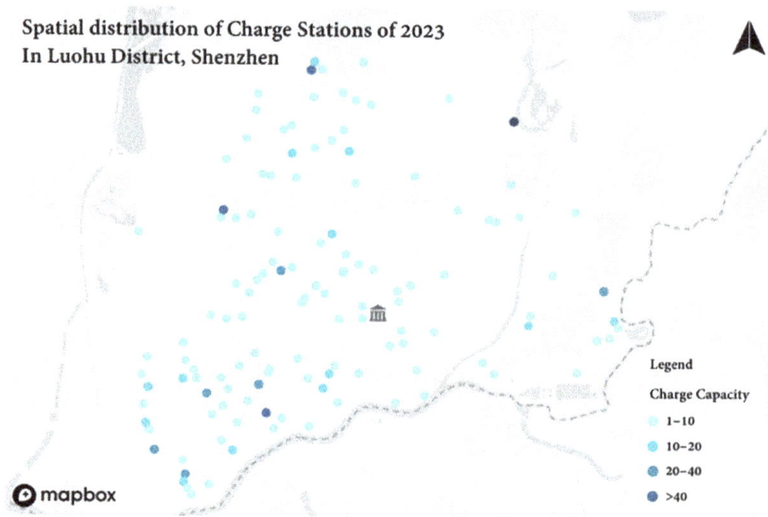

Figure 2. Charging capacity structure in Luohu District, Shenzhen.

3.2. *Multi-Agent-Based Model Specification for SAEVs' Charging and Parking*

The proposed multi-agent simulation model, based on Anylogic, contained four kinds of agents: Traveler, SAEV, Parking Lot, and Charging station. The whole system can be summarized in Figure 3, the description of all states defined for these four kinds of agents can be found in Table 5, and the transitions between these states are shown in Figure 4.

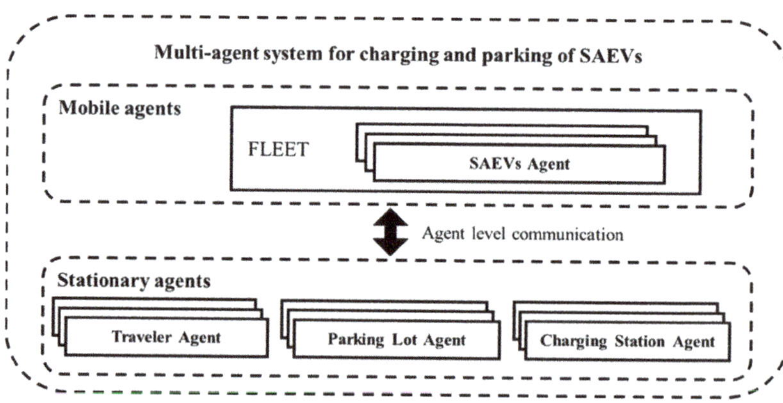

Figure 3. Overview of the proposed multi-agent simulation model.

Table 5. State description of the multi-agent simulation model.

Agent	State	Description
Traveler	WAIT TO DEPART	Initial state of a Traveler agent waits for departure.
	WAIT FOR SERVICE	The Traveler's request has been successfully assigned to a SAEV and it is waiting for the arrival of this specific SAEV.
	WAIT FOR MATCH	The Traveler's request cannot be assigned to a SAEV and it has to be waited until there is an available one.
	TO DESTINATION BY SAEV	The Traveler is now taken to its destination by a SAEV.
SAEV	IN THE INITIAL SPOT	Initial state of an SAEV agent where it waits for requests.
	DRIVING TO PICK-UP	The state that the SAEV is driving to the pick-up point of the matched request.
	AT PICK-UP POINT	The state that the SAEV has arrived at the pick-up point of the matched request.
	DRIVING TO DROP-OFF	The state that the SAEV is driving to the drop-off point of the matched request.
	AT DROP-OFF POINT	The state that the SAEV has arrived at the drop-off point of the matched request and checking the battery status.
	AVAILABLE TO SERVE	The state indicates that battery level is sufficient for the upcoming request.
	DRIVING TO TARGET PARKING LOT	The state indicates that the SAEV is driving to the target parking lot.
	AT PARKING LOT	The state indicates that the SAEV has arrived at its target parking lot.
	DRIVING TO CHARGE STATION	The state indicates that the SAEV is driving to a charging station.
	AT CHARGE STATION	The state indicates that the SAEV has arrived at the charging station.
Parking Lot	PARKING AVAILABLE	The state indicates that there is an available parking space in the parking lot.
	FULLY OCCUPIED	The state indicates that there is no available parking space in the parking lot.
Charging Station	CHARGING AVAILABLE	The state indicates that there is an available charging space in the charging station.
	FULLY OCCUPIED	The state indicates that there is no available charging space in the charging station.

3.2.1. Traveler Agent

For each *Traveler*, its initial state at the beginning of the simulation is called "WAIT TO DEPART". As the simulation proceeds, the *Traveler* generates a travel request based on the OD dataset, and the generated request will be instantly assigned to the nearest available SAEV. After the request has been successfully assigned, the Traveler changes its state to "WAIT FOR SERVICE" before the matched vehicle arrives. After the SAEV arrives and confirms the destination of this specific request, the *Traveler*, again, changes its state to "TO DESTINATION BY SAEV". If the request cannot be instantly assigned, in other words, all SAEVs are busy with requests that have been assigned to them before, the *Traveler* would update its state to "WAIT FOR MATCH" and the request will be stored into a collection called "Request list", which will automatically remove these requests once there is an available SAEV according to the First-In-First-Out principle. The time spent by the *Traveler* waiting for the arrival of a SAEV was called "Response time", which is an important indicator to evaluate the efficiency of the current transportation system. After the *Traveler* arrives at its destination, the SAEV completes its service and the travel request is satisfied.

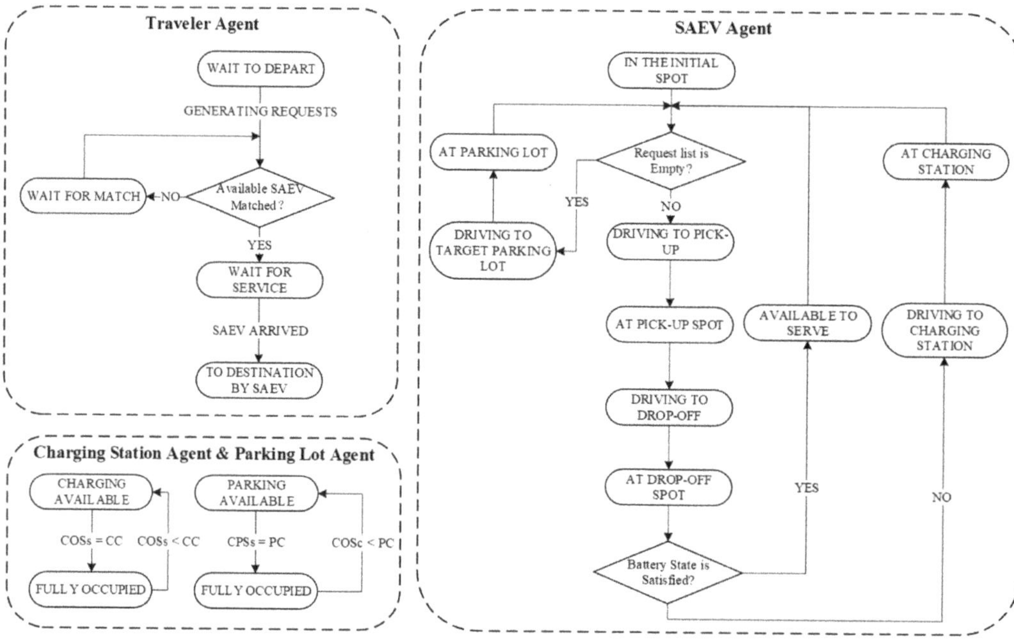

Figure 4. Individual agent actions in the multi-agent simulation model.

3.2.2. SAEV Agent

There are three phases for the *SAEV* agent, servicing phase, parking phase, and charging phase. At the beginning of the simulation, the *SAEV* are placed in the initial spot, a spot located at the center of the research area to indicate that all SAEVs start their daily work from here, and their state starts at "IN THE INITIAL SPOT".

The servicing phase begins after receiving the message "CHECK REQUEST QUEUE" from the system, the *SAEV* starts to check the collection called "Request list" which is used to store requests that need to be served. If the collection is empty, the parking phase begins, in order to reduce the cost of parking behavior and benefit the SAEV service provider economically, thanks to their self-driving ability, the *SAEV* would drive to PL_{target} which can be chosen using the following Formulas (1) and (2):

$$PL = \begin{cases} PL_{avail} & if\, CPSs < PC \\ PL_{full} & if\, CPSs = PC \end{cases} \tag{1}$$

$$PL_{target} = \min_{cost}\{PL_{avail}\} \tag{2}$$

where,

PL denotes the collection formed by all parking lots within the research area;

PL_{avail}, PL_{full} indicate two collections that contain available parking lots and fully occupied ones, respectively;

$CPSs$ refers to current parked spaces, PC refers to parking capacity;

PL_{target} denote the specific parking lot the *SAEV* would drive to;

C_{total} indicates the cost of choosing these parking lots as a target one at that specific simulation time, which can be calculated by Equation (3) where C_{pp} indicates the price of the parking lot, C_{ef} refers to the energy fee caused due to the travel to the parking lot [33] while C_{rt} represents the road toll charged driving there. l indicates the distance from the current location of the *SAEV* to the target parking lot, δ denotes the unit price of energy consumed by the SAEVs, in this case, the value is pre-set as 10, t represents the time spent to the parking lot and ε was the unit road toll with a pre-set value of 7:

$$C_{total} = C_{pp} + \underbrace{14.58 \times l \times \delta \times (l/t)^{-0.68}}_{C_{ef}} + \underbrace{\varepsilon \times l}_{C_{rt}} \tag{3}$$

Once the $SAEV$ starts to drive to PL_{target}, its state would switch to "DRIVING TO TARGET PARKING LOT", at the same time, the value of CPSs in PL_{target} would be increased by 1. At the time the $SAEV$ arrives at PL_{target}, its state updates to "AT PARKING LOT", the system will count this specific moment as the start of a parking demand and record the time when the $SAEV$ leaves the parking lot to accumulate the parking demand. But if the "Request list" is not empty, which means that there are requests that need to be served, then the servicing phase continues and the nearest available SAEVs (the $SAEV$ that has not been matched to a request or the $SAEV$ in the state called "AVAILABLE TO SERVE" to each of these requests will be assigned to them and update their state to "DRIVING TO PICK-UP". Once the $SAEV$ arrives at the specific pick-up point, its state update to "AT PICK-UP POINT", this is the state that confirms the destination of this specific request. "DRIVING TO DROP-OFF" is the state after the confirmation of the trip destination and by the time it arrived at the drop-off point, its state changes to "AT DROP-OFF POINT". At the drop-off point, the $SAEV$ will receive a message called "CHECK BATTERY STATE" from the system, and it will check its battery level, if the remaining power can cover the pre-set minimal range (in this study, 20%), its state will transfer to "AVAILABLE TO SERVE" to get ready for the next upcoming service and start checking the "Request list" again. But if the remaining power is lower than the minimal range, the charging phase begins. The $SAEV$ will drive to CS_{target} for charging and switch its state to "DRIVING TO CHARGING STATION". The CS_{target} can be chose using Formulas (4) and (5):

$$CS = \begin{cases} CS_{avail} & if\ COSs < CC \\ CS_{full} & if\ COSs = CC \end{cases} \tag{4}$$

$$CS_{target} = \begin{cases} \min\limits_{occ}\{CS_{avail}\} & if\ Requetlist\ is\ Empty \\ \min\limits_{dis}\{CS_{avail}\} & if\ Requestlist\ is\ not\ Empty \end{cases} \tag{5}$$

where,

CS denotes the collection formed by all charging stations within the research area;

CS_{avail}, CS_{full} indicate two collections that contain available charging stations and fully occupied ones, respectively;

$COSs$ refers to current occupied spaces, CC refers to charging capacity;

CS_{target} denotes the specific charging station the $SAEV$ would drive to; and,

occ and dis indicate the current occupied spaces of the charging station and the distance from the position of $SAEV$ to the charging station, respectively.

The above charging policy considers the relationship between the current requests and the availability of charging stations. In order to reduce the response time of traveler, when there is a request waiting to be served, the $SAEV$ would drive to the closest charging station to reduce the time spent on driving. However, if the request list is empty, the $SAEV$ would choose the charging station with the least occupied spaces to increase the possibility of an available charger when it arrives and increase the utilization rate of existing charging facilities.

Accordingly, the $COSs$ of CS_{target} would be increased by 1. When the $SAEV$ arrives and starts charging, its state changes to "AT CHARGING STATION" and similar to before, the system counts this specific moment as the start of a charging demand. The charging rate, *CR*, and the expected charged level, *ECL*, can be decided by Formulas (6) and (7), respectively:

$$CR = \begin{cases} GC & if\ \text{"Request list" is empty} \\ FC & if\ \text{"Request list" is not empty} \end{cases} \quad (6)$$

$$ECL = \begin{cases} TotalRange & if\ \text{"Request list" is empty} \\ AvailRange & if\ \text{"Request list" is not empty} \end{cases} \quad (7)$$

where,

GC, *FC* refer to the general charging rate and fast charging rate, respectively;

TotalRange indicates the *SAEV* would not leave the charging station until it is fully charged (100%) and, *AvailRange* indicates the *SAEV* would leave the charging station once it is charged to the pre-set value, which is 80% in this research.

As soon as the *SAEV* drives out of CS_{target}, its *COSs* would accordingly decrease by 1 and the system accumulates the value of this finished charging demand. The recharged *SAEV* will check the "Request list" after leaving CS_{target} and drive to serve if there are any requests waiting, otherwise, they make a parking choice again. The final state of *SAEV* is "AT PARKING LOT", indicating all requests have been served and all SAEVs are within satisfied battery level.

3.2.3. Charging Station Agent and Parking Lot Agent

There are only two states in each *Charging station* agent, called "CHARGE AVAILABLE" and "FULLY OCCUPIED", respectively. The transactions between these two states are decided by one parameter called charge capacity (CC) and one variable called COSs. When the value of COSs is equal to CC, which means that all charging spaces of the charging station are occupied, and the state changes to "FULLY OCCUPIED", when there is a vehicle leaving the charging station, the state changes to "FULLY OCCUPIED". Similar to the *Parking Lot* agent, the two states are called "PARKING AVAILABLE" and "FULLY OCCUPIED", as shown in the figure above. When the value of CPSs is less than PC, the parking lot agent updates its status to "PARKING AVAILABLE". When the value of CPSs is equal to PC, the state changes to "FULLY OCCUPIED". Apart from these parameters and variables that affect state changes, there are two variables for both these two kinds of agents called "charging time" and "parking time", respectively, to reflect the charging demand and parking demand at each charging station and parking lot.

3.3. *Simulated Scenarios and Experiment Setting*

Scenarios with different SAEVs' fleet sizes which in this case is equal to the ratio of the number of SAEVs to the number of total requests, vehicle range which refers to the driving capacity of a SAEV with 100% battery level, and charging rate are designed to evaluate their impact on urban parking demand, charging demand, VMT, and response time. Since SAEVs can provide car-sharing services, unlike the private conventional vehicle, there should be one vehicle for one request to serve the travel demand immediately, even a small penetration rate of SAEVs can serve all requests within a reasonable response time [1,7]. After several warm-up simulations, a 5% penetration rate for 13,380 trips (which has only 699 SAEVs) was chosen as the base case as the average response time can be maintained to a relatively short time between 5 to 8 min, and the value will constantly increase to 10% (1338 SAEVs) in 1% increments. Additionally, as the research area is not quite large, which only contains the traffic entities within 2 miles of the district center, we minimized the vehicle range accordingly. It was assumed that the vehicle range increased from 120 miles to 200 miles, in 20 mile increments. In order to simplify the calculation of SAEVs' charging time, mile/h is used as the unit of charging rate, which indicates the distance a SAEV can drive after pre-unit of time charging. The general charging rate varies from 30 mile/h through 50 mile/h, in 5 mile/h increments, while the fast charging rate varies from 60 mile/h through 100 mile/h, in 10 mile/h increments [6]. Table 6 provides a summary of all pre-set parameters. Additionally, in order to verify the effectiveness of the proposed charging policy in reducing the response time and decreasing the total VMT while increasing the utilization of the existing parking and charging facilities, comparison experiments were conducted where the SAEVs would choose the target charging station simply based on the distance.

Table 6. Summary of all simulated scenarios parameters.

Vehicle Range (mile)	120	140	**160**	180	200	NA
Charging rate (GC/FC) (mile/h)	30/60	35/70	**40/80**	45/90	50/100	NA
Fleet size (%)	**5**	6	7	8	9	10

Note: GC refers to general-charging, FC refers to fast-charging. Base case parameters are shown in bold.

Apart from the above parameters set, we assumed that:

- All SAEVs were fully charged at the beginning of the simulation;
- If the remaining power of a SAEV is not enough to cover the distance to the nearest charging station, it will transfer to the target charging station within 0.01 s as soon as it runs out of power;
- Due to the size of the research area and the fact that even in a scenario with the largest fleet size, there would be only 1338 SAEVs within the network, which is a relatively small number of vehicles, so the effect of road congestion on SAEVs' speed is not considered throughout the whole process, in other words, the SAEVs maintain a fixed speed throughout the whole simulation.

The output indicators of the simulation experiment include the parking demand, charging demand, total VMT, and average response time. The former two are counted in hours and would be accumulated through the whole simulation period. The spatial distribution of these two kinds of demand will be demonstrated on the map. At the same time, the total VMT will be divided into different parts to find out which part accounts for the most by counting the exact time SAEV arrived at different origins and destinations. The total VMT can be calculated by Formula (8):

$$VMT_{TOTAL} = VMT\_I + VMT\_E + VMT\_P + VMT\_CS \quad (8)$$

where,

VMT_{TOTAL} indicates the total VMT generated by a SAEV during the whole simulation period;

VMT_I, VMT_E, VMT_P, and VMT_CS indicate the VMT generated by a SAEV starting from the initial point, start point, end point, parking lot, and charging station respectively. They can be calculated by Formulas (9)–(12):

$$VMT\_I = VMT\_IS + VMT\_IP \quad (9)$$

$$VMT\_E = VMT\_ENS + VMT\_EP + VMT\_ECS \quad (10)$$

$$VMT\_P = VMT\_PS \quad (11)$$

$$VMT\_CS = VMT\_CSS + VMT\_CSP \quad (12)$$

where,

VMT_IS and VMT_IP denote the VMT generated by a SAEV starting from the initial point to one start point or to one parking lot, the latter one can only take place at the very beginning of the simulation as the number of requests then is less than the number of idle SAEVs;

VMT_ENS, *VMT_EP* and *VMT_ECS* indicate the VMT generated by a SAEV starting from one endpoint to the next start point, the parking lot and the charging station, respectively;

VMT_PS indicates the VMT generated by a SAEV starting from one parking lot to one start point;

VMT_CSS and *VMT_CSP* indicates the VMT generated by a SAEV starting from one charging station to one start point and to one parking lot, respectively.

The average response time can be calculated by Formula (13):

$$T = \frac{\sum\limits_{n=1}^{m} t_2^n - t_1^n}{m} \tag{13}$$

where,

t_1^n indicates the time the nth request was generated, t_2^n indicates the time the matched SAEV (for request n) arrived, and m indicates the number of the total travel demand within the whole system, in this case, the value of m is 13,380.

The simulation will stop when all requests have been served and all the SAEVs have been parked at the parking lots.

4. Results and Discussion

4.1. Parking Demand and Charging Demand

The results of parking demand and charging demand in each scenario are shown in Figures 5 and 6 separately.

(a) Different fleet size

(b) Different charging rate

(c) Different vehicle range

Figure 5. Overview of parking demand in various simulation scenarios.

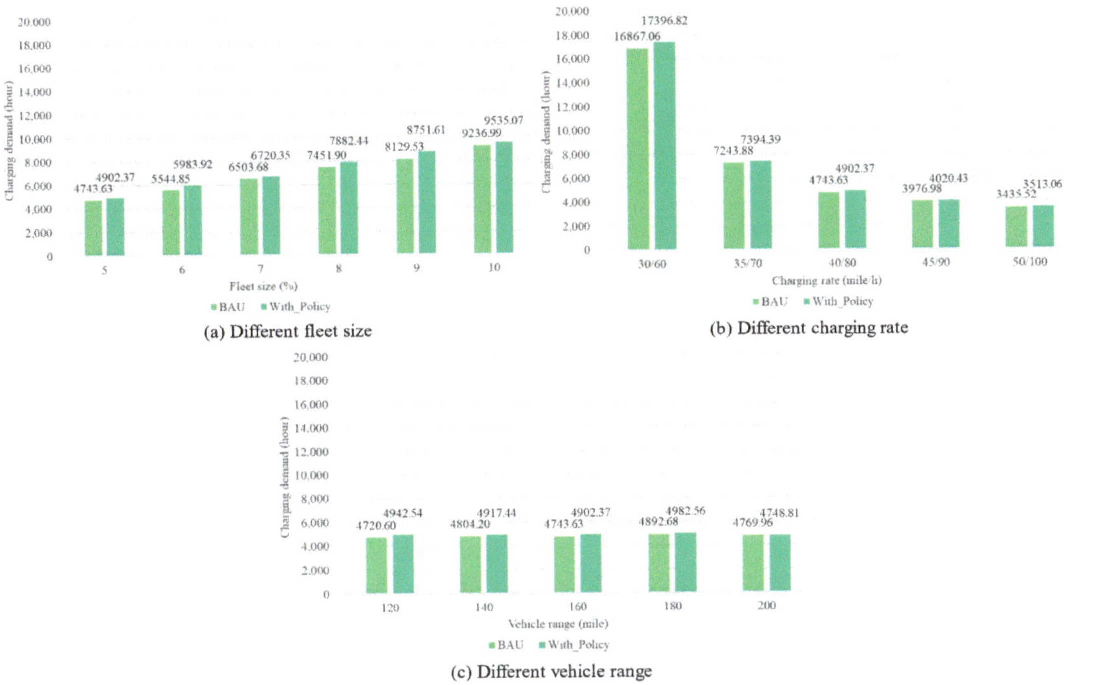

Figure 6. Overview of charging demand in various simulation scenarios.

According to Figures 5 and 6, as the fleet size continues to rise, the parking demand increases by nearly two times, from 547.47 h when the fleet size is only 5% (which is 669 SAEVs) to 1481.33 h with the maximized fleet size. The charging demand also saw an increase when the fleet size kept rising and the value doubled (from 4743 h to 9236 h). This is not unexpected as a larger fleet size means more SAEVs within the system are ready to serve and the travel request can be satisfied quicker, which increases the possibility of SAEVs becoming idle and choosing to park, and more SAEVs would inevitably increase the charging demand as the total number of SAEVs within the system that may need recharging has been increased, even though they may not drive to serve after charging. It is obvious that the charging demand in all scenarios (varying from 3436 h to 17,397 h) is much higher than the parking demand (376~2423 h). This is an expected result because compared with parking lots, charging stations can not only offer a place to park but also a place to recharge vehicles' batteries. Charging is a time-consuming process, which may also help these idle SAEVs that cannot find any requests at off-peak hours without doing nothing, in this case, recharging their batteries if necessary. For the car-sharing SAEVs, they can reduce their parking time by constantly serving different requests if there are any, for those that need recharging, the duration at the charging station can also be seen as the time waiting for the new requests, so it can be said that charging demand should be the one need to be concerned more compared with its parking counterpart after taking charging behavior and vehicle range into consideration.

Unlike the parking demand, which seems not to be effectively affected by the change in the charging rate, it is expected to see that the charging demand experienced a decrease as the charging rate increased (from 16,867 h to 3436 h). This is reasonable since a faster charging rate means less time spent in a charging station, which will shorten the time for SAEVs spent on charging and therefore increase the charging station's turnover efficiency and be seen as an effective way to reduce the time for SAEV to find an available charger. However, it is not possible to unlimitedly increase the charging rate since the reduction become much smaller and may not be an economical way to do so, even though this study did not place any limitations on this value, there should be some due to safety and power resource allocation considerations.

Increasing the vehicle range causes an increase in parking demand, as the vehicle range increases from 120 miles to 200 miles, the parking demand increases from 376 h to 698 h, while the charging demand keeps the value constant at 4800 h. The increase in parking demand may be because a larger vehicle range means a wider service range, a SAEV can serve more travel needs without being recharged, which makes some other SAEVs idle for some time, and consequently need to choose a parking lot. The constant charging demand may be because the total miles traveled to serve these requests are firm, so the change in vehicle range may only affect which SAEV is chosen to serve the traveler, which would not pose a threat to the total generated charging demand.

The proposed charging policy would increase the parking demand and charging demand in all scenarios, while the increase for charging is less noticeable (maximal value is 8%) than parking (from 49% to 70%). This may come from taking charging availability into consideration when choosing a charging station, all SAEVs within the system have a higher level of power, the possibility of charging a SAEV is not so common compared with the need to park as there may not be queued request and none of these idle SAEV were need to be charged and would be ready to serve request whenever there are any.

4.2. VMT

The VMT generated by empty driving due to SAEVs' self-driving ability has been considered a great waste of road resources and energy (electricity for SAEVs). So, it is necessary to clarify the origin of VMT and control it accordingly.

As shown in Figure 7, the total VMT rises as the fleet of SAEVs increases, but the increment is not significant which indicates that although more SAEVs on the road network will lead to more VMT, the impact is not significant. Also, as the vehicle range increases, the total VMT within the system increases significantly. The reason for this increase is the increase in the *VMT_ENS* part, possibly because increasing the vehicle range will inevitably expand the SAEVs' service range and increase the possibility of long-distance migration within its service range for request servicing. On the contrary, as the charging rate increases, the total VMT decreases significantly, mainly due to a significant decrease in the *VMT_ENS* part. This noticeable reduction may be because a fast charging rate brings less charging time, and the former requests served by relocated SAEVs which are far away from the start points can now be assigned to closer SAEVs which missed these requests due to the longer charging time they previously spent in the charging stations. This kind of situation occurs quite frequently based on the dramatic decrease in *VMT_ENS*.

The major part of the total VMT generated is *VMT_ENS*, while the others only accounted for small portions. The increase in vehicle range leads to higher values in *VMT_PS*, mainly because a wider service range brought by higher vehicle range enables SAEVs to serve more requests without recharging, while these not close enough to requests must drive to parking lots. A possible reason why *VMT_CSP* increases, as the fleet size becomes larger, may be the increase in the number of vehicles capable of providing services within the system reduces the possibility of demand queues, making it more likely for fully recharged SAEVs to drive to park once they leave the charging station. The large amount of *VMT_ENS* (nearly 90%) in all BAU (Business as usual) scenarios indicates that although relocating SAEVs to the next service immediately after they completed the previous one can increase the vehicle utilization rate, it can also increase the VMT within the entire road network. Since the effect of road congestion is not considered in this research, in reality, the more vehicles on the road, the more likely it is to cause congestion. Therefore, it is necessary to have a reasonable scheduling and allocation scheme of vehicles to reduce long-distance migration, thereby reducing excessive empty VMT, energy consumption, and road occupancy.

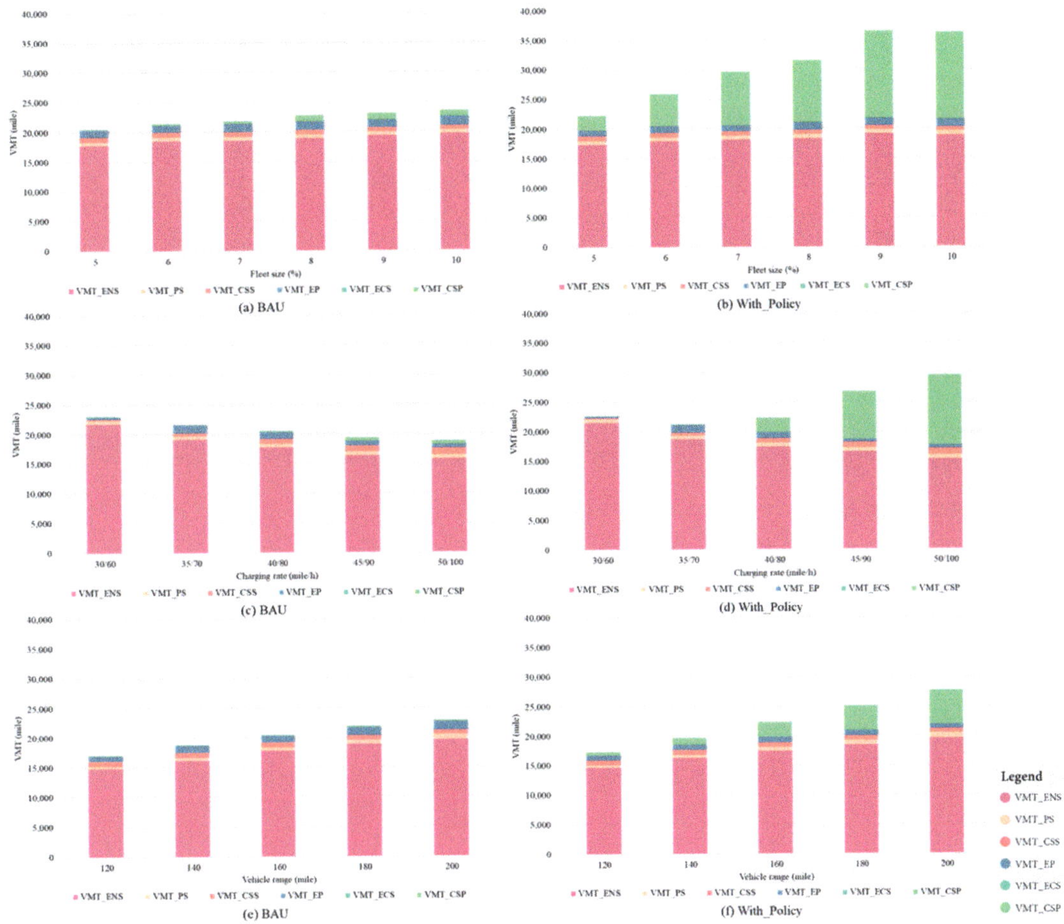

Figure 7. Overview of VMT in various simulation scenarios.

For scenarios with a charging policy, the total VMT is likely to show a rising trend as the fleet size and the vehicle range go up, similarly, even with some fluctuations, the total VMT also increase as the charging rate rises. The value of total VMT is much higher compared with BAU scenarios, mainly coming from the increase in the portion *VMT_CSP* part. This part of VMT also shows a positive relationship with fleet size, charging rate, and vehicle range. This is not unexpected because more SAEVs within the system means that there would be more vehicles ready to serve so during the charging process of some SAEVs, requests might be satisfied by other vehicles and these recharged vehicles may drive to park after finishing charging. A faster charging rate can reduce the time spent in the charging station and therefore enable the recharged SAEVs to return to the system quicker to serve the waited travel requests. Similarly, a larger vehicle range can ensure a longer time before the SAEV requires recharging, even if there were some of them below the expected battery level, others with the same vehicle range but not have been assigned to any requests can be the alternative when travelers need to be served, so after these recharged SAEVs out of the charging station, the only thing they have to do is drive to a parking lot.

4.3. *Average Response Time*

The average response time in different simulation scenarios can be found in Figure 8. As the fleet size increases, the average response time decreases from 7.99 min (5%) to 3.92 min (10%). This is reasonable as more SAEVs in the system lead to fewer orders being in the queue, therefore reducing the response time. However, it is worth noting that the reduction is becoming smaller which indicates that continuously increasing the number of vehicles in the transportation system can be an effective way to control the response time but the positive effect may be a threshold and an optimal value of the fleet size.

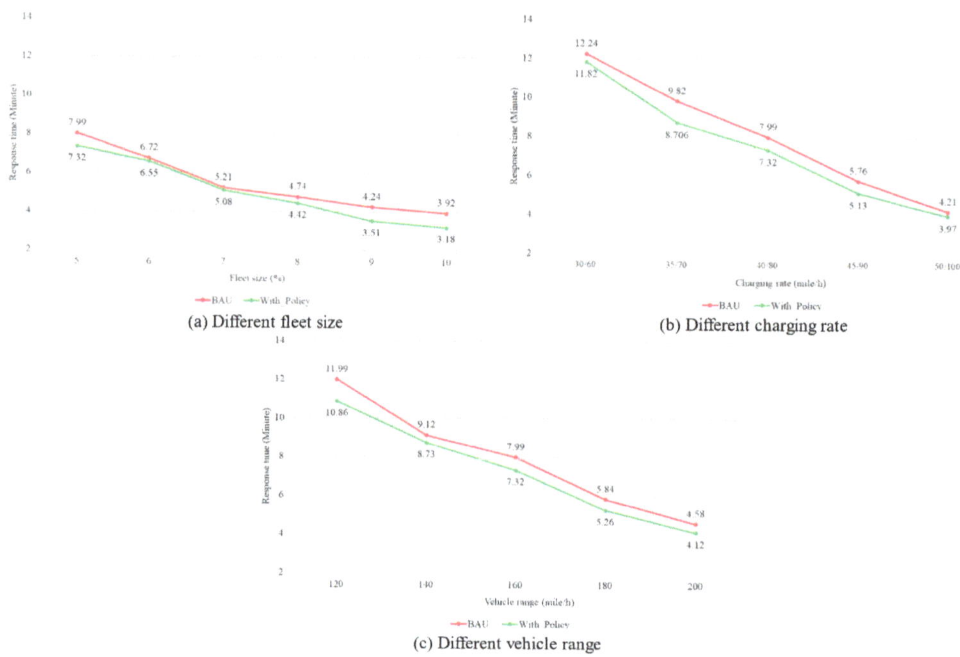

Figure 8. Overview of response time in various simulation scenarios.

As can be seen from the figure, the charging rate has a significant impact on the response time. The average response time at a charging rate of 30/60 miles/h (12.24 min) is almost 2 times longer than a charging rate of 50/100 miles/h (1.34 min). A faster charging rate can reduce the time SAEVs spend at charging stations and enable faster replenishment of available SAEVs into the service system. The change of vehicle range also shows a similar trend to the charging rate, the response time saw its peak value (which is 11.99 min) in a scenario with the smallest vehicle range, in this case, 120 miles, and as the vehicle range continue to rise, the response time reduces to 4.58 min. This can be explained from the point of view that before vehicles consume their entire battery capacity, vehicles with longer ranges can provide service with a shorter response time. After that, even a longer charging time would be required for SAEVs with longer vehicle range, the overall result of the simulation indicates that this would not affect the total response time within the whole simulation process, it is reasonable to say that vehicles with larger vehicle range can provide mobility service with a shorter response time. As it is not feasible to expect all SAEVs to complete all requests without recharging, therefore, to reduce response time, it is recommended to consider a combination of increasing the number of service vehicles, increasing the charging rate, and increasing the vehicle range.

It is also clear to see that in scenarios with the application of the proposed charging policy, the response time has been reduced to varying levels, from 2.5% to 18.9%. As the charging policy considers the current requests and the availability of the charging station, on one hand, during the peak hour, the SAEVs would choose the closest charging station to prioritize the charging demand so that they can finish the charging process as quickly as possible in order to serve the request if there are requests waiting to be satisfied. On the other hand, SAEVs would like to be easily charged as they would choose a charging station with minimal occupied spaces when in off-peak hours, this would reduce the time spent on driving to an available charging station and increase the number of ready-to-serve SAEVs in the traffic system. Both of these two kinds of charging behaviors were designed to fulfill the charging demand as soon as possible, the only different is the former one does this when it has to, and the latter one does this in advance to avoid poor battery state. Therefore, it is reasonable to see a reduction in response time after the implementation of the policy.

5. Conclusions and Future Work

SAEVs can provide car-sharing mobility services, reducing the total number of vehicles within the entire transportation system which will directly reduce the need for parking and reduce the pollution by using environmentally friendly electric power. But the vehicle range and their charging behavior can sometimes become barratries to further benefit the transportation system. This research proposed a multi-agent-based simulation model that considers both vehicle range and charging behavior of SAEVs and reflects their real-time battery level by monitoring their driving mileage. Based on a real dataset from the Luohu District in Shenzhen, various scenarios with different fleet sizes, charging rates, and vehicle ranges are explored to evaluate the impact of SAEVs on parking demand, charging demand, total VMT, and average response time and relationship between these indicators are revealed, and a charging policy consider the current requests and the availability of charging station are proposed and verified. The results indicate that after considering the charging behavior of SAEVs, the charging demand within the whole system is much more than the parking demand as SAEVs only park when their battery level is satisfied and there are no requests waiting to be served. Larger fleet size and longer vehicle range can lead to more parking demand while changing the charging rate would not influence the parking demand. A larger fleet size would result in more charging demand and a faster charging rate, which can dramatically reduce charging demand, while the change in vehicle range would not have an impact on the charging demand. The total VMT will increase as the fleet size becomes larger or the vehicle range increases and will decrease as the charging rate increases. Moreover, a large portion of VMT is generated by SAEVs relocating from the destination of the last service to the origin of the next request. This portion of VMT can be reduced by rational scheduling of SAEVs within the network. The average response time would decrease as the fleet size goes bigger, the charging rate goes faster and the vehicle range goes larger. But the reduction caused by fleet size changes is not as remarkable as the other two, which indicates that unlimitedly increasing the number of SAEVs in the system will not always shorten the response time as we expected. Therefore, to improve the quality of mobility service in terms of response time, it is reasonable to combine the change in fleet size, charging rate, and vehicle range. To verify the effectiveness of the proposed charging policy, the results of the comparison experiments show that when considering the current requests and the availability of charging stations, the response time would be further reduced by 2.5% to 18.9% in different scenarios, the parking demand would increase as there would be more SAEVs to be idle and choose to park, the charging demand would not be influenced much; however, the total VMT would increase because a great number of VMT is generated during the SAEVs driving to the parking lot after recharge.

The main contribution of this paper is threefold. Firstly, a multi-agent-based simulation model considering the vehicle range and charging behavior was proposed to reveal the relationship between fleet size, charging rate, vehicle range, parking demand, charging demand, VMT, and average response time.Secondly, the VMT has been divided into parts according to the different origins and destinations of SAEVs during the whole simulation process to clearly the portion of each part. Thirdly, the proposed charging policy considers the current request and the availability of charging stations was verified to be effective in reducing the response time.

However, there are some limitations of this research that should be addressed in future studies. First, the calculation of battery consumption and the recharging process should be more precise. Second, due to the poor performance of the used computer, the simulation experiment has narrowed down the research area to a specific scope, within a 2 mile radius of the Luohu District center, and the vehicle range has also been reduced accordingly. In future research, expanding the research area and using a more accurate vehicle range should be taken into consideration. Additionally, apart from fixed charging service providers such as charging stations mentioned here, a novel mobile charging approach provided by mobile charging vehicles should as be considered in future studies since the combination of these two can combine the advantages of both to serve SAEVs in dynamic and static ways [34,35], which may be the solution to a much more sustainable mobility system. Finally, this paper did not consider the impact of road congestion on vehicle speed, which should be incorporated into the transportation simulation model in future research.

Author Contributions: Conceptualization, Y.Z. and X.Y. (Xiaofei Ye); methodology, Y.Z.; software, Y.Z.; validation, Y.Z. and X.Y. (Xiaofei Ye); formal analysis, Y.Z.; resources, Y.Z.; data curation, Y.Z.; writing—original draft preparation, Y.Z.; writing—review and editing, Y.Z. and X.Y. (Xiaofei Ye); visualization, Y.Z.; supervision, X.Y. (Xiaofei Ye), X.Y. (Xingchen Yan), T.W., J.C. and P.Z.; funding acquisition, X.Y. (Xiaofei Ye), T.W. and P.Z. All authors have read and agreed to the published version of the manuscript.

Funding: This research was funded by the Fundamental Research Funds for the Provincial Universities of Zhejiang (SJLY2023009), Transportation Technology Plan Project of Ningbo, Zhejiang (202214), the National "111" Centre on Safety and Intelligent Operation of Sea Bridge (D21013), National Natural Science Foundation of China (Nos. 71971059, 52262047, 52302388, 52272334 and 61963011), the Natural Science Foundation of Jiangsu Province, China (no. BK20230853), the Specific Research Project of Guangxi for Research Bases and Talents [grant number AD20159035], in part by Guilin Key R&D Program [grant number 20210214-1], and Liuzhou Key R&D Program [grant number 2022AAA0103].

Data Availability Statement: Data used in this research can be found through these links provided in Section 3.1 Data description.

Acknowledgments: The authors thank their mentors who provided instructions on writing this paper.

Conflicts of Interest: The authors declare no conflict of interest.

References

1. Li, L.; Pantelidis, T.; Chow, J.Y.J.; Jabari, S.E. A real-time dispatching strategy for shared automated electric vehicles with performance guarantees. *Transp. Res. Part E Logist. Transp. Rev.* **2021**, *152*, 102392. [CrossRef]
2. Richter, M.A.; Hess, J.; Baur, C.; Stern, R. Exploring the Financial Implications of Operating a Shared Autonomous Electric Vehicle Fleet in Zurich. *J. Urban Mobil.* **2021**, *1*, 100001. [CrossRef]
3. Zhang, W.; Guhathakurta, S.; Fang, J.; Zhang, G. Exploring the impact of shared autonomous vehicles on urban parking demand: An agent-based simulation approach. *Sustain. Cities Soc.* **2015**, *19*, 34–45. [CrossRef]
4. Zhang, W.; Guhathakurta, S. Parking Spaces in the Age of Shared Autonomous Vehicles: How Much Parking Will We Need and Where? *Transp. Res. Rec. J. Transp. Res. Board* **2017**, *2651*, 80–91. [CrossRef]
5. Yan, Q.; Feng, T.; Timmermans, H. Private owners' propensity to engage in shared parking schemes under uncertainty: Comparison of alternate hybrid expected utility-regret-rejoice choice models. *Transp. Lett.* **2023**, *15*, 754–764. [CrossRef]
6. Chen, T.D.; Kockelman, K.M.; Hanna, J.P. Operations of a shared, autonomous, electric vehicle fleet: Implications of vehicle & charging infrastructure decisions. *Transp. Res. Part A Policy Pract.* **2016**, *94*, 243–254.
7. Loeb, B.; Kockelman, K.M.; Liu, J. Shared autonomous electric vehicle (SAEV) operations across the Austin, Texas network with charging infrastructure decisions. *Transp. Res. Part C Emerg. Technol.* **2018**, *89*, 222–233. [CrossRef]
8. Loeb, B.; Kockelman, K.M. Fleet performance and cost evaluation of a shared autonomous electric vehicle (SAEV) fleet: A case study for Austin, Texas. *Transp. Res. Part A Policy Pract.* **2019**, *121*, 374–385. [CrossRef]
9. Li, Y.; Li, X.; Jenn, A. Evaluating the emission benefits of shared autonomous electric vehicle fleets: A case study in California. *Appl. Energy* **2022**, *323*, 119638. [CrossRef]

10. Gonçalves Duarte Santos, G.; Birolini, S.; de Almeida Correia, G.H. A space–time-energy flow-based integer programming model to design and operate a regional shared automated electric vehicle (SAEV) system and corresponding charging network. *Transp. Res. Part C Emerg. Technol.* **2023**, *147*, 103997. [CrossRef]
11. Zhang, T.Z.; Chen, T.D. Smart charging management for shared autonomous electric vehicle fleets: A Puget Sound case study. *Transp. Res. Part D Transp. Environ.* **2020**, *78*, 102184. [CrossRef]
12. Vosooghi, R.; Puchinger, J.; Bischoff, J.; Jankovic, M.; Vouillon, A. Shared autonomous electric vehicle service performance: Assessing the impact of charging infrastructure. *Transp. Res. Part D Transp. Environ.* **2020**, *81*, 102283. [CrossRef]
13. Liao, Z.; Taiebat, M.; Xu, M. Shared autonomous electric vehicle fleets with vehicle-to-grid capability: Economic viability and environmental co-benefits. *Appl. Energy* **2021**, *302*, 117500. [CrossRef]
14. Nemoto, E.H.; Issaoui, R.; Korbee, D.; Jaroudi, I.; Fournier, G. How to measure the impacts of shared automated electric vehicles on urban mobility. *Transp. Res. Part D Transp. Environ.* **2021**, *93*, 102766. [CrossRef]
15. Dean, M.D.; Gurumurthy, K.M.; de Souza, F.; Auld, J.; Kockelman, K.M. Synergies between repositioning and charging strategies for shared autonomous electric vehicle fleets. *Transp. Res. Part D Transp. Environ.* **2022**, *108*, 103314. [CrossRef]
16. Grahn, R.; Qian, S.; Hendrickson, C. Environmental impacts of first-mile-last-mile systems with shared autonomous electric vehicles and ridehailing. *Transp. Res. Part D Transp. Environ.* **2023**, *118*, 103677. [CrossRef]
17. Fagnant, D.J.; Kockelman, K.M. The travel and environmental implications of shared autonomous vehicles, using agent-based model scenarios. *Transp. Res. Part C Emerg. Technol.* **2014**, *40*, 1–13. [CrossRef]
18. Harper, C.D.; Hendrickson, C.T.; Samaras, C. Exploring the Economic, Environmental, and Travel Implications of Changes in Parking Choices due to Driverless Vehicles: An Agent-Based Simulation Approach. *J. Urban Plan. Dev.* **2018**, *144*, 04018043. [CrossRef]
19. Zhong, H.; Li, W.; Burris, M.W.; Talebpour, A.; Sinha, K.C. Will autonomous vehicles change auto commuters' value of travel time? *Transp. Res. Part D Transp. Environ.* **2020**, *83*, 102303. [CrossRef]
20. Stinson, M.; Zou, B.; Briones, D.; Manjarrez, A.; Mohammadian, A. Vehicle ownership models for a sharing economy with autonomous vehicle considerations. *Transp. Lett.* **2023**, *15*, 1–17. [CrossRef]
21. Zhang, W.; Wang, K. Parking futures: Shared automated vehicles and parking demand reduction trajectories in Atlanta. *Land Use Policy* **2020**, *91*, 103963. [CrossRef]
22. Millard-Ball, A. The autonomous vehicle parking problem. *Transp. Policy* **2019**, *75*, 99–108. [CrossRef]
23. Kumakoshi, Y.; Hanabusa, H.; Oguchi, T. Impacts of shared autonomous vehicles: Tradeoff between parking demand reduction and congestion increase. *Transp. Res. Interdiscip. Perspect.* **2021**, *12*, 100482. [CrossRef]
24. Liu, Z.; Li, R.; Dai, J. Effects and feasibility of shared mobility with shared autonomous vehicles: An investigation based on data-driven modeling approach. *Transp. Res. Part A Policy Pract.* **2022**, *156*, 206–226. [CrossRef]
25. Winter, K.; Cats, O.; Martens, K.; van Arem, B. Relocating shared automated vehicles under parking constraints: Assessing the impact of different strategies for on-street parking. *Transportation* **2020**, *48*, 1931–1965. [CrossRef]
26. Kim, S.; Lee, U.; Lee, I.; Kang, N. Idle vehicle relocation strategy through deep learning for shared autonomous electric vehicle system optimization. *J. Clean. Prod.* **2022**, *333*, 130055. [CrossRef]
27. Ma, B.; Hu, D.; Wang, Y.; Sun, Q.; He, L.; Chen, X. Time-dependent Vehicle Routing Problem with Departure Time and Speed Optimization for Shared Autonomous Electric Vehicle Service. *Appl. Math. Model.* **2023**, *113*, 333–357. [CrossRef]
28. García-Magariño, I.; Palacios-Navarro, G.; Lacuesta, R.; Lloret, J. ABSCEV: An agent-based simulation framework about smart transportation for reducing waiting times in charging electric vehicles. *Comput. Netw.* **2018**, *138*, 119–135. [CrossRef]
29. Müller, J.; Straub, M.; Naqvi, A.; Richter, G.; Peer, S.; Rudloff, C. MATSim Model Vienna: Analyzing the Socioeconomic Impacts for Different Fleet Sizes and Pricing Schemes of Shared Autonomous Electric Vehicles. In Proceedings of the Transportation Research Board 100th Annual Meeting 2021, Washington, DC, USA, 5–29 January 2021.
30. De Wolf, D.; Diop, N.; Kilani, M. Environmental impacts of enlarging the market share of electric vehicles. *Environ. Econ. Policy Stud.* **2022**. [CrossRef]
31. Song, Y.; Zhao, H.; Luo, R.; Huang, L.; Zhang, Y.; Su, R. A sumo framework for deep reinforcement learning experiments solving electric vehicle charging dispatching problem. *arXiv* **2022**, arXiv:2209.02921.
32. Zhang, H.; Sheppard, C.J.R.; Lipman, T.E.; Zeng, T.; Moura, S.J. Charging infrastructure demands of shared-use autonomous electric vehicles in urban areas. *Transp. Res. Part D Transp. Environ.* **2020**, *78*, 102210. [CrossRef]
33. Gardner, L.M.; Duell, M.; Waller, S.T. A framework for evaluating the role of electric vehicles in transportation network infrastructure under travel demand variability. *Transp. Res. Part A Policy Pract.* **2013**, *49*, 76–90. [CrossRef]
34. Cui, S.; Yao, B.; Chen, G.; Zhu, C.; Yu, B. The multi-mode mobile charging service based on electric vehicle spatiotemporal distribution. *Energy* **2020**, *198*, 117302. [CrossRef]
35. Cui, S.; Ma, X.; Zhang, M.; Yu, B.; Yao, B. The parallel mobile charging service for free-floating shared electric vehicle clusters. *Transp. Res. Part E Logist. Transp. Rev.* **2022**, *160*, 102652. [CrossRef]

systems

MDPI

Article

Strategic Sensor Placement in Expansive Highway Networks: A Novel Framework for Maximizing Information Gain

Yunxiang Yang and Jidong J. Yang *

Smart Mobility and Infrastructure Laboratory, College of Engineering, University of Georgia, Athens, GA 30602, USA; yyang117@uga.edu
* Correspondence: jidong.yang@uga.edu

Abstract: Traffic sensors play a pivotal role in monitoring and assessing network-wide traffic conditions. However, the substantial costs associated with deploying an extensive sensor network across real-world highway systems can often prove prohibitive. Thus, the strategic selection of optimal sensor locations within budget and resource constraints becomes imperative, leading to the well-known Traffic Sensor Location Problem (TSLP). In this study, we introduce a novel framework to address the TSLP for large-scale highway networks, focusing on maximizing information gain in a joint vector space that comprehensively captures both network topology and segment-level features. To solve this optimization problem, we devised a genetic algorithm (GA) with penalty handling. Additionally, we developed a physics-guided random walk algorithm, which not only significantly reduces the search space but offers remarkable flexibility in striking a practical balance between computational load and the confidence of achieving global optimality. For illustration purposes, the proposed framework was applied to the Savannah highway network in Georgia. The results from our GA method align well with those from exhaustive research, but with significantly reduced computational time. By leveraging information theory and maximizing information gain in a low-dimensional vector space, the proposed framework permits parallel, scalable computation and offers considerable potential in the strategic planning and deployment of various sensors for expansive, real-world highway networks.

Keywords: Traffic Sensor Location Problem; topological embedding; information theory; Kullback–Leibler divergence; genetic algorithm; physics-guided random walk

Citation: Yang, Y.; Yang, J.J. Strategic Sensor Placement in Expansive Highway Networks: A Novel Framework for Maximizing Information Gain. *Systems* **2023**, *11*, 577. https://doi.org/10.3390/systems11120577

Academic Editor: Paolo Visconti

Received: 12 November 2023
Revised: 12 December 2023
Accepted: 14 December 2023
Published: 18 December 2023

1. Introduction

Traffic sensors represent a critical component in the ongoing evolution of Intelligent Transportation Systems (ITS) and provide essential data sources that can be leveraged by modern and emerging AI systems to revolutionize traffic management and the strategic planning of our vast highway networks. The significance of these sensors cannot be overstated as they offer crucial insights into traffic patterns, congestion levels, and dynamic road conditions, all of which are instrumental in optimizing traffic flow and enhancing overall road safety. However, as the role of traffic sensors in modern transportation infrastructure becomes increasingly important, so do the substantial challenges that come with their widespread deployment. These obstacles can range from the pragmatic constraints of limited budgets to the intricacies of selecting appropriate sensor types and dealing with the inherent possibility of sensor failures [1]. The quest to enhance the observability of traffic flow, a crucial objective within the realm of ITS, necessitates the creation of extensive sensor networks that can span vast geographical areas. This expansive network, while undeniably powerful in its capacity to capture comprehensive traffic data, brings with it the need for substantial initial investments for deployment and continuous, labor-intensive maintenance. These formidable financial and operational burdens pose a significant obstacle to the practical implementation of large traffic sensor networks.

It is in addressing these challenges that the Traffic Sensor Location Problem (TSLP) emerges as a critical focal point for transportation researchers and planners. This problem encapsulates the intricate process of strategically placing traffic sensors to maximize their utility while minimizing the financial and operational burdens associated with their maintenance and upkeep. Over the past few decades, extensive research has been devoted to tackling the TSLP, resulting in a diverse range of proposed methods. These approaches can be broadly classified into six categories [2]: O/D estimation [3,4], flow observability [5], link flow inference [6–8], path reconstruction [9], traffic surveillance [10], and travel time estimation [11]. Our proposed framework in this paper is related to flow observability and link low inference, while emphasizing the overall network coverage. Given the nature of the problem, which involves binary-choice decision for each candidate location, the most popular approaches center on constructing and solving mixed-integer linear or nonlinear programs. However, their practical utility in dealing with intricate, extensive real-world highway networks is frequently constrained by distinct challenges [12], such as the exceedingly high computational complexity associated with a large number of edges and nodes. For instance, with a network of n links, the TSLP search space is 2n for a single sensor. In large metropolitan regions, the highway network can rapidly expand to encompass tens of thousands of links, rendering the traditional approach unviable. In the context of this study, our objective is to harness data directly from regional transportation planning models and address the TSLP through a novel data-driven approach by leveraging information theory and maximizing information gain in a low-dimensional embedding (vector) space that jointly captures complex network topology and pertinent segment-level features. As such, the proposed framework is well suited for parallel, scalable computation with modern graphics processing units (GPUs) or tensor processing units (TPUs) for large-scale, real-world highway networks.

2. Methods

Our proposed framework is depicted in Figure 1. Firstly, we abstract the study highway network as a directional graph, where each segment is treated as a node. Then, we obtain the topological embedding of the graph in a vector space, where each node (segment) is represented by a dense vector in R^{n} while preserving the network topology. To enhance the topological embedding, we introduce additional dimensions (R^m) to account for segment-specific features such as AADT, functional class, and the number of lanes. This dimensional augmentation results in a joint vector space ($R^{(n+m)}$), highlighted by the shaded blue box in Figure 1. As a result, the inherent distance metric between two nodes (or segments) in this joint vector space incorporates both their topological relationship and segment attributes.

Subsequently, we model the node (segment) distribution of the network in this joint vector space using kernel density estimation (KDE), which is referred to as model distribution, denoted by $Q(x)$. On the other hand, the choice set of the proposed sensor locations (segments), together with existing sensor locations, represents the data distribution of the segments with a sensor, denoted by $P(x)$. By such construct, the TSLP is formulated as an optimization problem that minimizes the discrepancy between $P(x)$ and $Q(x)$, such as Kullback–Leibler (KL) divergence, with respect to the proposed sensor locations. Notably, this formulation presents a distinctive perspective compared to the common machine learning scenario, wherein the data distribution ($P(x)$) is known and the objective is to train the model ($Q(x)$) to align with the data. To effectively address this problem, we developed a customized genetic algorithm integrated with physics-guided random walks (PGRW) that operate from the existing sensor locations. This combination serves to effectively reduce the search space for determining the optimal locations for new sensors.

To set up the stage for coherent presentation, we delineate the process into six key components: (1) Graph Representation of Highway Network, (2) Topological Embedding of Graph, (3) Construction of Joint Vector Space, (4) Kernel Density Estimation, (5) Optimization Problem Formulation, and (6) Solution Algorithm. Throughout our discussion, we utilize the Savannah highway network as an illustrative example to facilitate a comprehensive understanding of the process.

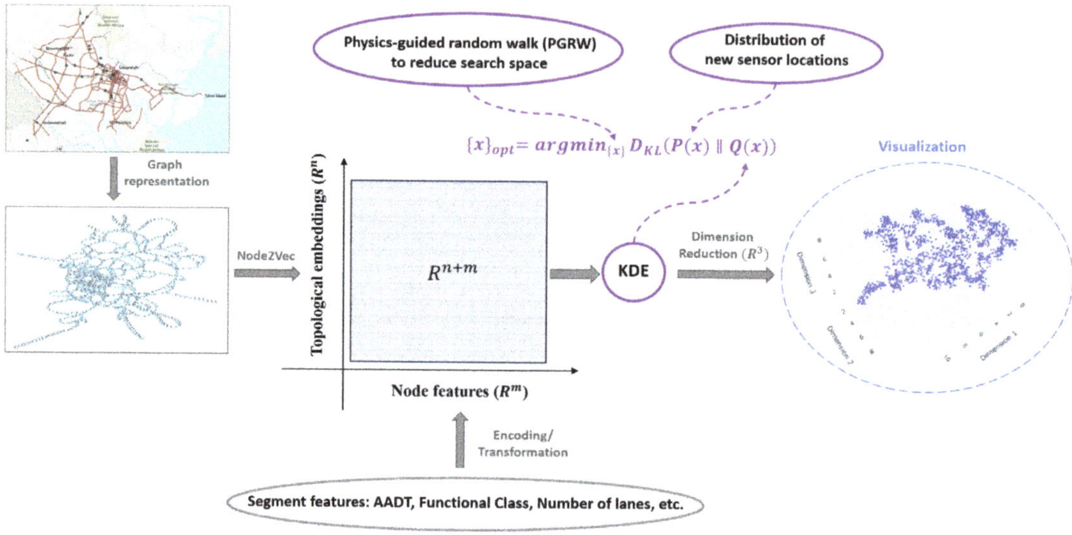

Figure 1. Illustration of the proposed framework.

Each of the six components is discussed in detail subsequently with a dedicated subsection.

2.1. Graph Representation of Highway Network

This initial step involves the representation of a highway network as a graph to capture its topology. For demonstration, the Savannah highway network is used, which consists of 1616 segments with different functional classes, such as interstate highways, arterials, collectors, and local roads. The GIS visualization of the network is shown in Figure 2. The green dots indicate the locations of the 26 existing Continuous Count Stations (CCSs). In this paper, sensor locations refer to CCSs.

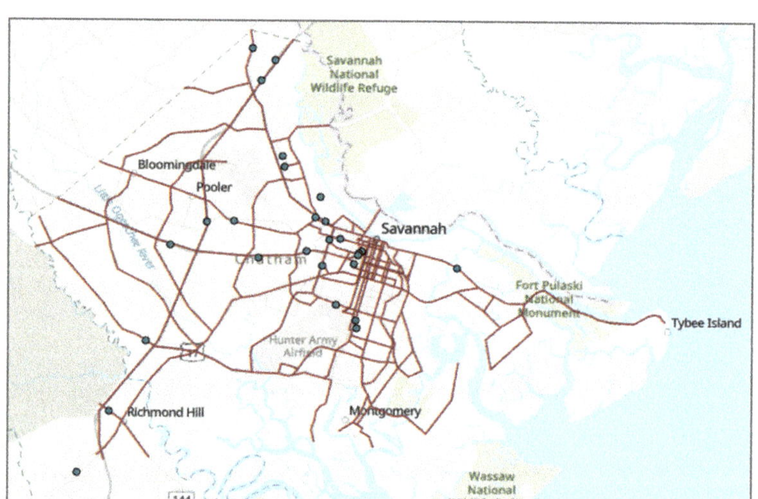

Figure 2. Visualization of the Savannah network and existing CCSs.

Both the undirected graph and the directed graph are constructed and shown in Figure 3. Each node in the graph represents a road segment in the original highway network. To capture the directional traffic flows, the directed graph is adopted in this study.

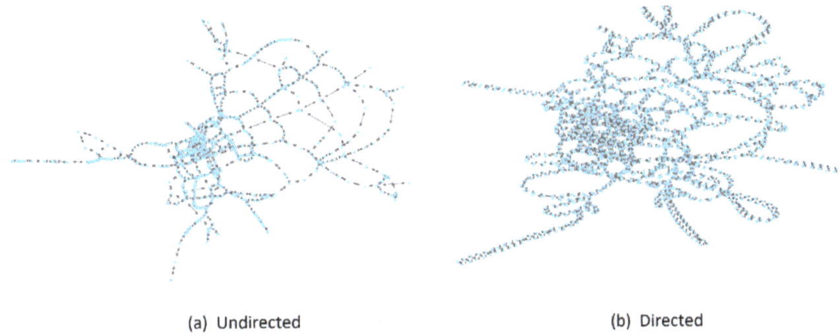

Figure 3. Graph representation of the Savannah highway network.

A straightforward way of assessing the overall topology of a highway network is to examine the distribution of the number of neighboring segments. As depicted in Figure 4, most segments demonstrate connections with two neighbors, followed by segments with four and three neighbors. The few segments with only one neighbor represent endpoints or *cul-de-sac* segments.

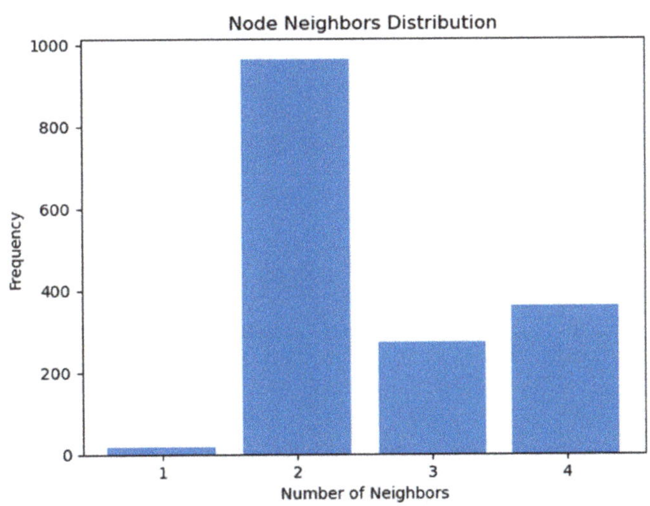

Figure 4. Distribution of the number of neighbors for the Savannah highway network.

2.2. Topological Embedding of Graphs

To effectively capture meaningful representations of the directed graph mentioned above, while simultaneously preserving its graph structure properties, we use the Node2Vec embedding method proposed by Grover and Leskovec [13], which is a scalable learning-based approach for embedding network topology. It harnesses the power of machine learning and embedding techniques to learn vector representations that intricately capture the nuanced relationships between nodes, rendering them highly useful for various downstream tasks in graph analysis. For our implementation, the adopted Node2Vec parameters are shown in Table 1.

Table 1. Parameters of Node2Vec algorithm.

Parameter	Value	Description
l_{wl}	30	Walk length, i.e., the number of nodes in each walk
N_{nw}	200	Number of walks per node
p	1	The likelihood of backtracking the walk and immediately revisiting a node in the random walk
q	1	The In-Out parameter q allows the traversal calculation to differentiate between inward and outward nodes
d_{dim}	8	The output Node2Vec embeddings dimension

Generally, the parameters p and q control the exploration and exploitation behavior when the Node2Vec algorithm starts sampling the graph. Node2Vec uses a 2nd-order random walk and guides the walking process by introducing a search bias α. Refer to Equation (1),

$$\pi_{vx} = \alpha_{pq}(t,\ x) \cdot w_{vx} \tag{1}$$

where π_{vx} denotes the unnormalized transition probability between nodes v and x and w_{vx} is the static edge weight. $\alpha_{pq}(t,\ x)$ is computed by Equation (2),

$$\alpha_{pq}(t,\ x) = \begin{cases} \frac{1}{p}\ if\ d_{tx} = 0 \\ 1\ if\ d_{tx} = 1 \\ \frac{1}{q}\ if\ d_{tx} = 2 \end{cases} \tag{2}$$

where d_{tx} denotes the shortest path distance between nodes t and x [13].

When considering the current node v in the context of Node2Vec, two important parameters, p and q, play a significant role in shaping the walker's behavior. Setting $0 < p < 1$ and $q > 1$ makes the walker more likely to exploit the visited nodes, favoring local exploration, while reducing the likelihood of exploring nodes further away. Conversely, when setting $p > 1$ and $0 < q < 1$, the walker becomes more inclined to visit nodes further away, promoting global exploration behavior. In our case, we intentionally set $p = q = 1$. The reason behind this choice is that our implementation of the Node2Vec model aims to encode the topology while maximally preserving the graph's structure without bias towards either local or global aspects.

For visualizing the embeddings produced by Node2Vec, we employ a 3D visualization technique using Uniform Manifold Approximation and Projection (UMAP). The UMAP has proven to be highly effective, offering competitive low-dimensional manifold representation while also preserving more of the global structure of the embeddings [14]. As depicted in Figure 5, the UMAP visualization demonstrates a balanced trade-off between local and global connectivity. The current CCS locations are shown as red dots, indicating a representative sampling of network segments in the embedding space.

In addition to topology embeddings, explicitly considering segment (node) features becomes important when determining the optimal sensor placement. In practical scenarios, state Departments of Transportation (DOTs) often prioritize achieving a well-balanced network coverage by strategically distributing sensors across various types of facilities and areas. To meet this requirement, we incorporate three important segment features, namely Total Volume, Lanes, and Functional Class, and combine them with the 8-dimensional Node2Vec embeddings, resulting in an expanded 11-dimensional joint feature space. This allows us to create a comprehensive representation that takes into account both the topological characteristics and the specific attributes of individual segments. Furthermore, we apply Min-Max scaling to all feature dimensions of the joint embedding space to ensure they share a consistent scale and contribute equally to the subsequent analysis and decision-making process.

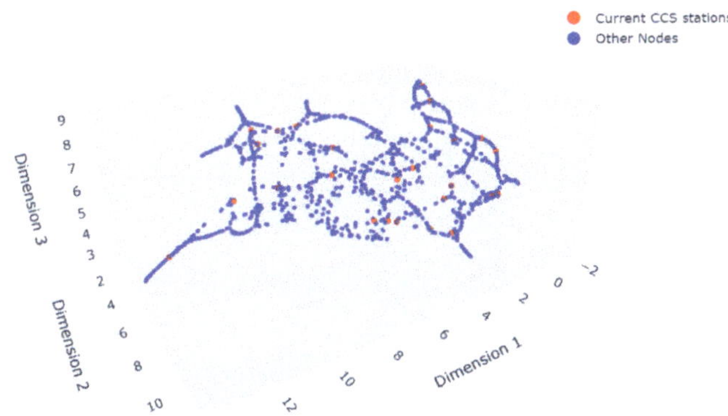

Figure 5. Three-dimensional UMAP visualization of Node2Vec embeddings.

2.3. Construction of Joint Vector Space

2.4. Kernel Density Estimation

Kernel density estimation (KDE) is a non-parametric method to estimate underlying distribution directly from data samples. Unlike the histogram, the KDE produces smooth estimate of the probability density function by using all sample data points and convincingly reveals multimodality [15]. Particularly, KDE imposes a kernel, a smooth and symmetric function, at each data point. The density estimation is derived by summing the contributions from these kernels. The choice of kernel and other parameters (e.g., bandwidth) can affect the smoothness of the estimated density. With a dataset of $\{x_1, x_2, \ldots, x_k\}$, the kernel density estimator can be computed by Equation (3),

$$\hat{f}_h(x) = \frac{1}{k}\sum_{i=1}^{k} K_h\,(x - x_i) = \frac{1}{kh}\sum_{i=1}^{k} K\left(\frac{x - x_i}{h}\right) \tag{3}$$

where $0 \le K_h(x - x_i) < \infty$ for all real x, and $\int_{-\infty}^{\infty} \hat{f}_h(x)\,dx = 1$. $K(.)$ is the kernel and $h > 0$ is a smoothing parameter, referred to as bandwidth. h controls the smoothness of the kernel. Improper selection of h can lead to one of two issues: over-smoothing or under-smoothing. Over-smoothing fails to capture the underlying structure of data, and eventually leads to an oversimplified representation of the true distribution, while under-smoothing captures noises in the data and leads to inaccurate representation of the data [16]. Figure 6 shows the 3D visualization of varying bandwidth values. The bandwidth of 0.4 (Figure 6a) indicates over-smoothing, while the bandwidth of 0.05 (Figure 6d) shows under-smoothing. The bandwidth of 0.2 (Figure 6b) is chosen to capture both local and global structures. Eventually, this KDE serves as the model distribution, $Q(x)$, for formulating our optimization problem in the following section.

2.5. Optimization Problem Formulation

Drawing inspiration from information theory, we formulate an optimization problem in the joint embedding space, guided by fundamental principles of information theory. Specifically, we would like our choice set of new sensor locations to maximize the information gain. This is equivalent to minimizing the Kullback–Leibler (KL) divergence between the data distribution, denoted as $P(x)$, and the model distribution, represented as $Q(x)$, through KDE. The KL divergence is also known as information divergence and relative entropy, measuring the dissimilarity between two distributions. The KL divergence is computed by Equation (4) for discrete probability distributions $P(x)$ and $Q(x)$, defined on the same sample space χ [17].

$$D_{KL}(P \parallel Q) = \sum_{x \epsilon \chi} P(x) log\left(\frac{P(x)}{Q(x)}\right) \tag{4}$$

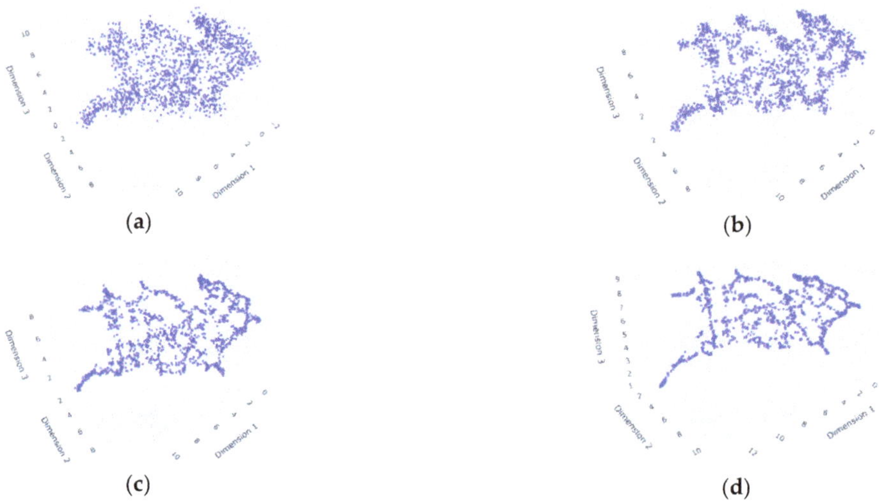

Figure 6. The KDE of the Savannah highway network on 3D UMAP embeddings: (**a**) bandwidth 0.4, (**b**) bandwidth 0.2, (**c**) bandwidth 0.1, (**d**) bandwidth 0.05.

The model distribution $Q(x)$ is estimated based on the outputs (e.g., the loaded network) of a travel demand model. The data distribution $P(x)$ represents the choice set of locations (segments) for the existing sensors (x_e) and the planned sensors (x_p). The objective is to select the planned sensor sites x_p to minimize the KL divergence between $P(x)$ and $Q(x)$, in the joint embedding space as previously described. Thus, the optimization problem is written in Equation (5),

$$\min_{x_p} \sum\nolimits_{\{x_e,\ x_p\}} P(x) log\left(\frac{P(x)}{Q(x)}\right), \qquad x_p \in \chi_{\text{pgrw}} \tag{5}$$

where χ_{pgrw} denotes the reduced search space by the PGRW algorithm, which is discussed in the next section.

2.6. *Solution Algorithm*

In this subsection, we introduce a solution algorithm, which consists of two components: (1) physics-guided random walk and (2) genetic algorithm. The former effectively reduces the search space, while the latter conducts search with penalty handling.

2.6.1. Physics-Guided Random Walk

Given the potentially large solution space for real-world highway networks, it is essential to reasonably narrow down the solution space first. For this purpose, we adopt a random walk-based approach. The idea is to design walkers to start from existing sensor locations, where traffic flows are known. Each walk is guided by traffic flows on the neighboring segments (nodes) subject to the network topology. Through a finite number of walk steps, each walker can cover a part of the network in the vicinity of existing sensors. By viewing traffic flow as a "diffusion process" across a highway network, the visited parts of network by the walkers are considered to be "inferable", which leave the remaining unvisited parts of the network to be explored for optimal sensor locations. In our implementation, we let multiple walkers start walking at the same time from the current sensor locations, which are 26 for the Savannah highway network.

To align with our intention on flow estimation, we guide all walkers with network traffic flow information, referred to as physics-guided random walk (PGRW) in this paper.

The probability of walking to each direct neighbor is computed by the softmax function of Annual Average Daily Traffic (AADT) over all neighboring nodes by Equation (6),

$$\sigma(z_i) = \frac{e^{z_i}}{\sum_{j=1}^{k} e^{z_j}}, \quad i, j = 1, \ldots, k \tag{6}$$

where k is the number of direct neighbors and z_i is AADT on segment i.

In the context of the PGRW algorithm, users have the ability to specify their preferred size for the reduced solution space when seeking optimal sensor locations. As an illustrative example, we opted for a target search space size of 100. The outcomes of the PGRW algorithm are visually presented in Figure 7, where distinct trajectories, distinguished by triangle markers in varying colors, depict the diverse paths taken by different walkers who started from the existing sensor locations. The presence of blue bubbles on the graph signifies the unexplored segments, which collectively compose the reduced space designated for the exploration of optimal sensor locations. It is important to underscore that the selection of the search space size confers a degree of flexibility, allowing users to strike a balance between computational resource usage and the confidence in achieving global optimality.

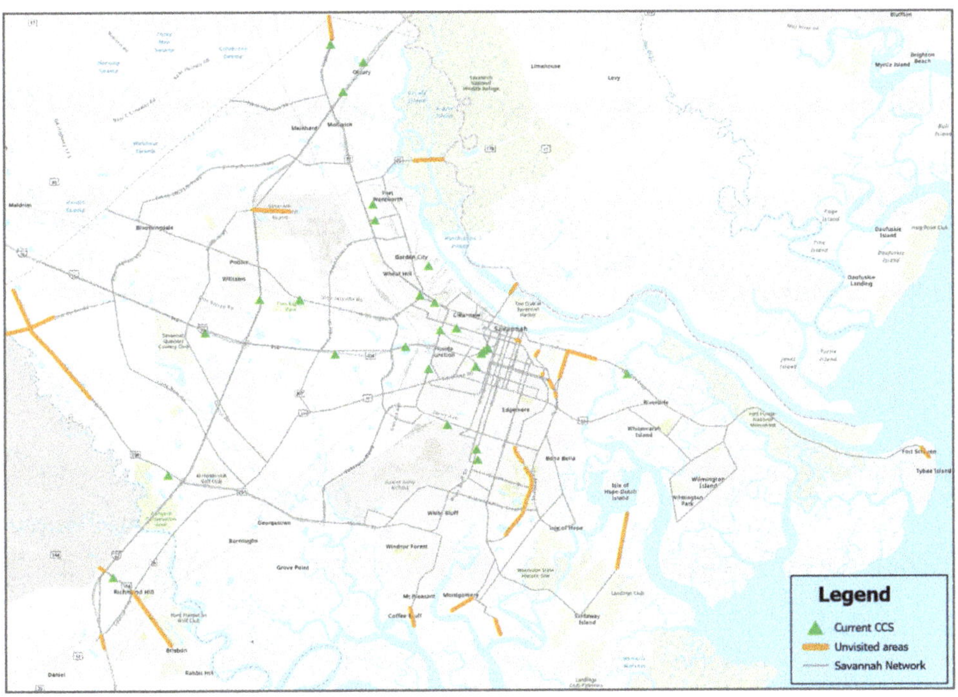

Figure 7. Reduced search space (indicated by segments in orange) by the PGRW.

2.6.2. Genetic Algorithm

In recent years, a multitude of researchers have proposed various metaheuristic algorithms to tackle real-world problems in engineering, economics, and management, among other fields [18]. Generally, these algorithms draw inspiration from three primary sources: biological evolution processes, physical laws, and swarm behavior [19,20]. Among population-based metaheuristics, a dominant and widely adopted subset, these algorithms iteratively search for optimal solutions by exploring a set of candidate solutions and leveraging population characteristics to guide the search [21]. An exemplary evolutionary algorithm within this family is the genetic algorithm (GA). In the GA, a population of candidate solutions (individuals) undergoes evolution through biological operators, including selection, crossover, and mutation. The process commences with a randomly generated population of individuals, and in each generation, every individual is evaluated by a fitness function. The individuals with higher fitness scores are retained from the current population and subsequently subjected to crossover and mutation to form a new generation. This iterative process continues until either the maximum fitness score is achieved or the maximum number of generations is reached.

In our setting, a gene is defined as a binary digit, taking the value of either 0 or 1. Each gene represents the selection status of a specific node or segment within the network, determining whether it is chosen for sensor placement. When a gene holds a value of 1, it signifies that the corresponding node or segment is selected for sensor placement, whereas a value of 0 indicates that it is not chosen. By this gene definition, an individual is represented as a binary string comprising candidate nodes or segments eligible for sensor placement. This binary string encapsulates the selection configuration of nodes or segments, with each gene in the string denoting the inclusion or exclusion of the corresponding node or segment for sensor deployment.

We define the fitness function as the reciprocal of *KL* divergence between model distribution $Q(x)$ and data distribution $P(x)$ as previously introduced.

It should be noted that the standard GA does not impose any restrictions on the number of "1" genes, which represent the segments selected for sensor placement in our setting. In real-world applications, the number of sensors to be installed is typically constrained by a limited budget or other practical considerations. Therefore, achieving a definitive number of sensors is often practically preferable. To ensure that the GA conforms to such a constraint, we employ a penalty trick, which involves pre-filtering candidate solutions that fail to meet the required number of "1" genes prior to fitness evaluation. Non-compliant solutions are assigned a substantial penalty with a low fitness score, effectively discouraging their continued participation in the evolutionary process. The penalty trick preserves the integrity of the GA's evolution process while respecting the desired constraint on the number of planned sensors. Algorithm 1 shows the implementation of our customized GA.

Algorithm 1. Pseudocode of GA with penalty trick.

Initialize population P with size N_p
For generation $i = 1$ to N_g:
For $j = 1$ to N_p:
If (the number of "1" gene in individual x_j) ! = (the desired number of sensors):
assign a low fitness score (e.g., 0.00001) to individual x_j.
Else: compute the fitness score for individual x_j.
End for
Select the best m individuals in the population P and save them as population, P_1
//crossover operation//
For $j = 1$ to $(N_p - m)$:
randomly select two individuals x_a and x_b from population P
generate x'_a and x'_b by crossover.
save x'_a and x'_b to population P_2
End for
//mutation operation//
For $j = 1$ to $(N_p - m)$:
select an individual x_j from P_2
apply mutation to obtain individual x''_j
replace x_j with x''_j in P_2
End for
update population $P = P_1 + P_2$
End for
Return the best individual, x_{best} in population P

3. Results

The optimal parameters for the GA were determined through a grid search and are summarized in Table 2. Notably, for the parent selection process, we employed the roulette wheel method.

Table 2. GA optimal parameters.

Parameter	Value
Number of generations	100
Number of parents mating	30
Population size	100
Gene space	[0, 1]
Parent selection	roulette wheel
Crossover	single point
Mutation	random
Mutation percent for genes	5

The GA is subsequently implemented using the optimal parameters in Table 2 with the reduced solution space previously generated by the PGRW (Figure 7). As depicted in Figure 8, the evolution process demonstrates convergence, with the fitness score approaching stability after approximately 130 evolutions.

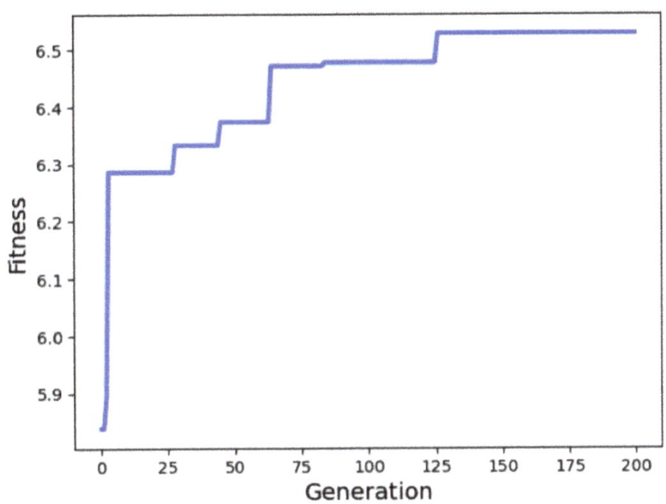

Figure 8. Evolution of GA with the optimal parameters.

For illustrative purposes, consider adding five new sensor sites to the Savanna highway network. Figure 9 shows the GA identified optimal locations, depicted by blue circles, for installing these new sensors. For comparison, the results derived from the exhaustive search (ES) method are also displayed in Figure 9, denoted by the red solid dots. Notably, four out of the five locations selected by the GA method align with those of the ES method. Upon closer inspection within the overlapping area in Figure 9, it becomes evident that the two divergent locations in the zoomed-in view correspond to the same direction of the same extended roadway section, while the matched location is for the other direction. Regarding computational efficiency, the GA method achieved convergence in 8 s, while the ES method took up to 20 h. These experiments were conducted on a laptop equipped with an Intel(R) Core(TM) i7-10870H CPU @ 2.20 GHz, 32 GB RAM, and NVIDIA GeForce RTX3070.

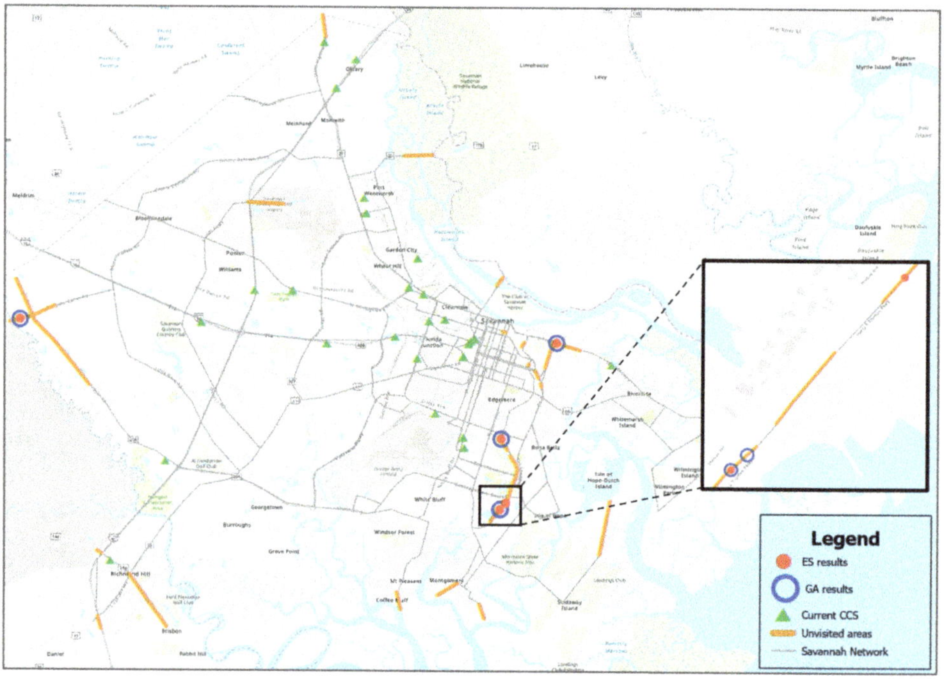

Figure 9. Visualization of GA results. (Note: orange segments denote the reduced search space by the PGRW, green triangles denote the existing sensor locations, blue circles indicate the locations selected by the GA method, and red dots are the locations from the ES method).

4. Discussion and Conclusions

In this study, we introduce a novel framework designed to tackle the TSLP. Inspired by information theory, our approach harnesses a distinctive fusion of network topology and segment-level features within a unified embedding space to identify optimal sensor locations, maximizing information gain. Unlike previous works, our proposed framework operates in a low-dimensional vector space, facilitating parallel and scalable computation using modern GPUs or TPUs, particularly for large-scale networks.

Employing a GA with penalty handling addresses the optimization challenge with desirable constraints, while the integration of the PGRW algorithm expedites the solution process. This dual approach not only significantly reduces the search space but also strikes a balance between computational load and the confidence of achieving global optimality. For illustration purposes, our framework has been applied to the Savannah highway network. The outcomes align well with those of exhaustive search, but with much faster convergence, demonstrating its potential for application in expansive real-world metropolitan or statewide highway networks.

Nonetheless, it is crucial to emphasize that while the GA serves as our primary solution method, our framework is inherently versatile and not bound to any specific solution algorithms. The pursuit of an optimal solution, as evidenced by the objective of minimizing KL divergence, is fundamentally shaped by the underlying network topology and the construction of the joint embedding space.

However, we acknowledge specific limitations and suggest avenues for future research. Our current investigation focused on three segment features, treating them equally in the joint embedding space. An essential direction for further exploration involves incorporating stakeholder preferences and policy guidance into the optimization framework, enabling a more nuanced consideration of diverse features and their varying contributions.

Additionally, a promising research approach entails formulating separate or interconnected optimization problems, especially for different sensor types with distinct purposes. This approach can facilitate a comprehensive planning and deployment strategy for large, hybrid sensor networks, tailored to meet specific objectives and stakeholder requirements.

Author Contributions: Conception and design: J.J.Y.; data processing: Y.Y.; analysis and interpretation of results: Y.Y. and J.J.Y.; draft manuscript preparation: Y.Y.; review and editing: J.J.Y.; visualization, Y.Y.; supervision, J.J.Y.; project administration, J.J.Y.; funding acquisition, J.J.Y. All authors have read and agreed to the published version of the manuscript.

Funding: This research was funded by Georgia Department of Transportation, grant number RP 20-28.

Data Availability Statement: Some or all of the data that support the findings of this study are available from the corresponding author upon reasonable request.

Acknowledgments: The work presented in this project is part of a research project (RP 20-28) sponsored by the Georgia Department of Transportation, United States. The contents of this paper reflect the views of the authors, who are solely responsible for the facts and accuracy of the data, opinions, and conclusions presented herein. The contents may not reflect the views of the funding agency or other individuals. The authors would like to acknowledge the financial support from the Georgia Department of Transportation in the United States.

Conflicts of Interest: The authors declare no conflict of interest.

References

1. Salari, M.; Kattan, L.; Lam, W.H.K.; Esfeh, M.A.; Fu, H. Modeling the effect of sensor failure on the location of counting sensors for origin-destination (OD) estimation. *Transp. Res. Part C Emerg. Technol.* **2021**, *132*, 103367. [CrossRef]
2. Owais, M. Traffic sensor location problem: Three decades of research. *Expert Syst. Appl.* **2022**, *208*, 118134. [CrossRef]
3. Mínguez, R.; Sánchez-Cambronero, S.; Castillo, E.; Jiménez, P. Optimal traffic plate scanning location for OD trip matrix and route estimation in road networks. *Transp. Res. Part B Methodol.* **2010**, *44*, 282–298. [CrossRef]
4. Fu, H.; Lam, W.H.; Shao, H.; Kattan, L.; Salari, M. Optimization of multi-type traffic sensor locations for estimation of multi-period origin-destination demands with covariance effects. *Transp. Res. Part E Logist. Transp. Rev.* **2022**, *157*, 102555. [CrossRef]
5. Castillo, E.; Conejo, A.J.; Menéndez, J.M.; Jiménez, P. The observability problem in traffic network models. *Comput. Aided Civ. Infrastruct. Eng.* **2008**, *23*, 208–222. [CrossRef]
6. Liu, Y.; Zhu, N.; Ma, S.; Jia, N. Traffic sensor location approach for flow inference. *IET Intell. Transp. Syst.* **2015**, *9*, 184–192. [CrossRef]
7. Morrison, D.R.; Martonosi, S.E. Characteristics of optimal solutions to the sensor location problem. *Ann. Oper. Res.* **2015**, *226*, 463–478. [CrossRef]
8. Ng, M. Synergistic sensor location for link flow inference without path enumeration: A node-based approach. *Transp. Res. Part B Methodol.* **2012**, *46*, 781–788. [CrossRef]
9. Gentili, M.; Mirchandani, P.B. Locating active sensors on traffic networks. *Ann. Oper. Res.* **2005**, *136*, 229–257. [CrossRef]
10. Owais, M. Location strategy for traffic emission remote sensing monitors to capture the violated emissions. *J. Adv. Transp.* **2019**, *2019*, 6520818. [CrossRef]
11. Edara, P.; Smith, B.; Guo, J.; Babiceanu, S.; McGhee, C. Methodology to identify optimal placement of point detectors for travel time estimation. *J. Transp. Eng.* **2011**, *137*, 155–173. [CrossRef]
12. Li, R.; Mehr, N.; Horowitz, R. Submodularity of optimal sensor placement for traffic networks. *Transp. Res. Part B Methodol.* **2023**, *171*, 29–43. [CrossRef]
13. Grover, A.; Leskovec, J. node2vec: Scalable Feature Learning for Networks. In Proceedings of the KDD'16: The 22nd ACM SIGKDD International Conference on Knowledge Discovery and Data Mining, ACM, San Francisco, CA, USA, 13–17 August 2016; pp. 855–864. [CrossRef]
14. McInnes, L.; Healy, J.; Melville, J. UMAP: Uniform Manifold Approximation and Projection for Dimension Reduction. *arXiv* **2020**, arXiv:1802.03426.
15. Węglarczyk, S. Kernel density estimation and its application. *ITM Web Conf.* **2018**, *23*, 00037. [CrossRef]
16. Kumar, V.; Chhabra, J.K.; Kumar, D. Parameter adaptive harmony search algorithm for unimodal and multimodal optimization problems. *J. Comput. Sci.* **2014**, *5*, 144–155. [CrossRef]
17. Van Erven, T.; Harremos, P. Rényi divergence and Kullback-Leibler divergence. *IEEE Trans. Inf. Theory* **2014**, *60*, 3797–3820. [CrossRef]
18. Bonabeau, E.; Theraulaz, G.; Dorigo, M. *Swarm Intelligence: From Natural to Artificial Systems*; Oxford University Press: Oxford, UK, 1999.

19. Katoch, S.; Chauhan, S.S.; Kumar, V. A review on genetic algorithm: Past, present, and future. *Multimed. Tools Appl.* **2021**, *80*, 8091–8126. [CrossRef]
20. He, T.; Wang, H.; Yoon, S.W. Comparison of Four Population-Based Meta-Heuristic Algorithms on Pick-and-Place Optimization. *Procedia Manuf.* **2018**, *17*, 944–951. [CrossRef]
21. Mirjalili, S. *Evolutionary Algorithms and Neural Networks, Studies in Computational Intelligence*; Springer: Berlin/Heidelberg, Germany, 2019. [CrossRef]

systems

MDPI

Article

Does Robotaxi Offer a Positive Travel Experience? A Study of the Key Factors That Influence Consumers' Use of the Robotaxi

Chun Yang [1], Chao Gu [2] and Wei Wei [3,*]

1 School of Design, Jiangnan University, Wuxi 214122, China; 8202201014@jiangnan.edu.cn
2 Academy of Arts & Design, Tsinghua University, Beijing 100084, China; cguamoy@my.honam.ac.kr
3 School of Textile Garment and Design, Changshu Institute of Technology, Changshu 215500, China
* Correspondence: doublewei@cslg.edu.cn

Abstract: Presently, robotaxi is being tested in cities such as Beijing, Changsha, Guangzhou, etc., and it remains a relatively new mode of transportation for consumers. Considering that robotaxi is a new mobility model, its popularity has an immediate impact on the function and efficiency of urban traffic, so further research on consumers' perceptions is necessary in order to improve their acceptance of robotaxi. In this study, we explored the behavioral intention of current users of robotaxi based on their performance expectancy, effort expectation, and perceived risk. Based on the results, it appears that performance expectations and effort expectations positively influence usage intentions, which indicates that improving travel efficiency and lowering the threshold for robotaxi use will assist consumers in accepting it. In terms of consumer behavior, perceived risk negatively impacts usage intention, meaning that personal safety, service quality, and travel experience are important factors. Performance expectancy and effort expectancy are positively correlated, indicating that improving travel efficiency and lowering thresholds are complementary.

Keywords: robotaxi; travel experience; behavioral intention

Citation: Yang, C.; Gu, C.; Wei, W. Does Robotaxi Offer a Positive Travel Experience? A Study of the Key Factors That Influence Consumers' Use of the Robotaxi. *Systems* **2023**, *11*, 559. https://doi.org/10.3390/systems11120559

Academic Editors: Mahyar Amirgholy and Jidong J. Yang

Received: 23 October 2023
Revised: 24 November 2023
Accepted: 28 November 2023
Published: 29 November 2023

1. Introduction

The self-driving automobile, also known as a driverless car or autonomous vehicle (AV), is able to sense the environment and navigate autonomously without the assistance of a driver [1]. The level of vehicle automation, as classified by the Society of Automotive Engineers (SAE) (2016), falls between L1 (driver assistance) and L5 (full driving automation) [2]. It is the goal of major manufacturers, such as Google, Tesla, and Huawei, to achieve L5 autonomous driving, which may be the final form of car driving.

In China, the popularity of intelligent vehicles is increasing, which makes autonomous driving more and more feasible. China's robotaxi market is expected to reach 15.8 billion yuan in 2022, according to China Robotaxi Industry Development Insights 2022. In the online car-hailing and traditional taxi markets, robotaxi has a penetration rate of 3%. During the period from 2024 to 2027, the market size will increase from 85 billion yuan to 139 billion yuan, and the penetration rate will increase from 15% to 22%. Market size is expected to reach 196.2 billion yuan by 2028, with a penetration rate of 30% [3]. A positive attitude towards unmanned vehicles is also demonstrated by the government and enterprises. Robotaxi is also a future trend and form of transportation.

A driverless vehicle may bring considerable benefits to road safety, flexibility, inclusivity, and sustainability in today's society, where vehicles are becoming more and more popular [4]. With China's large population base, daily commuters, tourists, and travelers, unmanned vehicles may have a bright future in this market. From the perspective of taxi companies, robotaxi can reduce the labor cost of taxi drivers and facilitate management, as well as serve as a new technology to attract consumers. From the perspective of consumers, robotaxi may be able to calculate the most efficient route more easily since it is autonomously driven, which may reduce communication issues with drivers. Female consumers may feel that robotaxi will be safer and have additional advantages. Robotaxi can impact the issues associated with security, trust, privacy, accountability, reliability, and transparency [5]. It is still a relatively new travel mode for users, even though robotaxi has been piloted in cities such as Beijing, Changsha, and Guangzhou. As a new mobility model, robotaxi's popularity has a certain effect on the function and efficiency of urban traffic, so further exploration of consumer perception is necessary in order to improve their acceptance of robotaxi.

This study aims to explore the perceptions of general consumers or potential consumers towards robotaxis and how these perceptions influence their behavioral intention to ultimately use robotaxis by investigating China's robotaxi users. This study employs variables such as performance expectancy, service expectancy, and perceived risk to establish a hypothetical model to examine the factors influencing consumers' usage of robotaxis. This research perspective is relatively novel in the current field and aids businesses in deepening their understanding of the underlying motivations of consumers. At a time when urban transportation faces challenges, the promotion of autonomous vehicles becomes particularly crucial.

Additionally, the results of this study can provide governments, businesses, or investors with an in-depth analysis of the robotaxi market outlook from a consumer perspective, offering valuable decision-making direction. Our research can aid in understanding the development of China's robotaxi market and delve deeper into key factors such as user needs, technological advancement, and market prospects, providing profound insights and robust support for the future development of robotaxis.

2. Theoretical Framework

In order to effectively evaluate consumers' willingness to use robotaxi and establish consumers' relevant cognition, this study builds a research framework by selecting relevant constructs as the assessment content involved in the study through literature review and discussion.

2.1. Performance Expectancy and Effort Expectancy

The subjects of performance expectation and effort expectation are generally discussed together in UTAUT [6] or UTAUT2 [7]. In terms of performance expectancy, it is the degree to which an individual believes that the system will contribute to the improvement of their job performance [6]. The effort expectancy of a system relates to how easy it is for individuals to use it [7]. According to the definition, performance expectancy and effort expectancy refer to the degree to which consumers accept new products and new technologies, i.e., the degree to which they are easy to use and useful [8]. According to this study, performance expectancy refers to the degree to which robotaxi's travel efficiency meets expectations, while effort expectancy refers to the ease with which it can be used. The study by Kanwaldeep et al. considers performance expectancy when investigating key factors for consumer adoption of autonomous vehicles [9]. According to their study, self-driving cars will perform better than manual-driven vehicles [10]. In addition, effort expectancy has implications for consumers' behavioral intentions regarding autonomous vehicles [11]. Unmanned vehicles are more likely to be adopted by consumers when their effort expectations are lower [12]. In this study, we focus on performance expectancy and effort expectancy because robotaxi is still in the exploratory stage in China, and there are only a few cities currently being piloted. In addition to conducting research from the viewpoint of unmanned vehicle technology, it is important to look at consumer views, perceptions, and acceptance of robotaxi. While there are similar factors in UTAUT, such as facilitating conditions and social influence, there is a lack of social influence conditions because we are analyzing subjective cognition and robotaxi has not yet established a large-scale presence in China. Due to this, other constructs of UTAUT are not taken into account in this study.

2.2. Perceived Risk

First, perceived risk is associated with psychology research, which relates to consumers' expectations of negative outcomes when purchasing a particular product [13]. Because driverless cars are a new technology undergoing development, it is understandable that consumers are concerned about them. Unmanned vehicles may cause consumers' concerns due to online reports [14], media perceptions [15], etc. In addition, consumers may have concerns about hacker attacks, vehicles being remotely controlled, and autonomous driving being disrupted by emergencies [9,16]. Therefore, unmanned vehicles still require improvements in order to be reliable in the face of small probability events [17]. The perceived risk in this study refers to the likelihood that consumers perceive risks associated with robotaxi's service quality, safety, and travel experience. As consumers weigh risks and benefits before making a final decision [18], the lower the perceived risks, the higher the perceived advantages of unmanned vehicles [18]. Additionally, the less risk consumers perceive, the more likely they are to adopt robotaxi.

2.3. Behavioral Intention

The behavioral intention indicator is an important component of consumer research, since it indicates the likelihood of the consumer taking a particular action [19]. Consumer behavior has also been extensively discussed in the field of unmanned vehicles. Shirley et al. discuss how value orientation, media attention, and scientific knowledge influence Singaporeans' behavioral intention to use self-driving cars [20]. Ghasri and Vij investigated the influence of media comments and social influence on consumer behavioral intentions, particularly in relation to the distinction between different natural attributes of consumers [21]. Kaur and Rampersad discuss the aspects of security and privacy that consumers are most concerned about [9]. A reduction or elimination of consumer concerns will result in an increase in consumer behavioral intention. According to this study, the consumer's behavioral intention represents the degree of willingness to ride a robotaxi. An individual's behavioral intention is positively influenced by positive intentions and negatively influenced by negative intentions. As a result, in this study, the consumers' intentions to adopt robotaxi may be influenced by their PE and EE, and at the same time, they may be negatively affected by their PR.

3. Research Method and Hypothesis

Based on the research purpose, this study draws up a research structure and integrates and analyzes the theoretical basis of relevant literature in order to carry out research methods and develop a research plan for implementation. This chapter explores the relevant influencing factors that affect consumers' adoption of robotaxi based on literature research, establishes hypotheses, and uses quantitative questionnaires and scale survey methods to test these hypotheses. Following the reliability analysis and item analysis of the questionnaires, a correlation analysis was performed using a structural equation model. In addition to providing specific practice recommendations, the factors that impact the consumer the most were analyzed.

3.1. Proposed Theoretical Model and Research Hypothesis

Considering the above discussion, this study proposes the following hypotheses and constructs a model based on the hypothetical relationship. The model includes four constructs, performance expectancy (PE), effort expectancy (EE), perceived risk (PR), and behavioral intention (BI), as well as five related research hypotheses (Figure 1).

Hypothesis 1 (H1). *Performance expectancy is significantly positively correlated with the consumers' behavioral intention of robotaxi.*

Hypothesis 2 (H2). *Performance expectancy and consumers' perceived risk for robotaxi are significantly negatively correlated.*

Hypothesis 3 (H3). *Effort expectancy is significantly positively correlated with the consumers' behavioral intention of robotaxi.*

Hypothesis 4 (H4). *Effort expectancy and consumers' perceived risk for robotaxi are significantly negatively correlated.*

Hypothesis 5 (H5). *Perceived risk is significantly negatively correlated with the consumers' behavioral intention of robotaxi.*

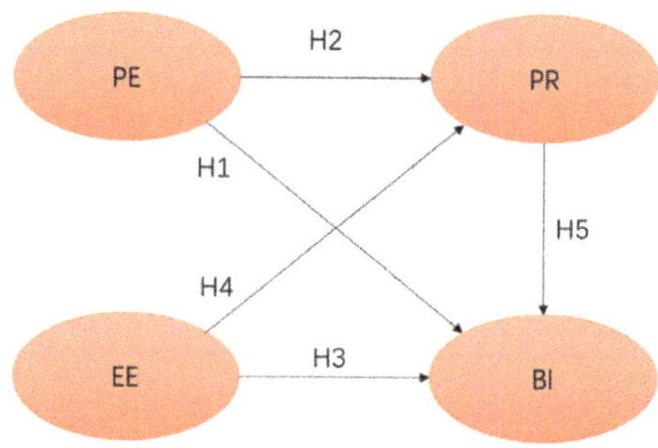

Figure 1. The model developed in this study.

3.2. *Definitions of Research Variables*

According to the research topic and literature relevant to this study, the items of the questionnaire were designed. In Table 1, we present the definition of variable operability and the reference source for the scale.

Table 1. Reference sources for variables and items.

Variable	Operational Definition	Reference
Performance expectancy	The degree to which robotaxi is expected to improve travel efficiency.	[22]
Effort expectancy	The degree to which robotaxi is easy to use.	[22]
Perceived risk	The degree of the consumers' perception of the risk associated with quality of service, safety, and travel experience.	[23]
Behavioral intention	The degree of the consumers' intention to use robotaxi.	[22]

4. Results

This chapter focuses primarily on the research objectives of two stages: the distribution and collection of questionnaires, and the subsequent quantitative analysis of the data. A detailed description of the calculation process and statistical results is provided below:

4.1. *Descriptive Analysis of Demographic Variables*

From February to May 2022, a Chinese online questionnaire was distributed through the Questionnaire Star network platform to citizens in Beijing, Changsha, Guangzhou, and other cities with robotaxi pilot programs. As robotaxi is still in the pilot stage, consumers in these major cities, particularly those who have experienced autonomous taxis, constitute our primary target for investigation. Additionally, given the preliminary nature of this study, there are no extensive restrictions imposed on the consumer demographic.

In the survey, all subjects click on a link to view the survey description, while they also voluntarily answer the research questions and can launch the survey at any time. In this regard, all subjects are fully informed. After completing the questionnaire and scale, in order to express our gratitude, the subjects will receive a bonus of 15 RMB and a lottery on the platform.

With the exception of demographic variables, all variables were rated using a seven-point Likert scale (1. Strongly disagree; 2. Disagree; 3. Slightly disagree; 4. Neutral; 5. Slightly agree; 6. Agree; 7. Strongly agree). In this study, 750 questionnaires were sent out, and a total of 640 valid samples were obtained, which was in line with Jackson's estimated parameter-to-sample ratio, greater than 1:10 [24]. The distribution of demographic variables in this study is shown in Table 2 based on the statistical analysis of the data collected from the valid questionnaire.

Table 2. Definitions of variable operability and reference scales.

Sample	Category	Number	Percentage (%)
Gender	Male	317	49.531
	Female	323	50.469
Age	Under 22	53	8.281
	23–30	251	39.219
	31–40	220	34.375
	41–50	54	8.438
	Over 51	62	9.688
Marital status	Married	444	69.375
	Not Married	196	30.625
Income	Under 4000	102	15.937
	4001–6000	137	21.406
	6001–12,000	260	40.625
	12,001–18,000	102	15.937
	Over 18,001	39	6.094
Education background	The junior high school level and below	10	1.563
	Secondary school or high school	42	6.563
	College or university	488	76.250
	A graduate degree or higher	200	15.625
Occupation	Civil servant	73	11.406
	Clerk	253	39.531
	Worker	88	13.750
	Public Service Agencies	80	12.500
	Student	71	11.094
	Self-employed	75	11.719
Area	Eastern Region	402	62.813
	Central Region	110	17.188
	West Region	96	15.000
	Northeast Region	31	4.844
	Hong Kong, Macao, and Taiwan	1	0.156

4.2. Reliability Analysis

This study recruited IBM SPSS 24 software to conduct reliability and validity analysis. A reliability analysis was conducted on the questionnaire in order to remove unstable questions to ensure reliability and discrimination. In Table 3, Cronbach's α value of each facet is greater than 0.6, and the CITC (corrected item-to-total correlation) is greater than 0.4, indicating high confidence for all constructs [25]. Moreover, deleting any item will result in a lower aspect of Cronbach's α than the current result, indicating that the item should not be deleted. The comprehensive data show that the data is reliable and can be used for further analysis.

Table 3. Results of the reliability analysis.

Dimension	Item	CITC	Cronbach's α after Item Deletion	Cronbach's α
PE	PE1	0.633	0.775	0.814
	PE2	0.689	0.718	
	PE3	0.671	0.737	
EE	EE1	0.634	0.762	0.812
	EE2	0.643	0.758	
	EE3	0.604	0.776	
	EE4	0.640	0.759	
PR	PR1	0.658	0.715	0.800
	PR2	0.702	0.666	
	PR3	0.581	0.791	
BI	BI1	0.553	0.742	0.769
	BI2	0.647	0.639	
	BI3	0.612	0.680	

4.3. Exploratory Factor Analysis

An exploratory factor analysis was conducted to test the validity of the questionnaire in this study. The method of analysis used in the calculation process is principal component analysis (PCA). In addition, factor rotation was performed using the varimax method. Table 4 illustrates the results. The KMO (Kaiser–Meyer–Olkin) value of all constructs is greater than 0.5 and the significance of Bartlett's Sphere test is less than 0.05, indicating that the data meet the criteria for factor analysis [26,27]. A further analysis shows that the commonality of each item exceeds 0.5 and the factor loading that contributes to its construct exceeds 0.6. This suggests that the construct has good validity [28]. In the extraction of new factors, all items belonging to each construct are included. An eigenvalue greater than 1 can only be extracted from one new factor belonging to each construct [29], indicating that it is a good single construct factor [30].

Table 4. Exploratory factor analysis results.

Construct	Item	KMO	Bartlett's Sphere Test	Commonality	Factor Loading	Eigenvalue	Total Variation Explained
PE	PE1	0.712	0.000	0.695	0.834	2.186	72.852%
	PE2			0.755	0.869		
	PE3			0.736	0.858		
EE	EE1	0.788	0.000	0.645	0.803	2.558	63.962%
	EE2			0.655	0.809		
	EE3			0.610	0.781		
	EE4			0.650	0.806		
PR	PR1	0.691	0.000	0.730	0.855	2.146	71.527%
	PR2			0.772	0.879		
	PR3			0.643	0.802		
BI	BI1	0.686	0.000	0.625	0.791	2.054	68.467%
	BI2			0.734	0.856		
	BI3			0.695	0.834		

4.4. Measurement Model

This study used IBM AMOS 22 software for structural equation model analysis. Because a large number of studies have used AMOS for analysis, AMOS is proven to be a reliable structural equation modeling software. In Figure 2, all latent variables are correlated and satisfy the path analysis premise. Moreover, all fit points in this model meet the recommended criteria, as shown in Table 5. This indicates that the first-order confirmatory factor analysis (CFA) model is well-fitted [31].

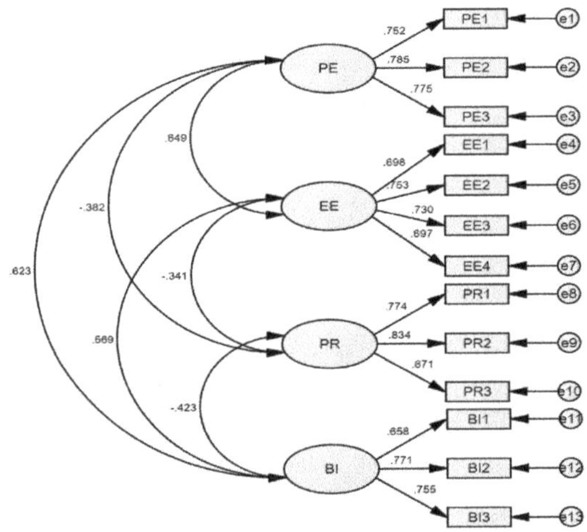

Figure 2. CFA model.

Table 5. Model fitting index comparison results of CFA.

Common Indices	χ^2/df	RMSEA	GFI	AGFI	NFI	CFI	SRMR
Judgment criteria	<3	<0.08	>0.9	>0.9	>0.9	>0.9	<0.08
CFA value	2.633	0.051	0.963	0.943	0.953	0.970	0.040
CCLFM value	2.678	0.051	0.970	0.942	0.953	0.970	0.040

Note: χ^2/df: normed Chi-square; RMSEA: root-mean-square-error approximation; GFI: goodness-of-fit index; AGFI: adjusted goodness-of-fit index; NFI: normative fit index; CFI: comparative fit index; SRMR: standardized root-mean-square residual.

Table 6 presents the results of convergent validity. Each item in the first-order CFA model has a factor loading greater than 0.5. There was a significant correlation between coefficient estimates and standard errors, $t > 1.96$, $p < 0.05$, as indicated in the fit index. Each construct's combined reliability (CR) exceeds 0.6 [32], and the average variance extracted (AVE) exceeds the base value of 0.36 [33]. According to the results of this study, the questionnaire data show good convergent validity.

Table 6. Convergent validity of CFA.

Construct	Items	Factor Loading	t Value	SE	*p* Value	SMC	AVE	CR
PE	PE1	0.752	20.548	0.163	0.001	0.565	0.594	0.815
	PE2	0.785	21.749	0.167	0.001	0.617		
	PE3	0.775	21.374	0.166	0.001	0.601		
EE	EE1	0.698	18.561	0.178	0.001	0.487	0.518	0.811
	EE2	0.753	20.506	0.161	0.001	0.566		
	EE3	0.730	19.697	0.166	0.001	0.533		
	EE4	0.697	18.542	0.172	0.001	0.486		
PR	PR1	0.774	20.600	0.200	0.001	0.599	0.582	0.805
	PR2	0.834	22.475	0.187	0.001	0.696		
	PR3	0.671	17.472	0.184	0.001	0.451		
BI	BI1	0.658	16.854	0.171	0.001	0.433	0.532	0.773
	BI2	0.771	20.433	0.178	0.001	0.595		
	BI3	0.755	19.916	0.163	0.001	0.570		

Note: SE: standard error; SMC: square multiple correlation; CR: composite reliability; AVE: average variance extracted.

The discriminant validity is based on the work of Fornell and Larcker [31]. The model is considered discriminant if the square root of the AVE for each facet is greater than the correlation coefficient between the facets. In this study, all diagonal values are greater than the values outside the diagonal, and therefore all aspects of this study have good discriminant validity (Table 7). This study shows that each construct has good discriminant validity.

Table 7. Discriminant validity.

	PE	EE	PR	BI
PE	0.771			
EE	0.523 *	0.720		
PR	−0.324 *	−0.286 *	0.763	
BI	0.501 *	0.447 *	−0.351 *	0.729

* The level of significance is 0.05.

4.5. Structural Equation Model

In this study, we used the research of Jackson et al. [34], Kline [35], Schumacker [36], and Hu and Bentler [37], as well as other scholars. To evaluate the fit of the structural model, multiple indicators (MLχ^2, DF, χ^2/DF, RMSEA, SRMR, AGFI, CFI, NFI, GFI) were selected. As shown in Table 8, the research constructs are measured according to the research assumptions and models. In addition, all standard model fit evaluation indicators satisfy the independent level and combination rule of recommended fit, which indicates that the structural model has a good fit. According to the study, the theoretical framework assumed is consistent with the actual survey results. The path coefficients are shown in Figure 3.

Table 8. Adaptability of SEM.

Common Indices	χ^2/df	RMSEA	GFI	AGFI	NFI	CFI	SRMR
Judgment criteria	<3	<0.08	>0.9	>0.9	>0.9	>0.9	<0.08
Value	2.633	0.051	0.963	0.943	0.953	0.970	0.040

Note: χ^2/df: normed chi-square; RMSEA: root-mean-square-error approximation; GFI: goodness-of-fit index; AGFI: adjusted goodness-of-fit index; NFI: normative fit index; CFI: comparative fit index; SRMR: standardized root-mean-square residual.

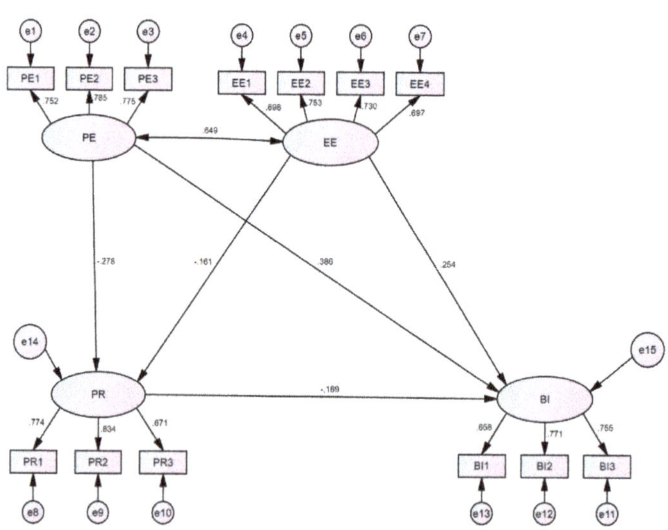

Figure 3. Structural equation model.

According to Liao and Hu's research, this study tests the path effect in the model [38]. The test standard was set at * $p < 0.1$, ** $p < 0.05$, and *** $p < 0.001$. As can be seen from Table 9, when PR is the dependent variable, PE has a direct negative effect on PR ($p = 0.001$, $\beta = -0.278$), and EE also has a direct negative effect on PR ($p = 0.029$, $\beta = -0.161$). When BI is the dependent variable, PR negatively affects BI ($p = 0.001$, $\beta = -0.189$), while PE positively affects BI ($p = 0.001$, $\beta = 0.386$). There is a positive indirect effect of PE on BI ($p = 0.001$, $\beta = 0.052$) and a positive impact in total ($p = 0.001$, $\beta = 0.438$). There is a direct positive effect of EE on BI ($p = 0.001$, $\beta = 0.254$), as well as an indirect positive effect ($p = 0.016$, $\beta = 0.030$), with a total positive effect ($p = 0.001$, $\beta = 0.285$).

Table 9. Direct and indirect effects.

Path	Direct Effect		Indirect Effect		Total Effect	
	β	B-C Sig.	β	B-C Sig.	β	B-S Sig.
PE→PR	−0.278	0.001 *	/	/	−0.278	0.001 *
EE→PR	−0.161	0.029 *	/	/	−0.161	0.029 *
PE→BI	0.386	0.001 *	0.052	0.001*	0.438	0.001 *
EE→BI	0.254	0.001 *	0.030	0.016*	0.285	0.001 *
PR→BI	−0.189	0.001 *	/	/	−0.189	0.001 *

* The level of significance is 0.05.

4.6. Analyzing Moderated Variables

The study further examined the moderating effect of gender as a moderator variable on each pathway, as shown in Table 10. According to the results, only EE had a significant moderating effect on PR when gender was used as a moderator variable.

Table 10. Results of mediation effect.

Moderating Variable	IV	→	DV	CMIN	*p*
Gender	PE	→	PR	1.313	0.252
	EE	→	PR	5.560	0.018 *
	PE	→	BI	1.717	0.190
	EE	→	BI	1.129	0.288
	PR	→	BI	3.492	0.062

* The level of significance is 0.05. Note: IV: independent variable; DV: dependent variable; CMIN: chi-square.

Further, as shown in Table 11, we compared the path coefficients that moderated the effects of variables. When it comes to perceiving PR, only female consumers are sensitive to EE, whereas men are not.

Table 11. Comparison between path coefficients with significant moderating effects.

Moderating Variable		Path	β	*p*
Gender	Male	EE→PR	−0.006	0.953
	Female		−0.337	0.002 *

* The level of significance is 0.05.

5. Discussion

The results of the empirical analysis provide some key findings, which are discussed below.

The results of this study demonstrate that PE and EE have positive effects on BI(H1, H3), which is consistent with the results of previous studies [9,39]. It has been demonstrated that consumers' expectations of whether or not to use robotaxi depend on their perceptions of its travel efficiency and ease of use—robotaxi's commitment to meeting consumers' commuting and travel needs in a cost-effective and efficient manner [40]. In this regard, both researchers and related service organizations must take effective measures to improve the PE of potential users in order to increase the degree of acceptance or use of the product [39]. In addition to improving travel efficiency, consumers also value robotaxi's ease of use. In order to obtain relatively high benefits, consumers must spend the least amount of time, money, and effort (including time and economic costs) [41]. Therefore, it is necessary to work closely with all relevant organizations and departments to develop an easier-to-use travel form, similar to the MaaS system, and to optimize the robotaxi experience.

It has been demonstrated that PE and EE negatively affect PR (H2, H4). Therefore, consumers perceive a lower risk when PE and EE are higher. Based on the interpretation of this study, consumers' perceptions of robotaxis include service quality [42], safety [9], travel experience, etc. [43]. The higher the robotaxi property, the lower the perceived risk. As a result, consumers' judgment of PR is influenced by the positive perception of their peers represented by PE and EE. This study found that PR negatively impacted BI(H5). Thus, the higher the PR, the lower the BI of the consumer. A number of studies have confirmed the relationship between risks and BI in the context of autonomous vehicles [43,44]. Therefore, it is necessary to take measures to reduce the risk perception of consumers, thereby improving BI, such as providing timely customer service communication and ensuring the arrival of spare vehicles in the event of a vehicle failure.

It is interesting to note that PE has a direct influence coefficient of 0.386 and a total influence coefficient of 0.438 on BI. EE has a direct influence coefficient of 0.254 on BI, and a total influence coefficient of 0.285. It has been demonstrated that representing PR as a variable intermediary improves the perception of BI by consumers from PE and EE. The reason for this may be that robotaxi already has a certain degree of perfection in its form, function, and service in the current environment and conditions. Some of the concerns of consumers have been addressed by manufacturers and service providers. Consequently, consumers perceive less risk than they expect, which facilitates BI.

Furthermore, this study examined whether gender, as a moderator variable, affects different pathways. When gender was used as a moderator variable, it only had a moderating effect on the pathway from EE to PR. Further analysis reveals that only female consumers have sensitive perceptions. In line with previous research, female consumers may be more sensitive to public relations [45]. The above phenomenon may be attributed to the fact that female consumers are less comfortable with new technologies than male consumers, and they have a greater sense of self-protection and insecurity.

6. Managerial Implication

The contribution of this study lies in deconstructing the impact of performance expectancy, effort expectancy, perceived risk, and gender on users' intention to use robotaxi, and in providing relevant decision-making directions or suggestions for industry decision-makers.

Firstly, we identified that performance expectancy plays a key role in the model of consumer behavioral intention to use robotaxis. Manufacturers and service providers should adopt the latest technologies and intelligent dispatch systems to optimize service design, reduce consumers' time costs, and improve overall service efficiency. Additionally, to enhance the convenience experience for users, manufacturers and service providers need to consider offering personalized services and integrated service chains, such as with Mobility-as-a-Service systems. This approach can reduce the learning curve for new users and increase user coverage, making it easier for consumers to use robotaxis. Such measures would help enhance effort expectancy, fostering a positive anticipation of the service among users.

To alleviate consumer concerns about the unknown risks associated with robotaxi services, manufacturers and service providers should offer detailed and transparent information, such as disclosures on safety measures and vehicle maintenance details. Improving transparency in information will help reduce users' perceived risk, enhance their positive perception of the service, and provide them with a sufficient sense of security. Of course, manufacturers and service providers should also establish and implement stricter safety standards to further enhance the reliability and credibility of the overall service.

Additionally, the study on gender differences in this research reveals the high sensitivity of female consumers to perceived risk. Manufacturers and service providers can adopt differentiated marketing strategies, strengthen safety measures for female consumers, and provide more detailed information to alleviate their safety concerns, encouraging more female consumers to accept and trust robotaxi services. To increase service acceptance, manufacturers and service providers can also educate and inform consumers about the safety and convenience of the service, thereby strengthening positive expectations. Especially for female consumers, targeted advertising that emphasizes the safety of the service can be used to increase their acceptance.

7. Conclusions and Implications

In this study, we examined the factors that influence consumers' adoption of robotaxi as a mode of transportation, specifically consumers' perception of five constructs: PE, EE, PR, and BI. We used structural equation modeling to estimate the relationship between constructs and examined the moderating effect of gender. Our results can provide a certain degree of reference value for consumers, practitioners, and government agencies.

Based on the results, all assumptions are valid. PR stimulated PE and EE effects on BI through its role as a mediator. Gender as a moderator variable affects the effect of EE on PR. According to the hypotheses presented in this study, PE and EE have a positive impact on BI. This positive impact occurs as a result of consumers learning about robotaxi's related systems, processes, and collection of related information, and then making a decision accordingly. Furthermore, the negative factors of robotaxi may also negatively affect consumers' expectations, thereby affecting BI.

The academic contribution of this study is the deconstruction of UTAUT and the inclusion of PR as a mediating variable in PE, EE, and BI. Based on the results, it proves that the four constructs are causally related, which lays a certain foundation for further research related to unmanned driving and has a certain value as a reference for future researchers.

This study has a number of limitations that may suggest directions for future research:

1. Although all aspects of this study are related, there may be some latent variables or second-order aspects that have not been explored and discussed. To enhance the model's explanatory power and improve its performance, researchers can add new facets, including second-order facets. As an example, the influence or interference of supplementary news media on consumers.
2. There are different responses of men and women to the use of robotaxi, and subsequent research will be able to investigate further into the internal reasons for these differences. In addition, it may also be able to conduct detailed research, analysis, or discussion on other different attributes associated with consumers.
3. Structural equation modeling has been used as the primary method of analysis and research of quantitative research papers. In the future, qualitative research (expert interviews, field investigations, etc.) may be added to complement quantitative data to convey deeper meanings.
4. Considering the focus of the study on Chinese consumers, researchers may be able to compare Chinese and foreign consumers in the future and facilitate the coordination of the different conditions on a global scale in the future.

Author Contributions: Conceptualization, C.Y., C.G. and W.W.; methodology, C.Y. and C.G.; formal analysis, C.G. and W.W.; data curation, C.G.; writing—original draft preparation, C.Y.; writing—review and editing, C.Y. All authors have read and agreed to the published version of the manuscript.

Funding: This work was supported by Jiangsu Education Department (Philosophy and Social Sciences Projects of Higher Education, grant number 2022SJYB1508).

Institutional Review Board Statement: Not applicable.

Informed Consent Statement: Not applicable.

Data Availability Statement: Data are contained within the article.

Acknowledgments: We are grateful to Jie Sun for her work. We also thank the anonymous reviewers who provided valuable comments on the manuscript.

Conflicts of Interest: The authors declare no conflict of interest.

References

1. Du, H.; Zhu, G.; Zheng, J. Why travelers trust and accept self-driving cars: An empirical study. *Travel Behav. Soc.* **2021**, *22*, 1–9. [CrossRef]
2. Hamburger, Y.A.; Sela, Y.; Kaufman, S.; Wellingstein, T.; Stein, N.; Sivan, J. Personality and the autonomous vehicle: Overcoming psychological barriers to the driverless car. *Technol. Soc.* **2022**, *69*, 101971. [CrossRef]
3. Microcomputer. Driverless Overtaking on a Curve in China. Available online: https://baijiahao.baidu.com/s?id=1743095020214119806&wfr=spider&for=pc (accessed on 6 May 2022).
4. González-González, E.; Nogués, S.; Stead, D. Automated vehicles and the city of tomorrow: A backcasting approach. *Cities* **2019**, *94*, 153–160. [CrossRef]
5. Fagnant, D.J.; Kockelman, K. Preparing a nation for autonomous vehicles: Opportunities, barriers and policy recommendations. *Transp. Res. Part A Policy Pract.* **2015**, *77*, 167–181. [CrossRef]
6. Venkatesh, V.; Morris, M.G.; Davis, G.B.; Davis, F.D. User acceptance of information technology: Toward a unified view. *MIS Q.* **2003**, *27*, 425–478. [CrossRef]
7. Ain, N.; Kaur, K.; Waheed, M. The influence of learning value on learning management system use: An extension of utaut2. *Inf. Dev.* **2015**, *32*, 1306–1321. [CrossRef]
8. Nordhoff, S.; Louw, T.; Innamaa, S.; Lehtonen, E.; Beuster, A.; Torrao, G.; Bjorvatn, A.; Kessel, T.; Malin, F.; Happee, R.; et al. Using the utaut2 model to explain public acceptance of conditionally automated (l3) cars: A questionnaire study among 9,118 car drivers from eight european countries. *Transp. Res. Part F Traffic Psychol. Behav.* **2020**, *74*, 280–297. [CrossRef]
9. Kaur, K.; Rampersad, G. Trust in driverless cars: Investigating key factors influencing the adoption of driverless cars. *J. Eng. Technol. Manag.* **2018**, *48*, 87–96. [CrossRef]
10. Kyriakidis, M.; Happee, R.; de Winter, J.C.F. Public opinion on automated driving: Results of an international questionnaire among 5000 respondents. *Transp. Res. Part F Traffic Psychol. Behav.* **2015**, *32*, 127–140. [CrossRef]
11. Nordhoff, S.; Malmsten, V.; van Arem, B.; Liu, P.; Happee, R. A structural equation modeling approach for the acceptance of driverless automated shuttles based on constructs from the unified theory of acceptance and use of technology and the diffusion of innovation theory. *Transp. Res. Part F Traffic Psychol. Behav.* **2021**, *78*, 58–73. [CrossRef]
12. van der Waal, N.E.; de Wit, J.; Bol, N.; Ebbers, W.; Hooft, L.; Metting, E.; van der Laan, L.N. Predictors of contact tracing app adoption: Integrating the utaut, hbm and contextual factors. *Technol. Soc.* **2022**, *71*, 102101. [CrossRef]
13. Dunn, M.G.; Murphy, P.E.; Skelly, G.U. Research note: The influence of perceived risk on brand preference for supermarket products. *J. Retail.* **1986**, *62*, 204–216.
14. Wakabayashi, D. Self-Driving Uber Car Kills Arizona Pedestrian, Where Robots Roam. Available online: https://www.nytimes.com/2018/03/19/technology/uber-driverless-fatality.html (accessed on 15 July 2022).
15. Anania, E.C.; Rice, S.; Walters, N.W.; Pierce, M.; Winter, S.R.; Milner, M.N. The effects of positive and negative information on consumers' willingness to ride in a driverless vehicle. *Transp. Policy* **2018**, *72*, 218–224. [CrossRef]
16. Ring, T. Connected cars—The next targe tfor hackers. *Netw. Secur.* **2015**, *2015*, 11–16. [CrossRef]
17. Waldrop, M.M. Autonomous vehicles: No drivers required. *Nature* **2015**, *518*, 20–23. [CrossRef]
18. Wang, S.; Wang, J.; Lin, S.; Li, J. Public perceptions and acceptance of nuclear energy in china: The role of public knowledge, perceived benefit, perceived risk and public engagement. *Energy Policy* **2019**, *126*, 352–360. [CrossRef]
19. Fishbein, M.; Ajzen, I. Belief, attitude, intention, and behavior: An introduction to theory and research. *Philos. Rhetor.* **1977**, *6*, 244–245.
20. Ho, S.S.; Leow, V.J.X.; Leung, Y.W. Driving without the brain? Effects of value predispositions, media attention, and science knowledge on public willingness to use driverless cars in singapore. *Transp. Res. Part F Traffic Psychol. Behav.* **2020**, *71*, 49–61. [CrossRef]
21. Ghasri, M.; Vij, A. The potential impact of media commentary and social influence on consumer preferences for driverless cars. *Transp. Res. Part C Emerg. Technol.* **2021**, *127*, 103132. [CrossRef]
22. Venkatesh, V.; Thong, J.Y.L.; Xu, X. Consumer acceptance and use of information technology: Extending the unified theory of acceptance and use of technology. *MIS Q.* **2012**, *36*, 157–178. [CrossRef]
23. Abbasi, G.A.; Kumaravelu, J.; Goh, Y.-N.; Dara Singh, K.S. Understanding the intention to revisit a destination by expanding the theory of planned behaviour (tpb). *Span. J. Mark.-ESIC* **2021**, *25*, 282–311. [CrossRef]
24. Jackson, D.L. Revisiting sample size and number of parameter estimates: Some support for the n:Q hypothesis. *Struct. Equ. Model. Multidiscip. J.* **2003**, *10*, 128–141. [CrossRef]

25. Zijlmans, E.A.; Tijmstra, J.; Van der Ark, L.A.; Sijtsma, K. Item-score reliability as a selection tool in test construction. *Front. Psychol.* **2019**, *9*, 2298. [CrossRef] [PubMed]
26. Kaiser, H.F. An index of factorial simplicity. *Psychometrika* **1974**, *39*, 31–36. [CrossRef]
27. Norusis, M. *SPSS Professional Statistics*; Prentice-Hall: Upper Sadler River, NJ, USA, 1998.
28. Osborne, J.W. *Best Practices in Quantitative Methods*; Sage: Thousand Oaks, CA, USA, 2008.
29. Harman, H. *Modern Factor Analysis*; University of Chicago Press: Chicago, IL, USA, 1960.
30. Kohli, A.K.; Shervani, T.A.; Challagalla, G.N. Learning and performance orientation of salespeople: The role of supervisors. *J. Mark. Res.* **1998**, *35*, 263–274. [CrossRef]
31. Fornell, C.; Larcker, D.F. Evaluating structural equation models with unobservable variables and measurement error. *J. Mark. Res.* **1981**, *18*, 39–50. [CrossRef]
32. Liang, A.R.-D.; Lim, W.M. Exploring the online buying behavior of specialty food shoppers. *Int. J. Hosp. Manag.* **2011**, *30*, 855–865. [CrossRef]
33. Fernandes, D.W.; Moori, R.G.; Filho, V.A.V. Logistic service quality as a mediator between logistics capabilities and customer satisfaction. *Rev. Gest.* **2018**, *25*, 358–372. [CrossRef]
34. Jackson, D.L.; Gillaspy Jr, J.A.; Purc-Stephenson, R. Reporting practices in confirmatory factor analysis: An overview and some recommendations. *Psychol. Methods* **2009**, *14*, 6. [CrossRef]
35. Kline, R.B. *Principles and Practice of Structural Equation Modeling*, 4th ed.; Guilford Publications: New York, NY, USA, 2015.
36. Whittaker, T.A. *A Beginner's Guide to Structural Equation Modeling*; Taylor & Francis: Abingdon, UK, 2011.
37. Hu, L.t.; Bentler, P.M. Cutoff criteria for fit indexes in covariance structure analysis: Conventional criteria versus new alternatives. *Struct. Equ. Model. Multidiscip. J.* **1999**, *6*, 1–55. [CrossRef]
38. Liao, H.-L.; Lu, H.-P. The role of experience and innovation characteristics in the adoption and continued use of e-learning websites. *Comput. Educ.* **2008**, *51*, 1405–1416. [CrossRef]
39. Bernhard, C.; Oberfeld, D.; Hoffmann, C.; Weismüller, D.; Hecht, H. User acceptance of automated public transport: Valence of an autonomous minibus experience. *Transp. Res. Part F Traffic Psychol. Behav.* **2020**, *70*, 109–123. [CrossRef]
40. Madigan, R.; Louw, T.; Wilbrink, M.; Schieben, A.; Merat, N. What influences the decision to use automated public transport? Using utaut to understand public acceptance of automated road transport systems. *Transp. Res. Part F Traffic Psychol. Behav.* **2017**, *50*, 55–64. [CrossRef]
41. Smyth, J.; Chen, H.; Donzella, V.; Woodman, R. Public acceptance of driver state monitoring for automated vehicles: Applying the utaut framework. *Transp. Res. Part F Traffic Psychol. Behav.* **2021**, *83*, 179–191. [CrossRef]
42. Papadima, G.; Genitsaris, E.; Karagiotas, I.; Naniopoulos, A.; Nalmpantis, D. Investigation of acceptance of driverless buses in the city of trikala and optimization of the service using conjoint analysis. *Util. Policy* **2020**, *62*, 100994. [CrossRef]
43. Jing, P.; Du, L.; Chen, Y.; Shi, Y.; Zhan, F.; Xie, J. Factors that influence parents' intentions of using autonomous vehicles to transport children to and from school. *Accid. Anal. Prev.* **2021**, *152*, 105991. [CrossRef]
44. Potoglou, D.; Whittle, C.; Tsouros, I.; Whitmarsh, L. Consumer intentions for alternative fuelled and autonomous vehicles: A segmentation analysis across six countries. *Transp. Res. Part D Transp. Environ.* **2020**, *79*, 102243. [CrossRef]
45. Rice, S.; Winter, S.R. Do gender and age affect willingness to ride in driverless vehicles: If so, then why? *Technol. Soc.* **2019**, *58*, 101145. [CrossRef]

systems

MDPI

Article

Estimating Benefits of Microtransit for Social Determinants of Health: A Social Return on Investment System Dynamics Model

Mohammad Maleki and Janille Smith-Colin *

Department of Civil and Environmental Engineering, Southern Methodist University, P.O. Box 750340, Dallas, TX 75275-0340, USA; mmaleki@smu.edu
* Correspondence: jsmithcolin@smu.edu

Abstract: Lack of transportation services in low-income communities greatly affects people's health and well-being, creating barriers to social determinants of health (SDOH). One potential solution that has gained the attention of US decision-makers in recent years is microtransit, a transportation intervention aimed at addressing this issue. Despite promising results from prior microtransit implementation, the extent to which these programs deliver social benefits remains uncertain. This study presents a novel model called Social Return on Investment System Dynamics (SROISD) to forecast the social benefits of a microtransit program in Holmes County, Mississippi. The SROISD model identifies the scope and key stakeholders, maps outcomes, and gives outcomes a value. A causal loop diagram is developed next based on mapped outcomes and a literature review, thereby conceptualizing the processes through which social benefits are gained from the microtransit program. Three stock and flow diagrams are then created from the causal loop diagram to formulate the system and produce results. Outcomes mapped relative to three SDOH areas (1) accessing healthcare, (2) accessing employment, and (3) social participation indicate an overall positive return from investing in microtransit within the low-income community of interest. Additionally, ridesharing demonstrates a significant positive correlation with the SROI ratio. These findings offer support for the advantages of investing in microtransit. Additionally, the SROISD methodology offers decisionmakers a dynamically responsive approach that integrates traditional return on investment methodologies with system dynamics to explore social benefits across a variety of impact categories.

Keywords: microtransit; transportation intervention; social determinants of health; social return on investment; system dynamics; social benefits

Citation: Maleki, M.; Smith-Colin, J. Estimating Benefits of Microtransit for Social Determinants of Health: A Social Return on Investment System Dynamics Model. *Systems* **2023**, *11*, 538. https://doi.org/10.3390/systems11110538

Academic Editors: Mahyar Amirgholy and Jidong J. Yang

Received: 20 September 2023
Revised: 28 October 2023
Accepted: 31 October 2023
Published: 4 November 2023

1. Introduction

In recent years, there has been a growing recognition of the significant importance of social determinants of health (SDOH), as defined by the World Health Organization (WHO), as non-medical factors that influence health outcomes [1]. These factors are commonly categorized into five primary domains: health access and quality, education access and quality, social and community context, economic stability, and neighborhood and built environment [2]. While the influence of medical care on health is undeniable, research indicates that medical care alone accounts for only about 10–15 percent of population health outcomes, with social determinants playing a more substantial role, contributing to 50–60 percent of overall health outcomes [3].

Transportation, often considered a subcategory of the built environment SDOH categories, is an important factor contributing to population health. It is thought of by many as a social determinant of health [4,5], while also supporting access and mobility to almost all other SDOH categories (e.g., transportation access to healthcare, education, and employment) [6]. In low-income areas, where economic disparities and limited resources prevail, reliable and accessible transportation systems are even more vital for providing access to employment, healthcare, and other SDOH. Limited transportation access in a region may result in a cycle of poverty where residents are unable to fully participate in economic and social opportunities such as steady employment, leading to economic instability and thus creating transportation mobility barriers [7].

Recognizing the pivotal role of transportation as a social determinant of health, several innovative solutions have been employed to effectively address these challenges. One such solution is the implementation of microtransit programs, which have shown promise in mitigating barriers related to SDOH [8–10]. As defined by the Federal Transit Administration, Microtransit is a technology-enabled, multi-passenger transportation service that operates on dynamically generated routes [11]. Rossetti et al., further describe microtransit as a variety of on-demand transportation services that provide shared rides within a designated service area, typically utilizing vehicles such as vans, minivans, or microbuses [12]. In contrast to conventional fixed-route public transportation systems, microtransit programs are intended to be responsive to passenger demand and can change their routes and schedules in real-time. Users crowdsource rides by using a smartphone app or phone call provided by the private operator to make requests for rides [13].

Furthermore, microtransit has the potential to boost community cohesion and economic development, making it a vital tool for promoting sustainability and enhancing livability. Microtransit contributes to sustainability by addressing several sustainable development goals (SDG) described by the United Nations [14]. It promotes Good Health and Well-Being (SDG 3) by providing convenient and affordable transportation options that provide access to healthcare and reduce air pollution [15,16]. It supports Decent Work and Economic Growth (SDG 8) by creating job opportunities in the transportation sector [17]. Microtransit provides affordable access to employment opportunities and regular access to healthcare, helping to reduce poverty (SDG 1) by improving economic prospects for individuals and families, ultimately making urban areas more sustainable and livable.

Although prior experiences with several microtransit programs have demonstrated varying degrees of success [10], there remains a lack of comprehensive understanding regarding the advantages of investing in such transportation interventions. There is a need for an approach that can assess the benefits and social returns of microtransit interventions across different populations, as well as the impacts of budget availability on service quality.

Studies show that Social Return on Investment (SROI) is a valuable approach for comprehensively evaluating or forecasting social value created through investments in funded activities [18]. Given the potential social, economic, and environmental impacts of microtransit programs on SDOH outcomes, utilizing SROI becomes particularly relevant in assessing the effectiveness of such programs. However, the costs and benefits of a microtransit program accrue over several years, are influenced by several factors, and change over time. For instance, the program's ridership can fluctuate due to changes in service quality, changes in car affordability, shifts in the cost of rides, and more. Moreover, these programs struggle with financial solvency and economic sustainability, in many instances limited by a monetary valuation of costs and benefits that fails to consider the full scope of societal value created. These realities create a need for a system of valuation that can dynamically track changes in system characteristics and provide a monetary valuation of costs and social benefits across broad categories. Hence, an adaptive approach is essential to precisely forecast the SROI arising from microtransit programs as they evolve over time.

In response to this matter, this study introduces an innovative approach called Social Return on Investment System Dynamics (SROISD), which pioneers the application of a system dynamics (SD)-based framework to forecast SROI and thus overall social value from investments. It is proposed that SROISD can serve as a highly valuable tool for forecasting future returns from programs whose costs and social benefits accrue and change dynamically. The ability to capture intricate, interconnected relationships within a system, simulate various scenarios, and offer insights into how changes over time can affect outcomes are just a few advantages that SD modeling offers. In order to provide a thorough understanding of how investments in social programs can yield long-term social value by accounting for dynamic interactions and dependencies, SROISD makes use of these advantages. This enables both policymakers and stakeholders to make more informed decisions.

The rest of this paper is organized as follows: In Section 2, a comprehensive review of existing literature related to microtransit programs, Social Return on Investment (SROI), and System Dynamics is provided. Section 3 delves into the specifics of the case study and the applied methodological framework. The system conceptualization is addressed in Section 4, followed by model formulation in Section 5. Section 6 is dedicated to model validation, while Section 7 encompasses the presentation of results, including scenario analysis, sensitivity analysis, and policy implications. Finally, Section 8 presents the conclusions.

2. Literature

2.1. *Characteristics of Microtransit Programs*

The term "microtransit" first appeared in 2014 to describe a brand-new class of transportation services provided by private organizations, including VIA, Bridj, and Chariot [13]. While conventional transit services are typically operated by public agencies, follow fixed routes and schedules, and provide more targeted coverage of densely populated areas, microtransit offers more flexible, demand-responsive services, often using smaller vehicles or ridesharing platforms. Microtransit programs are known for their adaptability, as routes and schedules can be adjusted in real-time to respond to passenger demand. These programs have also been designed to address the specific SDOH needs of targeted populations within a community and typically provide door-to-door or curb-to-curb service [19].

Microtransit programs are designed to cater to a diverse range of users with specific transportation needs. These users often include individuals who may lack access to traditional fixed-route transit or require more flexible transportation options. Microtransit services play a pivotal role in enhancing transportation equity by serving various groups, such as shift workers, low-income individuals, the elderly, disabled, and underserved communities. For example, Transportation Disadvantaged Late Shift (TD Late Shift) [17] offered by Pinellas Suncoast Transit Authority (PSTA) offers service to individuals with jobs that either begin or end between 9 p.m. and 6 a.m. The service specifically serves those who have no other means of transportation and have annual incomes of no greater than 150% of the federal poverty level. Another notable example is Rides to Wellness (R2W), operated by the Mass Transportation Authority (MTA) [16]. R2W targets the elderly, disabled, and transportation-disadvantaged community members of Flint, Michigan. It also extends its services to residents of areas not previously served by fixed-route transit, ensuring improved transportation access for underserved populations. Additionally, GoLink service in Dallas serves as a critical component of microtransit programs by providing access to areas that were previously not served by fixed-route public transit. It plays a vital role in offering first-mile access to fixed-route transit stations for all residents, including people who work in areas such as Inland Port Dallas [20]. Additional information on potential passengers of microtransit programs can be found in Table 1.

In addition to the programs in the US, there are examples of community-based microtransit programs in other countries. In the UK, the Dial-a-Community Bus in Maud, Aberdeenshire, is a charitable microtransit service combating isolation for vulnerable community members [21]. In Germany, Sprinti operates in the Hannover region, focusing on improving public transport accessibility, particularly through first- and last-mile solutions [22]. Additionally, TransLink in Queensland, Australia, offers a flexible local transport program connecting people to public transport networks, shopping, healthcare, and employment opportunities through shared and pre-booked services [23].

Microtransit programs, like all other funded activities, involve both costs and benefits. Understanding and evaluating these costs and benefits are fundamental to effective transportation planning and resource allocation. The costs incurred and social benefits delivered by microtransit can be assessed using performance measures [10], as is customary in transportation planning and performance measurement [24,25]. Table 1 presents the characteristics, target population, and performance measures used to capture the impacts of various microtransit programs across the United States, as published in work by the American Public Transportation Association (APTA) [10]. Performance measures will play a pivotal role in subsequent phases of this paper by facilitating the estimation of the benefits yielded by microtransit.

Table 1. Characteristics of microtransit programs in the U.S.

Program	Agency	Target Population	General Characteristics	Performance Measures
Transportation Disadvantaged Late Shift (TD Late Shift) [10,17]	Pinellas Suncoast Transit Authority (PSTA)	- People with jobs that begin or end between 9 p.m. and 6 a.m. - Have no other means of transportation, including family and friends - With annual incomes of no greater than 150% of the federal poverty level	- Launched in 2016 - Door-to-door service to and from work - Total of 25 trips a month	- Monthly savings in operating costs - Average time gained for personal and leisure purposes per person - Number of new job opportunities or work shifts - Average number of people served per month - Number of jobs that program has provided - Passenger Satisfaction
Regional Transportation Commission FlexRIDE (RTC FlexRide) [10,26]	Regional Transportation Commission (RTC)	- People in unincorporated areas of Washoe County	- Launched in 2019 - Rides booked using the RTC smartphone app or via phone call - Funded with local sales tax dollars - Curb-to-curb rides to jobs, education, medical services, and grocery stores	- Monthly savings in operating costs - Monthly amount of economic activity gained - Average decrease in time spent on daily commutes to and from work, education, medical centers, and grocery stores per person. - Average number of people served per month - Number of jobs accessed by the program
Rides to Wellness (R2W) [10,16]	Mass Transportation Authority (MTA)	- The elderly, disabled, or transportation-disadvantaged community members of Flint, Michigan.	- Launched in 2016 - Same-day door-to-door microtransit service for eligible residents - Mainly for non-emergency medical transportation (NEMT), but can also be booked for accessing grocery stores, beauty salons, farmers markets, pharmacies, government agencies, non-profits, laundromats, or senior centers	- Average monthly decrease in no-show appointments - Average number of people served per month - Number of jobs accessed by the program - Average monthly cost savings when accessing medical centers
RideKC Microtransit [10,27]	Johnson County Government	- Residents of areas not previously served by fixed-route transit	- Launched in 2019 - Curb-to-Curb rides - Operates from 6 a.m. to 8 p.m. - Rides booked via the smartphone app, website, and phone call	- Average monthly cost savings for mobility needs - Average number of people served per month - Number of jobs that program has created

2.2. Social Return on Investment

Social Return on Investment (SROI) is a valuable approach and framework for comprehensively evaluating or forecasting social value created through investments in funded activities [18,28]. The concept of SROI was pioneered in the late 1990s by the Roberts Enterprise Development Fund (REDF) in the United States [29,30] and was later tested by the New Economics Foundation (NEF) in the United Kingdom [31–33]. Early descriptions of SROI methodology imply that the approach initially developed from common methodologies for evaluating investments in business and finance allows nonprofit sector returns/payoffs to be defined in broader social terms [34]. SROI is often defined as a stakeholder-informed cost-benefit analysis (CBA) or a Return on Investment (ROI) approach that takes a broader perspective of returns by integrating social benefits in addition to project revenues [28,35]. The main objective of SROI is to estimate costs and benefits resulting from an investment, whether social, economic, or environmental, in monetary values [20], with a focus on non-traded, non-market products. SROI requires the participation of stakeholders in the estimation of financial proxies and the evaluation of social value created by organizations [36].

SROI analysis is classified as two different types: evaluative SROI, which examines past outcomes, and forecast SROI, which predicts the social value created when planned future outcomes are achieved [37]. SROI analysis comprises six distinct phases: (1) defining this study scope and identifying stakeholders; (2) mapping outcomes; (3) collecting data on the outcomes and assigning a value to them; (4) establishing impacts; (5) computing the SROI ratio; and (6) calculating, reporting, and validating the SROI measure [37]. SROI has been used to evaluate the impact of funded activities on SDOH [38–40] including transportation. For instance, SROI has been used to evaluate the impact of modifying vehicles for use by people with disabilities [32]. Results showed a return ranging from \$2.78 to \$17.32 for every dollar invested in vehicle modifications. SROI has also been used to measure the social value created by investing in risk-based transportation asset management systems in the state of Iowa [41]. Although there are studies that evaluate the monetary benefits of transportation programs through approaches such as CBA [42,43] and ROI [44], studies utilizing SROI to measure the social value created by transportation interventions remain few [45]. Considering the potential for societal gains, including in SDOH, that can result from investing in transportation services, SROI is an appropriate approach to capture social, economic, and environmental returns from such investments.

2.3. System Dynamics

System dynamics (SD) is a powerful tool that enables comprehensive analysis and modeling of complex, dynamic systems, providing valuable insights into the interdependencies and feedback loops that drive their behavior. Transportation systems, in particular, are often complex systems, involving various stakeholders and components that interact and influence one another, making them an ideal context for leveraging the power of SD [46]. SD originated in the mid-1950s through the pioneering work of Professor Jay W. Forrester at the Massachusetts Institute of Technology [47]. While SD initially found early applications in business management [48], many research papers have applied SD over the last several decades to fields including transportation-related issues [46,49]. SD has also been extensively used in transportation and health [50]. Causal loop diagrams (CLD) are used in SD to establish dynamic hypotheses, which serve as the foundation for constructing quantitative stock and flow diagrams (SFD). These SFDs then enable the simulation of a system's behavior, facilitating the analysis and understanding of dynamic problems within the system.

3. Data and Methods

3.1. Case Study Context

This study forecasts SROI to estimate the social value generated from the implementation of microtransit in Holmes County, Mississippi, which is recognized as one of the lowest-income areas in the nation. A rural county with a population of nearly 17,000 individuals, of which 84 percent identify as Black or African American, the county faces significant socio-economic challenges, with a median income of $16,311 per year and 42 percent living below the poverty line. Additionally, Holmes County has limited public transit service available to residents. Transportation-related barriers to employment, healthcare, healthy food, and education are thus prevalent due to limited transit service. To address these challenges, a free-ride microtransit program was launched in fall 2021 by Feonix Mobility Rising, a non-profit impact organization, offering on-demand, door-to-door rides to local points of interest with the cooperation of other transit agencies in the city. Financial support for the program was offered by a major health insurance provider acting as the payer. Riders could book trips online or through the call center of the program, and transportation requests were fulfilled using taxis and wheelchair-accessible vehicles operated by volunteer drivers or a vehicle provided by one of the local transit agencies. The microtransit program was delivered as a collaboration between the local transit agencies and Feonix, the non-profit operator. This integrated, on-demand service thus filled in gaps in the existing rural transit service. While limited radio and newspaper advertisements were in place, the program primarily relied on a community resource coordinator for promotion, with further word-of-mouth within the community being a crucial rider attraction mechanism. The pilot program operated from September through December 2021, produced 373 rides, and served 61 individuals. Locations accessed by riders included employment, healthcare facilities, and other destinations.

3.2. SROISD Framework

Social Return on Investment System Dynamics (SROISD), introduced in this paper, is a model that calculates SROI through an SD structure. As the name implies, SROISD integrates SROI steps with stages of SD and, informed by external data, allows interested parties to calculate SROI. Figure 1 presents the SROISD framework and steps used to calculate SROI. As shown in Figure 1, characteristics of microtransit programs, data from service provider teams, and literature will be used to establish scope and identify key stakeholders (SROI stage 1), map outcomes (SROI stage 2), and value outcomes (SROI stage 3). The results of these three stages, along with external data, will be used to develop two main stages of the SD model: system conceptualization, or causal loop diagrams (CLD), and model formulation, or stock and flow diagrams (SFD).

The primary objective of CLDs is to visually represent the key entities that influence the interventions being modeled and the beneficiaries of the system. On the other hand, SFDs aim to quantify these entities by assigning values or equations to the variables within the CLD. This allows for a computational simulation of the system, enabling a quantitative understanding of how it evolves over time. The simulation enables us to predict the social benefits and costs of the program, thereby allowing us to calculate SROI ratios for a specific time frame, as represented by the SROI calculation in Figure 1. Based on the categories of dynamic problems introduced by Hovmand [51], the problem modeled in this case analysis is a dynamic learning problem. In this learning problem, we aim to identify factors, such as microtransit program characteristics, stakeholder needs, geographic context, and service provider constraints, that contribute to fluctuations in the SROI ratio.

One of the needs of any pilot program is to demonstrate long-term sustainability and financial viability. This case study analysis applying the SROISD framework demonstrates how to dynamically capture returns from investment in such pilot programs, ultimately supporting future program decisions. Through this model, we can assess various scenarios and policy implications, evaluating their impact on the social return of the microtransit program. The sections that follow the SROISD framework are described in more detail, specifically using the Holmes County microtransit pilot program as a case example of the framework's application.

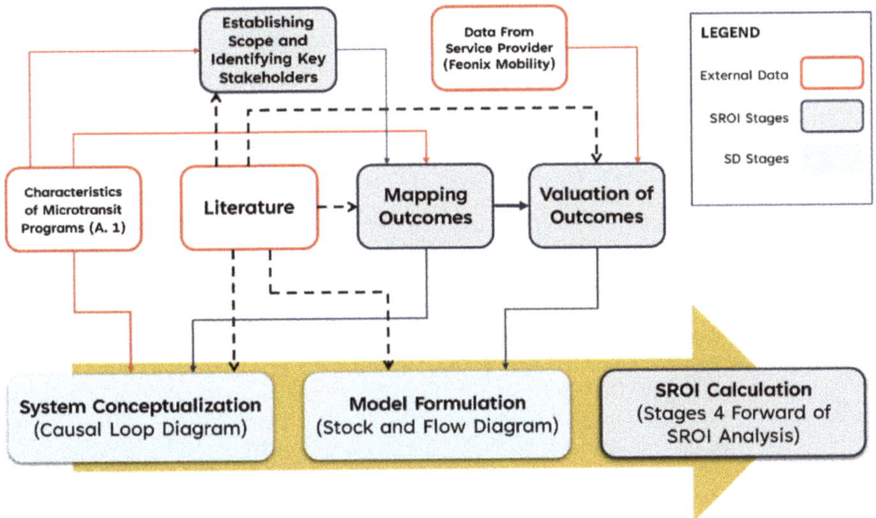

Figure 1. SROISD framework.

4. System Conceptualization

According to the SROISD framework (Figure 1), the first stage of the model is system conceptualization. System conceptualization is informed by many steps, including establishing the scope, identifying key stakeholders, and mapping outcomes. External data from the literature and other key partners, such as transportation service providers, may also be relevant. The main output from system conceptualization is a causal loop diagram.

4.1. Defining Scope and Stakeholders

The first step of the SROISD framework is defining the analysis scope and identifying key stakeholders. The scope of this study is focused on assessing the social value derived from an on-demand microtransit program implemented to address transportation-related barriers to SDOH in Holmes County. Stakeholders were thus defined as riders (i.e., the target population), healthcare providers, transportation service providers and their staff, volunteers (i.e., drivers), and the Holmes County community at large.

4.2. Mapping Social Outcomes

This step maps the outcomes of the microtransit program. In this project, outcomes are mapped relative to three SDOH areas: (1) accessing healthcare, (2) accessing employment, and (3) social participation. For example, in the healthcare access category, on-demand microtransit has the potential to decrease the overall rate of missed medical appointments within a community [16]. Reduced rates of missed medical appointments can enhance health status, which in turn can reduce the frequency of emergency department (ED) visits [52–55]. Consequently, an increase in the number of medical visits and a decrease in missed appointments can yield financial benefits for healthcare providers. The microtransit intervention also has the potential to create social benefits related to job access. Outcomes include expanded access to new job opportunities as well as increased accessibility to jobs with varied working hours (i.e., work shifts), for example, early AM and late PM hours. Additionally, previous studies highlight the pivotal role that transportation plays in fostering social inclusion among individuals [56,57]. In areas characterized by limited car ownership, such as Holmes County, having access to a reliable transportation system can significantly enhance social inclusion. Change in social isolation is thus mapped as an outcome in this study. Consequently, this study identifies the following social outcomes of the microtransit program: decreased ED visits, increased medical appointments at healthcare centers, income gains, enhanced social inclusion, and improved mental health. These outcomes will be integrated into the CLD to show connections and feedback between key outcomes and system variables such as program characteristics and stakeholder needs. Outcomes mapped and monetary valuations assigned are by no means intended to be exhaustive but instead illustrative of the monetary valuations and variables included in this specific scenario development.

4.3. Causal Loop Diagram

After problem identification, the first step of every SD analysis involves formulating dynamic hypotheses, which utilize qualitative methods to create CLDs [58]. This step integrates mapped outcomes into the model as variables, along with other relevant factors specific to the problem, to create a CLD that conceptualizes the system as shown in the SROISD model (Figure 1). A CLD is comprised of two main components: variables and connectors. Connectors represent direct relationships between variables. Additionally, polarity (positive or negative) is used to signify the effect of one variable on another. A positive connector indicates that the connected variables change in the same direction, while a negative connector indicates the opposite. Frequently observed within a CLD are reinforcing loops, denoted as (R), and balancing loops, denoted as (B). Balancing and reinforcing loops, as their names suggest, are used to simultaneously erode the system and grow the system, thus preventing infinite growth.

Transportation systems consist of many variables and parameters, leading to the creation of an SFD that becomes excessively large. In order to tackle this challenge, Ercan et al. [59] propose the utilization of smaller subdivisions, known as subsections, within the overall model. Figure 2 shows the CLD, which is divided into three modules or subsections: Social Benefits, Costs, and Service Operations. The CLD also shows the relationship between these three modules and the target variable, SROI, which is distinct from these modules and is calculated by dividing social benefits by the program's costs at each instance.

4.3.1. Social Benefits Module

The social benefits module (i.e., the bottom module) consists of all the variables that contribute to benefits gained by stakeholders. The variables on the right side of the module were identified in the mapping outcomes step discussed in Section 4.2. For example, an increased number of "riders to opportunities" increases "medical visits", which decreases "ED visits", which increases "social benefits" gained. These benefits can be broadly categorized into three categories: access to healthcare, access to employment, and access to social activities. Late4 in the model formulation stage, proxy values will be assigned to each social benefit category.

On the left side of the social benefits module (R1), which is the word-of-mouth reinforcing loop, are the dynamics that lead to a change in the number of riders (i.e., target population) within the system. This loop is powered by word-of-mouth, or the passing of information about the pilot program and its available services from person to person. When the microtransit program first launches, only a small proportion of individuals are aware of the program, while many individuals who need the service are unaware of its existence. Those who are aware increase awareness by telling uninformed individuals about the program. People who become aware become potential riders, and ultimately some join the program, and the number of users increases. In the R1 loop, the 'awareness-raising capacity' represents the maximum number of individuals that can be reached by a single person. As information spreads about the program, there is 'increasing awareness' which creates 'potential riders,' leading to 'riders joining the program,' which increases the number of riders accessing opportunities as shown in Figure 2. These riders then create and receive social benefits, as further identified in the module.

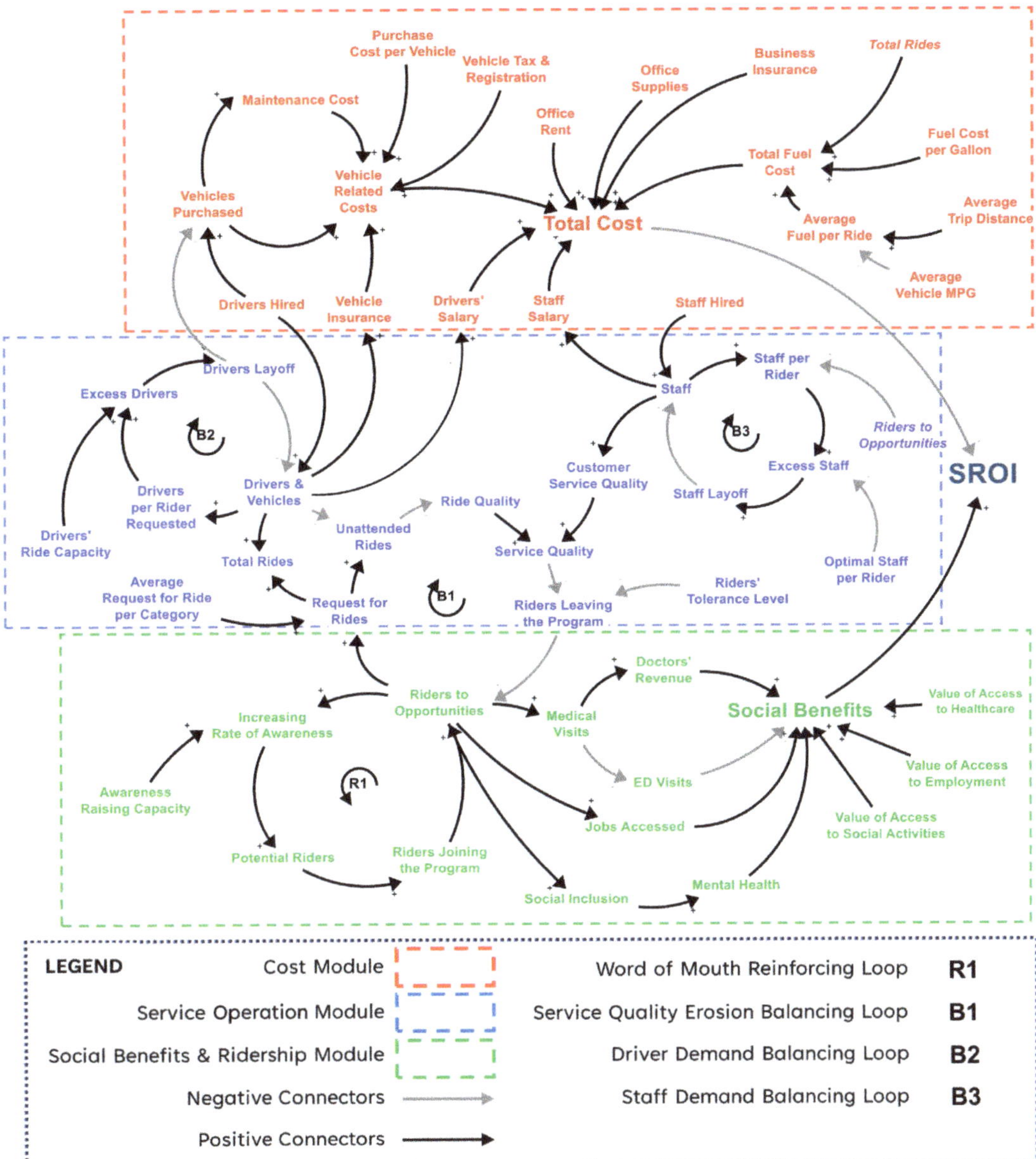

Figure 2. Casual Loop Diagram.

4.3.2. Cost Module

The cost module consists of variables contributing to the total monetary costs of operating the microtransit service; these include vehicle-related costs such as vehicle purchase expenses, vehicle insurance, maintenance, registration, and annual taxes; office-related costs such as building rent and office supplies; salary-related costs such as driver and staff salaries; and fuel-related costs. Many of these costs are dynamic, meaning that they change based on patterns of related variables. For example, fuel costs depend on the quantity and average distance of rides, the miles per gallon (MPG) rating of vehicles, and the fuel price per gallon, while driver and staff costs are calculated based on the number of people hired and salaries. The costs included in the cost module were identified from the literature and based on data provided to the research team by the transportation provider operating the microtransit service. As shown in Figure 2, the total value of the model is captured in terms of total cost.

4.3.3. Service Operation Module

The final module of the CLD is the service operation module. The service operation module consists of a service quality erosion balancing loop, a driver demand balancing loop, and a staff demand balancing loop. Service quality, which is the target variable of the module, contributes directly to riders leaving the program and may impact riders joining the program. The microtransit program's ability to offer rides is limited and reliant on the availability of drivers and cars. If ride requests surpass the program's capacity, some requests go unattended, resulting in decreased ride quality and overall reduced service quality. Likewise, if the number of users exceeds the capacity of the customer service staff, unanswered calls and long wait times for feedback decrease customer service quality. Ultimately, a decrease in service quality, encompassing both ride quality and customer service quality, and driver capacity, causes some users to exit the program.

All three balancing loops of the CLD are located in the service operation module. B1 is the service quality erosion loop, showing that as more users join the program, service quality is likely to be lowered, leading some users to leave the program. B2 and B3 loops, related to driver and staff demand, work similarly. If driver supply exceeds demand, some drivers are let go for balance. B3 mirrors this for staff. Decisions about staff and driver surplus are measured using an optimal rider-staff ratio and a driver capacity measure. A higher staff-to-rider ratio results in more excess staff, and increased driver capacity results in more excess drivers. Finally, the rider tolerance variable informs the acceptance of service quality issues. At a 100% tolerance level, no one leaves the system, even with low service quality. Whereas, with a 10% tolerance level, small changes in service quality can lead to riders exiting the program.

Finally, it should be noted that the CLD model designed for microtransit programs can be adapted to a larger public transit system, such as trains and large buses with fixed routes. The flexible nature of the model can support several adjustments to vehicle types, workforce categories, and broader benefit and cost parameters. For example, large public transit systems encompass various vehicle types and require an expanded workforce, including drivers, maintenance crews, cleaning staff, and more. The social benefits module can also be broadened to address the larger ridership and diverse service areas related to these systems. Additionally, the cost module can consider the increased expenses associated with maintaining a more extensive and varied fleet of vehicles. Infrastructure maintenance and a more diverse workforce, encompassing driver salaries, maintenance personnel, and cleaning staff, can also be integrated into a model for larger transit systems. In the service operation module, balancing loops can manage numerous drivers, staff, and service quality factors on a larger scale, and parameters can be adapted to reflect the complexities and dynamics of a large public transit system. The model can thus be expanded and adjusted to align with the diverse nature and scope of these comprehensive transportation networks.

5. Model Formulation

The model formulation uses SFDs to simulate changes in the system. As shown in Figure 1, the model formulation stage of SD is informed by the SROI step "valuation of outcomes," which relies heavily on the literature to define proxy values that monetize outcomes and, where available, also relies on real-world data, in this case provided by the transportation service provider.

5.1. Valuation of Outcomes

The primary focus of the outcome valuation step is assigning monetary values to the benefit categories identified in the Social Benefits module of Figure 2, including access to healthcare, access to employment, and access to social activities.

The SROISD framework assigns proxy values either through direct valuation or replacement valuation [32]. Direct valuation calculates the tangible benefits stakeholders would gain from an outcome, while replacement valuation assesses the costs that would be incurred without the outcome. For example, when assessing the benefit of transportation access to employment, direct valuation quantifies the total increase in income resulting from securing employment. In contrast, replacement valuation for a funded mental health improvement activity estimates the cost of mental health classes or therapy expenses that would result if mental health services were not covered through a proposed intervention. Proxy values are used in the stock and flow simulation of the social benefits module (see Figure 2) to forecast total annual social benefits gained. Table 2 shows the proxy values assigned to social benefits categories (i.e., access to healthcare, access to employment, and access to social activities). Table 2 shows the variables related to each social benefit category, measures of effectiveness (mostly identified in the literature), valuation methods used, and annual and total proxy values in dollars.

For example, the healthcare access social benefit category is represented by variables including decreased ED visits and increased medical visits, as shown in Figure 2. Based on the literature, an appropriate measure of effectiveness for medical visits is the average cost of a doctor's appointment. This benefit accrues in the system because medical centers, which have been identified as system stakeholders, benefit from the increased number of rides to healthcare in terms of payments made to them. The assumption is that medical centers see an increased number of visits of 1/month or 12/year at an average cost of $450 per visit for a total annual valuation of $5400, as shown in Table 2.

The other access to healthcare variable captured in Table 2 is the number of ED visits avoided per person. Existing literature on the impacts of preventative care shows that, on average, primary care visits can reduce ED visits by approximately 0.34 visits per year [55,60,61]. As such, using replacement valuation, the measure of effectiveness is formulated as the total number of ED visits avoided per year (0.34) multiplied by the average cost of an ED visit, subtracting the cost of monthly primary care appointments in a year.

In the employment access category, direct valuation is used to monetize income gained using data for the average salary in Holmes County, which is estimated at $40,701. This benefit is gained when a person goes to work 260 days in a year (i.e., total number of working days per year [62]). To calculate the benefit per ride, $40,701 is divided by 520, accounting for total rides to and from employment per person in a year, which is calculated at $78.27. Similarly, for healthcare access and social inclusion benefits, the annual proxy values are divided by 24 (monthly rides to and from healthcare) and 156 (seeing friends or relatives once or twice a month [63]), as detailed in Table 2.

The output of the model formulation is stock and flow diagrams. These are discussed in more detail in the section that follows.

Table 2. Proxy Values of Outcomes.

Social Benefit Category	Variables	Measure of Effectiveness	Valuation Method	Annual Proxy Value	Total Proxy Value Per Ride	Source
Healthcare access	- Decreased ER visits. - Decreased ambulance use	[((average cost of ED visits + average cost of ambulance rides) × 0.34) − (average cost of medical appointments × 12)]/24	Replacement valuation	530 + 980 − (12 × 24.04 [2]) = $225	5625/24 = $234.375	[64–67]
	Increased medical visits to healthcare centers	Average cost of doctor's appointments × 12	Direct valuation	450 × 12 = $5400		[64–67]
Employment access	Income gained	Average state salary	Direct valuation	$40,701	40,701/520 = $78.27	[45,68]
Social inclusion	Social participation	Value of seeing friends and relatives once or twice a week − value of seeing friends and relatives once or twice a month	Replacement valuation	£12,000 in 2003; after exchange to USD and adjusted for inflation [1]: $24,824	26,557/156 = $170.23	[32,63,69]
	Improved mental health	Cost of Medicare part b deductible + (average cost therapy sessions Mississippi × 12 × co pay rate of therapy visits) + (cost of 12 months antidepressants × co pay rate of drugs)	Replacement valuation	$1733		[32,70–73]

[1] Inflation is calculated based on [74] through www.measuringworth.com/exchange/. (accessed on 13 November 2022). [2] Calculations are based on values from the State of Mississippi.

5.2. Stock and Flow Diagrams

This step uses proxy values to forecast total annual social benefits using stock and flow diagrams. As was conducted in the CLDs, the SFD is divided into three subsections or modules (see Figure 2). For simplicity, the SFD for the social benefits category "access to employment" is discussed.

SFDs show how quantities accumulate (stocks) and change (flows) over time in a system. While CLDs conceptualize the system, SFDs convert entities of the system into constants and variables, assign values to these variables, use equations to define relationships, simulate the system, and allow for changes over time to be observed.

5.2.1. Social Benefits Module SFD

Figure 3 illustrates the SFD for the "employment" social benefits module. Three key stocks are shown: "Unaware People in Need of Rides", "Potential Riders to Employment", and "Riders to Employment". The diagram encompasses three flows: "Riders to Employment Becoming Aware", "Riders to Employment Joining", and "Riders to Employment Leaving". A comprehensive elaboration on all stocks, flows, and associated variables within the diagram is presented as follows.

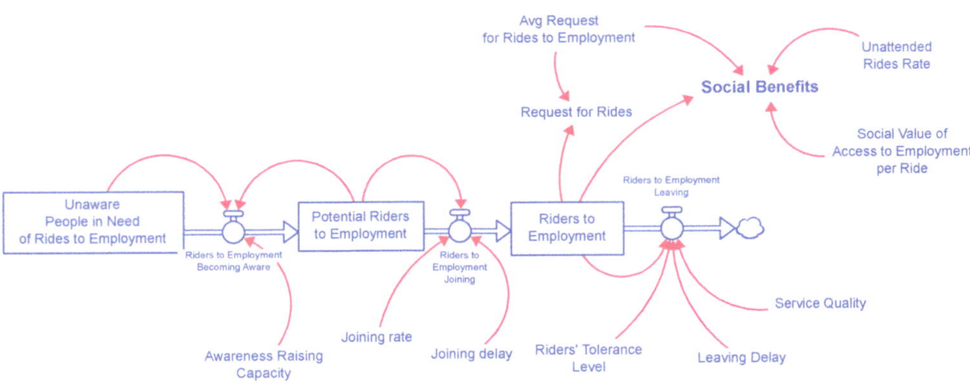

Figure 3. Stock and flow diagram for the social benefits module.

As shown in Figure 3, the social benefits variable is a function of the "unattended rides rate", "requests for rides to employment", and the "social benefit gained per ride to employment" variables. Social benefits are calculated using the following equation:

$$SB = (1 - \lambda)\sum_{i=1}^{3} R_i V_i \overline{n}_i \quad (1)$$

where SB is social benefits, λ is unattended rides rate, which is between 0 and 1, R is quantity of riders to opportunities, V is proxy value of access to opportunities, $\overline{n}$ is average number of requests for rides to opportunities per person, i = 1 represents employment; i = 2 is healthcare; and i = 3 is social activities. The proxy value, or monetized social benefit, of access to employment is identified in Table 2. The components of the social benefits SFD are discussed in more detail below.

Unattended Rides Rate

In Figure 3, The "unattended rides rate" represents the number of ride requests that go unfulfilled due to limitations in driver capacity. The unattended rides rate ensures that benefits accrue based on completed rides, not those that drivers fail to complete. The summation of benefits is thus reduced by this factor, as the appropriate reason for multiplying the summary of benefits by $(1 - \lambda)$ is that social benefits will only be obtained from rides completed and not those that are unattended. The following equation is used to calculate the unattended ride rate (λ):

$$\lambda = \frac{U}{Req} \tag{2}$$

where λ is the unattended ride rate, U is the number of unattended rides, and Req is the total number of requests for rides per year.

Methods for calculating unattended rides (U) and total requests for rides (Req) are described in the service operations module below.

Number of Riders

The next step includes determining the number of riders across three ride categories (see Figure 3). The process of individuals becoming users of the microtransit program does not happen instantly but takes some time. Therefore, in the SFD, three stocks are utilized to represent this process and account for the associated delays. Initially, individuals are in need of rides but are unaware of the program's existence (stock 1: Unaware people in need of rides to employment). Then, they become acquainted with the program through word of mouth and transform into potential riders (stock 2: Potential riders to employment). Finally, they make a decision on whether or not to join the program and become riders (stock 3: Riders to employment), as shown in Figure 3.

Individuals in Need of Rides

The next step is to determine the number of riders in three categories. The process involves a gradual transition of persons who need rides, becoming aware of the microtransit program through word of mouth, and finally deciding to join as riders. Three stocks represent this process, with associated delays.

The first step uses census data to identify the total number of individuals in urgent need of transportation services across three ride categories. Unemployed individuals aged 16–65 without ambulatory difficulties living in zero-vehicle households and employed people who work beyond walking distance from their residence were considered in need of rides to employment [75–78]. To determine the number of individuals in need of rides to healthcare, this paper considers those in zero-vehicle households, aged above 65, and those between 16 and 65 with a disability [76,79]. Furthermore, residents who live in zero-vehicle households can benefit greatly from the social cohesion provided by the microtransit program [79]. Notably, populations from zero-vehicle households reflect a conservative estimate of microtransit program use, potentially leading to increased social benefits.

The Word-of-Mouth Process

The word-of-mouth process involves potential riders (individuals aware of the program) informing uninformed individuals about it (see Figure 3). It is important to note that a person can only inform those whom they interact with. McCormick et al., [80,81], found that, on average, each person interacts with 600 individuals. As such, this study estimates "Awareness Raising Capacity" at 3.6%, which is equal to 600 people in Holmes County. After becoming aware of the program, potential riders will decide whether or not to join.

Decision to Join or Leave the Program

The joining rate represents the proportion of potential riders who choose to participate. In Holmes County, where transportation options are limited, it is assumed that half of all potential riders will join the program within an average delay of three months. Notably, different values for the above variables were tested during the sensitivity analysis to assess their impact on the system's performance and to check the sensitivity of the model to assumptions, as described in the sensitivity analysis section below.

The service quality variable, assessed within the service module, plays a crucial role in influencing the number of users who decide to leave the service. When there is a decline in service quality, individuals are less likely to stay in the program, based on their tolerance level. Both service quality and rider tolerance level are measured on a scale ranging from 0 to 1. Moreover, it is assumed that a continuous period of 9 months with declining service quality is required for individuals to decide about leaving the program. The equation used to determine riders' departure from the service is as follows:

$$RD = \frac{R_i(1 - TL)(1 - Q)}{LD} \tag{3}$$

where RD is riders' departure, R is number of riders, LD is leaving delay, TL is riders' tolerance level, and Q is service quality.

According to the equation, a tolerance level of 1 (the maximum tolerance) results in 0 riders leaving. Other factors like relocation or job loss are not considered in this equation, which focuses solely on riders' decisions based on service quality.

Finally, the request for rides variable is calculated using the following equation:

$$Req = \sum_{i=1}^{3} R_i \overline{n}_i \tag{4}$$

where Req is request for rides, R is number of riders, and $\overline{n}$ is average number of requests for rides to opportunities per person.

Equation (3) above states that the total number of requests for rides that are submitted to the program each year is calculated by multiplying the number of riders within each distinct category—namely healthcare; employment; and social participation—by the corresponding average request count for that specific category.

5.2.2. Service Operations Module SFD

Figure 4 shows the SFD for the service operations module. The main stocks of this SFD are "Drivers", which represents the number of drivers in the program, and "Staff", which shows the number of customer service staff. The 4 flows are "Hiring New Drivers", "Drivers Leaving the Program", "Hiring New Staff", and "Staff Layoff".

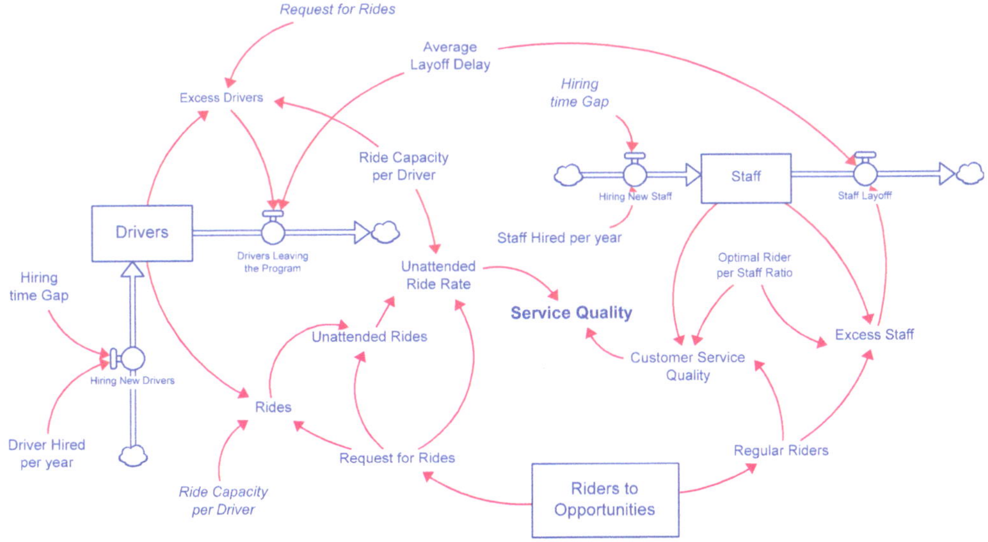

Figure 4. Stock and flow diagram for the service module.

The target variable of the module is service quality, combining the unattended ride rate and customer service quality. Drivers in the program are estimated to have an average ride capacity of 2600 per year, or approximately 10 rides per day across 260 working days. Hence, certain ride requests go unattended as a result of driver ride capacity reaching its limit. Calculating the number of these unattended rides necessitates the application of a non-linear function. To accomplish this, the IF, THEN, and ELSE functions within Stella were used, as represented by the piecewise function below. The approach for computing unattended rides is outlined in Equation (5):

$$f(U) = \begin{cases} Req - (D \times C), & if \ \frac{Req}{D \times C} > 1 \\ 0, & if \ \frac{Req}{D \times C} \leq 1 \end{cases} \quad (5)$$

where U is unattended rides, D is the number of drivers, and C is the ride capacity per driver.

Conversely, the number of rides can be calculated simply by subtracting unattended rides from requests for rides.

According to the B2 loop in the CLD (Figure 2), the program keeps the number of drivers at an optimum level to minimize the cost. Therefore, it is needed first to find the number of excess drivers using the following piecewise function:

$$f(S) = \begin{cases} \left\lfloor Staff - \frac{Reg}{OR} \right\rfloor, & if \ Staff > \frac{Reg}{OR} \\ 0, & if \ Staff \leq \frac{Reg}{OR} \end{cases} \quad (6)$$

where S is excess staff, Reg is regular riders, and OR is optimal rider per staff. *Staff* leaving the program is also calculated by multiplying excess staff by the average layoff delay.

The variable customer service quality, which is between 0 and 1, is calculated using the following:

$$f(CQ) = \begin{cases} 1, & if \ \frac{Staff \times OR}{Reg} \geq 1 \\ \frac{Staff \times OR}{Reg}, & if \ \frac{Staff \times OR}{Reg} < 1 \end{cases} \quad (7)$$

Finally, the service quality, which is the average of attended rides and customer service quality, is calculated as follows:

$$Q = \frac{CQ + (1 - \lambda)}{2} \quad (8)$$

5.2.3. Cost Module SFD

Figure 5 illustrates the SFD for the cost module. The stocks and flows in this SFD are explained in Service Operations Module SFD (Section 5.2.2).

Vehicle-related costs are a major expense category for microtransit. In this model, vehicle-related costs are calculated as follows, and all the variables are explained in detail in the subsequent section (fuel costs are addressed separately):

$$\text{Vehicle related cos ts} = (\text{Vehicle Purchased} \times \ (1 + \text{Tax}) \times \ \text{Price}) + \text{Cars} \times (\text{Insurance} + \text{Maintenance} + \text{Registration}) \quad (9)$$

In this case, it is assumed that the microtransit program purchases one vehicle per driver hired, unless there are spare vehicles available. Spare vehicles become available once a driver leaves the program. Therefore, the number of vehicles purchased each year is equal to the number of drivers hired minus the number of drivers leaving the program. Each vehicle in Mississippi incurs a 5% purchase tax [82]. On average, each vehicle has an annual insurance cost of $1471 [83] and an annual registration cost of $14 [84]. Average maintenance costs are estimated at $506 per year [85], including repairs and oil changes.

Other important costs considered in this analysis were personnel costs, office-related costs, and fuel costs. According to Salary.com, the median driver wage and median salary of a Customer Service Representative in Mississippi were estimated at $32,786 and $31,216, respectively [86,87]. Office-related costs were estimated at $36,000 per year, which is the average annual cost to rent a 1500 sqft class A office space in Mississippi [88]. Finally, fuel costs were determined by multiplying the average fuel consumed per ride, the number of rides, and the average fuel cost per gallon. Average fuel use per ride was estimated to be a function of average vehicle MPG (i.e., 31.7 mpg) [89] and average trip distance, which based on Holmes County data (i.e., reference data) were 6.3 miles [90]. Full details about variables included in the Cost Module, including equations, properties, units, and corresponding values, can be found in Table S1.

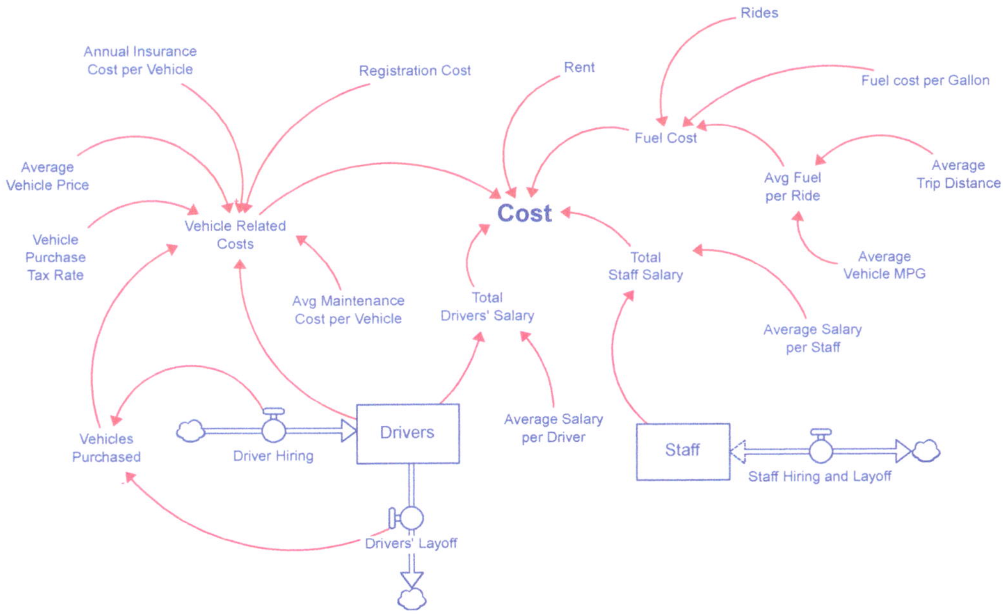

Figure 5. Stock and flow diagram for the cost module.

6. Model Validation

According to Sterman [58], all models, whether mental or formal representations, are simplified versions of the real world. In the field of SD, model validation remains a critical stage in the development process to ensure that constructed models accurately represent the structure and behavior of complex systems [91]. Model validation consists of several stages and plays a vital role in confirming the accuracy and reliability of models.

This study conducts multiple tests, drawing from existing literature and extending methods employed in prior research [58,91,92], including structure verification, parameter verification, dimensional consistency, and a behavior reproduction test. The initial tests focused on structure and parameter verification. The structure verification test ensured that all parameters and relationships within the model were representative of the context being investigated, which is a microtransit program in a low-income community. During the structure verification test, variables and relationships outlined in the CLD (Figure 2) were also checked to ensure that they represented characteristics of microtransit programs across the US (Table 1) as well as data provided by the microtransit program operator in Holmes County. Finally, variables were checked against findings from the literature, as they are cited in Section 4. In the parameter verification test, all values of variables used in the model were checked to ensure they were based on existing literature, which is all cited in Section 5.2 (Stock and Flow Diagram). Next, a dimensional consistency test was used to check each variable's units; this check ensured dimensional equivalence and consistency. Table S1 provides a detailed overview of the verified units.

The final test conducted was a behavior reproduction test, where model results were compared to historical data provided by the microtransit program operator. To accomplish this, the values of constants, such as the number of drivers, driver salaries, staff salaries, etc., are adjusted based on the data at the time of the program. For example, since the drivers in the Holmes County program were unpaid volunteers, the average driver salary was set to zero in the model. Subsequently, a one-way Analysis of Variance (ANOVA) test is conducted, and the R-squared metric is measured to determine if there are statistically significant differences in means between the results of the model and historical data.

Figure 6 shows a comparison between model output and real data gathered during the fall 2021 piloting of the microtransit program in Holmes County, including F-statistics, p-values, and R-squared values. Results indicate that p-values for both results are higher than the confidence value (α) of 0.05 and R-squared values are above 0.8, meaning that there is no statistically significant difference between the results of the model and historical data.

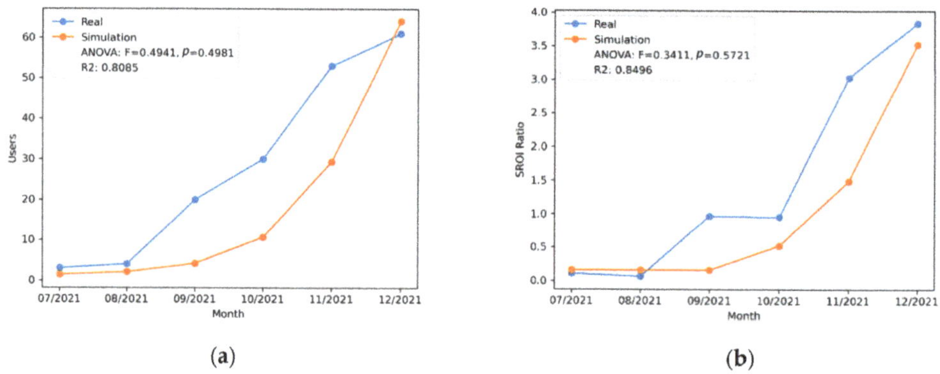

Figure 6. Results of the Behavior Reproduction Test: (**a**) Number of riders; (**b**) SROI ratio.

7. Results and Discussion

7.1. Scenario Analysis

The microtransit program's social benefits depend on completed rides, which are influenced by its capacity, primarily determined by hiring drivers and staff. To assess different hiring options, 5 scenarios are formulated: scenario 1—hires 2 drivers and 2 staff per year; with the number of drivers and staff hired increasing by 2 in each subsequent scenario. Scenario 5 hires 10 drivers and 10 staff per year. Stella software was used to simulate each of the five scenarios and observe the corresponding changes in the *SROI* ratio. The target variable of the overall model was calculated as follows:

$$SROI = \frac{Present\ Value\ of\ Benefits}{Present\ Value\ of\ Investments} \tag{10}$$

In the SROISD model, the present value of benefits is represented as "Social Benefits" (the target variable of the social benefits module), and the present value of investments is represented as "Costs" (the target variable of the cost module). Scenario analysis outcomes and results are discussed below.

Dynamic changes in the model's primary output variable, SROI, are shown in Figure 7, which shows SROI ratios for each scenario spanning a 10-year period. The graph demonstrates a clear trend wherein the SROI ratio rises as the program allocates more funds towards the recruitment of drivers and staff. These findings indicate that increasing investment in human resources leads to higher social returns. An analysis of the budget allocated to the program each year reveals that the SROI ratio can vary from 4 in the initial year to exceeding 6 by the end of the ten-year timeframe. This implies that for every $1 invested in the program, a social benefit of $4 to $6 can be realized. These results highlight the program's potential for generating significant social value and offer a strong rationale for allocating resources towards the recruitment of drivers and staff to maximize social returns. Nevertheless, it can be observed that in the initial year, Scenario 1 produces the highest SROI ratio. This implies that the hiring of two drivers and staff suffices for the initial phase. However, as the user base expands, a greater number of drivers and staff members become needed.

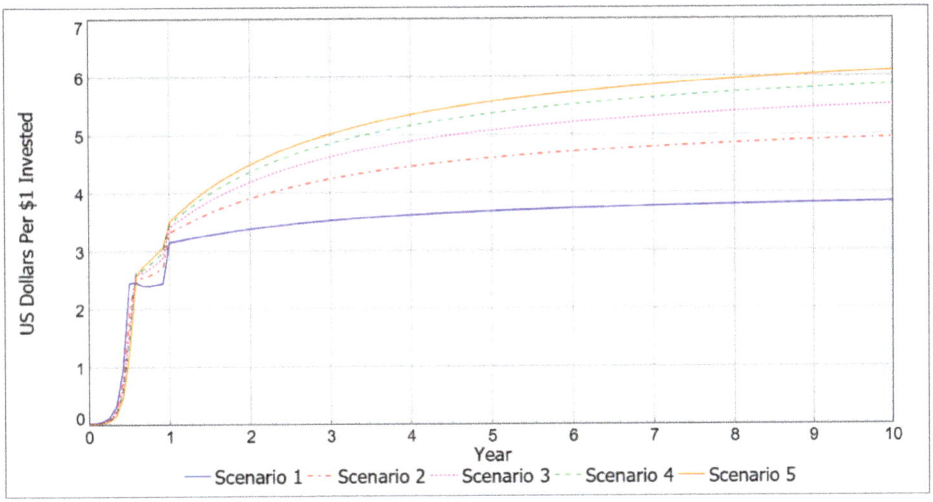

Figure 7. SROI Ratios per Scenario.

Figures 8 and 9 show trends in the number of unattended rides and service quality across each scenario. The findings show that hiring 10 drivers and 10 staff leads to a higher SROI ratio in the span of 10 years. As the number of drivers and staff increases, customer service quality improves, and the rate of unattended rides decreases.

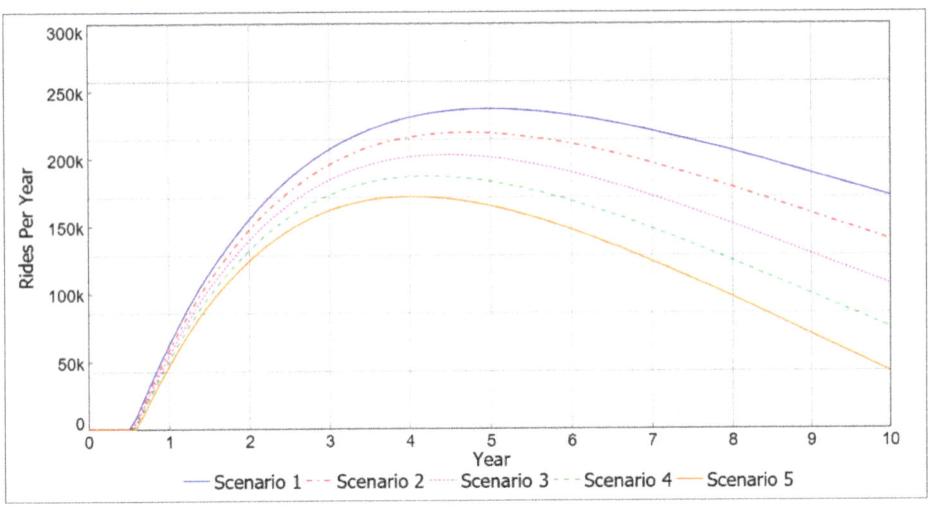

Figure 8. Unattended Rides per Scenario.

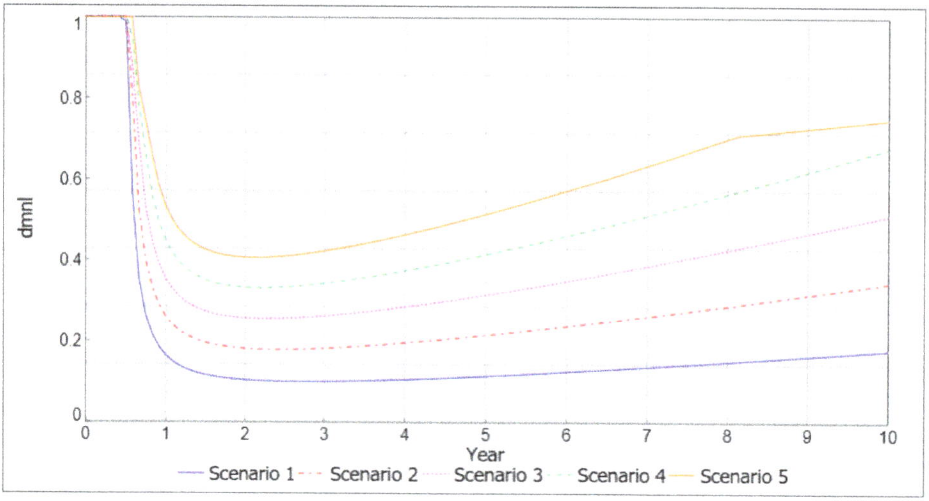

Figure 9. Service Quality per Scenario.

Figure 8 highlights a noteworthy trend where the number of unattended rides shows an increase during years 3 to 6. This rise can be attributed to the growth in the number of requests for rides, which outpaces the available number of drivers hired during this period. However, as the recruitment of drivers gradually catches up with the increasing demand, the number of unattended rides starts to fall.

These results underscore the importance of maintaining an adequate workforce to meet customer demand and ensure high service quality. Increasing the number of drivers in line with the growth in ridership is thus a likely strategy for effectively tackling the problem of unattended rides and improving the customer experience.

7.2. Sensitivity Analysis

Sensitivity analysis plays a crucial role in both SROI and SD studies. According to Nicholls et al. [37], sensitivity analysis is a key step in calculating SROI, allowing for the identification of proxy values that have the most significant impact on model output. In SD simulations, sensitivity analysis serves as a valuable tool for assessing the reliability of conclusions given uncertainties in the assumptions made during the system conceptualization and model formulation phases [58].

The SROISD model focuses on the SROI ratio as the primary output variable. The sensitivity analysis is thus performed by varying the SROI ratio. Figure 10 depicts the results of a sensitivity analysis performed using Stella software and plotted with the Matplotlib library in Python. Each line represents the relationship between variable changes (x-axis) and their corresponding impact on the SROI ratio (y-axis). Steeper line slopes indicate greater sensitivity of the SROI ratio to the variable being tested.

The SROI ratio is most sensitive to the following three variables: average salary per staff member, average salary per driver, and ride capacity per driver. As shown in Figure 10, decreasing average staff salary and average salary per driver by 80% leads to a around 125% increase in SROI after 10 years. On the other hand, a 20% increase in ride capacity per driver (from 10 to 12 rides per day) leads to a 22% increase in SROI. Additionally, as shown in Figure 10, sensitivity analysis of the social values (i.e., proxy values) of all benefit categories revealed that the social value of access to employment has the strongest impact on SROI. An 80% increase in social value of access to employment boosts the SROI ratio by around 40%, while an 80% decrease in social value of access to employment reduces SROI by 45%. Access to social activities shows relatively similar results. SROI is least sensitive to the social value of access to healthcare.

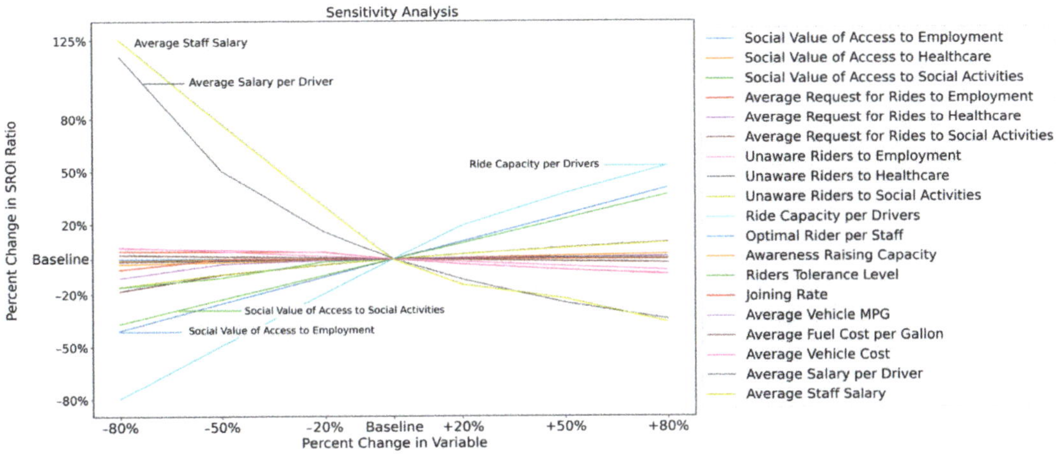

Figure 10. Sensitivity Analysis Results.

7.3. Policy Analysis and Implications

The SROISD model has several implications for transportation program runners. This model equips decision-makers and program administrators with a powerful tool to make informed financial and operational decisions. It allows them to not only assess the overall impact and returns of a microtransit program and other similar shared mobility services but also provides the flexibility to determine the magnitude and timing of resource allocations. The application of the SROISD approach to microtransit, as a proxy and example of transportation in general, showcases the adaptability of the methodology and its potential use for understanding various transportation programs by simply calibrating variables and adjusting relevant parameters. Decision-makers can use the model to optimize resource allocation decisions based on budget constraints and priorities. They can make data-driven decisions on when and where to invest, whether it is in expanding the fleet of vehicles, hiring additional staff, or incentivizing drivers. This level of precision in resource allocation is vital to ensuring that limited resources are utilized efficiently and social benefits are maximized.

Furthermore, the SROISD model offers insights into program dynamics over time, enables transportation decision makers to evaluate long-term sustainability and effectiveness, and identifies key points for potential adjustment and refinement. This flexibility not only benefits decision makers but also community members and other stakeholders, as the data driven, adaptive approach provided by SROISD expands the understanding of microtransit program reach and success.

Based on the results of our model, several policy considerations become apparent. First, the model demonstrates that policymakers may consider implementing optimization measures for driver and staff hiring to improve the program's SROI over time. However, it is essential to carefully evaluate the potential trade-offs between cost savings, service quality, and financial earnings. Secondly, according to sensitivity analysis (Section 7.2), the model highlights the significance of ride capacity per driver in influencing both the SROI ratio and service quality. Policymakers should explore strategies to optimize ride capacity per driver, for example, through driver training and route planning, as these actions significantly impact driver capacity and subsequently SROI, as shown in Figure 10. Additionally, policymakers may consider ride sharing as an approach to increasing driver capacity, as discussed below.

It can be concluded from the results of the sensitivity analysis (Figure 10) that incorporating ride-sharing initiatives can have a positive impact on both the SROI ratio and the service quality of the microtransit program. By encouraging ridesharing among passengers with similar routes or destinations, policymakers can optimize vehicle capacity and reduce operational costs. Particularly for access to employment services, which have the highest average request for rides, a ride share of at least 2 people in one vehicle results in a 50% decrease in requests for rides to employment and an increase of approximately 5% and 13% in the SROI ratio and service quality, respectively.

In fact, by incorporating a ridesharing variable into the model, we can further explore the effects of ridesharing on the SROI ratio. This variable is multiplied by the rides requested and divided by the social benefit variable. When we set the ridesharing variable at different values, such as 0.5, 0.4, 0.3, 0.2, or 0.1, it represents the average proportion of rides that are shared among two individuals: 100%, 80%, 60%, 40%, or 10%, respectively. We conducted tests using different values for the ridesharing variable, and the results are presented in Figures 11 and 12. These findings demonstrate a positive relationship between ridesharing and the SROI ratio, providing further evidence of the beneficial impacts of ridesharing.

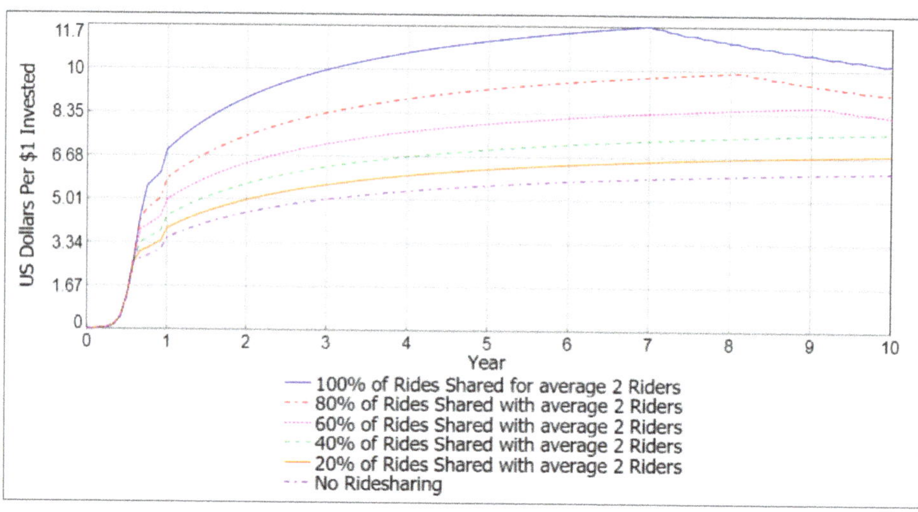

Figure 11. Effect of Ridesharing on SROI.

Policymakers can explore various strategies to promote ride sharing, such as implementing technology solutions that facilitate matchmaking between passengers with compatible travel plans. By leveraging the power of ridesharing, the microtransit program can achieve higher SROI and enhanced service quality, further benefiting the low-income area and its residents.

Finally, efforts to increase the average request for rides to employment can enhance the overall program impact. Policy interventions may involve targeted campaigns, partnerships with local employers, or improved accessibility to employment centers. By considering these strategies, decision-makers can make informed choices to maximize the social benefits of the microtransit program in the low-income area while balancing the effects of salary reductions on staff and drivers.

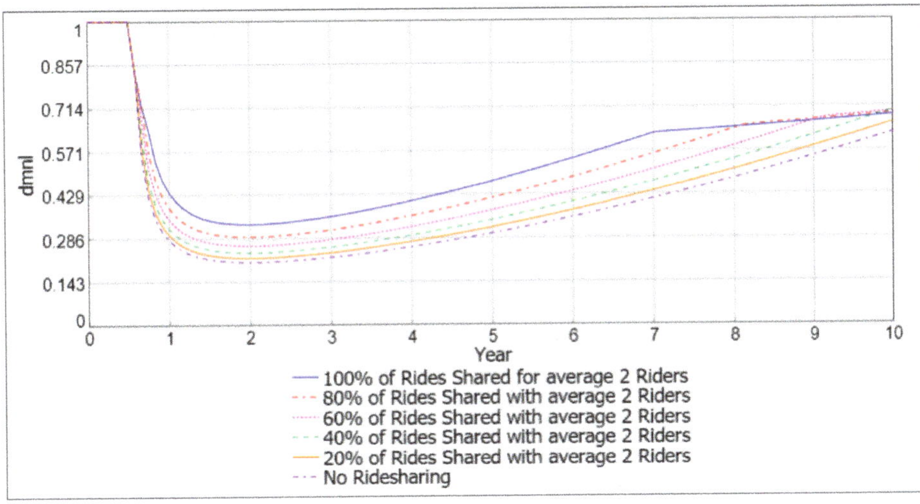

Figure 12. Effect of Ridesharing on Service Quality.

8. Conclusions

Reliable and accessible transportation plays a crucial role in low-income areas by connecting communities to essential resources. Transportation is not only an SDOH but also helps to address barriers that people may face in accessing other SDOH categories such as healthcare, employment, and education. Transportation availability thus not only improves personal mobility but can also help break the cycle of poverty by providing access to opportunities. Previous studies have highlighted the importance of microtransit programs as reliable transportation solutions that aim to address barriers to SDOH. However, there remains a need for an approach to effectively measure the long-term social and economic impact of microtransit and other on-demand transit systems. SROI is a comprehensive method that takes into account the perspectives of stakeholders and considers both the social benefits and project revenues, going beyond the traditional ROI approach. However, microtransit costs are not typically incurred upfront, and both costs and social benefits derived from such programs are subject to change over time due to the effects of several factors. This study presents an innovative SD-based model called SROISD, which offers a 10-year forecast of the SROI for a microtransit program in a low-income area. The model takes into consideration various factors that influence the costs and social benefits generated by such a program, as well as the complex interactions between them. To illustrate the application of the model, results from a case study analysis in Holmes County, MS, are reported.

Results of this case analysis suggest that microtransit, depending on the amount of money invested, can offer a social return, or SROI, where social benefits gained from the program outweigh costs by approximately 4 to 6 times. The model further suggests that an increase in the number of rides that one driver can accommodate per day has a significant impact on the SROI ratio over time, offering support for the concept of ridesharing in microtransit. Results of this case analysis further indicate that potential benefits derived from microtransit are notably higher than costs when considering access to SDOH categories (e.g., healthcare, employment, and social activities) and impacts on stakeholders, including riders, healthcare providers, and transportation operators.

Overall, the results underscore the considerable positive impact of microtransit on enhancing access to crucial services, promoting sustainability, and fostering social inclusion. Furthermore, these findings provide invaluable insights for decision-makers seeking to optimize resource allocation over the long term. The model empowers them to make data-driven decisions, determining both the degree and timing of resources allocated to the microtransit program. Such strategic flexibility is vital to ensuring that limited resources are utilized efficiently while maximizing the program's social benefits. Additionally, the model offers a dynamic view of the program over time, allowing decision-makers to assess its sustainability and effectiveness, which, in turn, supports the long-term success of the program.

This study contributes to the existing knowledge of SROI and SD by applying SD to forecast SROI, providing a dynamic, forward-looking perspective. SROI stands as a superior metric for evaluating the impact of transportation programs on SDOH when compared to traditional ROI approaches. Unlike conventional ROI, which primarily focuses on monetary returns, SROI encompasses a broader spectrum of social benefits and costs, resulting in a more comprehensive evaluation of social impact. Furthermore, SROISD sets itself apart by offering substantial advantages in comparison to other studies that have employed SROI. While SROI methodologies have been utilized for forecasting, they are often static in nature and assume upfront costs and immediate returns. In contrast, the SROISD model is useful in the context of transportation programs, where costs dynamically evolve over time. The dynamic modeling capabilities of SD allow for the accurate tracking of both costs and social returns, offering decision-makers a clearer understanding of the intricate financial and social dynamics at play.

Moreover, SROISD introduces the pivotal capability of forecasting, a feature that is challenging in traditional SROI methodologies, particularly when it comes to transportation programs with evolving costs and benefits. While SROI has been used for forecasting in certain contexts, it assumes upfront costs and does not align well with transportation programs, where costs and benefits change dynamically over time. In contrast, the SROISD model excels in this regard, providing the necessary flexibility for forecasting within such complex and evolving systems.

This study, despite its significant contributions, had several limitations that warrant further investigation. Firstly, while a microtransit program can benefit multiple stakeholders, this study focused only on those who would experience significant benefits (i.e., riders to employment, riders to healthcare, riders to social activities, and healthcare providers), leaving room to explore the potential advantages for other stakeholders in future research. Additionally, it is important to note that microtransit programs can have environmental impacts that should be considered in future work, and the associated technology costs need closer examination to provide a comprehensive understanding of the program's overall implications. Secondly, the evaluation of transportation interventions typically involves quality of life surveys like health-related quality of life (HRQOL) and measures such as the quality-adjusted life-year (QALY). However, as this study aimed to forecast social benefits, the feasibility of utilizing measures like QALY was limited, as surveys needed to be conducted pre- and post-program, necessitating the reliance on alternative measures such as the number of ED visits avoided, incomes gained, benefits gained from seeing friends and relatives, etc. Future investigations could explore the inclusion of QALY and HRQOL to provide a more comprehensive evaluation. Moreover, this study focused on three access categories—health; employment; and social activities—to demonstrate the methodological application of the SROISD model; disregarding other potential social benefits that microtransit programs offer; such as access to education, food, etc. Future studies should expand the analysis to encompass a broader range of service categories to capture the full scope of social benefits provided by microtransit programs. Additionally, variables such as the leaving delay for riders and the average layoff delay for drivers and staff rely on researcher assumptions due to challenges faced in their precise measurement. These assumed variables underwent additional calibration to align them with the reference data. Furthermore, during sensitivity analysis, delay variables were checked to determine their impacts on model output. Findings indicated a limited impact on the SROI output variable.

To further develop the SROISD model, future work can include additional stakeholders and aim to capture a wider range of social benefits. Such expanded analysis would provide more comprehensive insights into the social advantages associated with microtransit programs. Additionally, conducting a survey with stakeholders would be instrumental in exploring their viewpoints and gaining a better understanding of the social benefits they perceive. Furthermore, recent advancements in microtransit programs have seen the incorporation of technologies such as in-vehicle cameras. Future work should consider the costs and benefits associated with integrating these technologies into the analysis. This could involve evaluating how these technologies impact safety, service quality, and operational efficiency, as well as assessing the potential privacy and data security implications. Moreover, it is crucial to acknowledge that the emergence of new ridesharing modes, such as robotaxis, has the potential to alter outcomes related to SROI. These innovations may reduce driver salary costs, decrease unattended rides, and increase trip chaining possibilities, further improving service operation efficiencies. As technologies advance, population needs change, and opportunities expand, there will be an ongoing need to evaluate SROI and the impact of new emerging factors.

Supplementary Materials: The following supporting information can be downloaded at: https://www.mdpi.com/article/10.3390/systems11110538/s1, Table S1. Description of Variables.

Author Contributions: Conceptualization, M.M. and J.S.-C.; Data curation, M.M. and J.S.-C.; Formal analysis, M.M. and J.S.-C.; Methodology, M.M. and J.S.-C.; Resources, J.S.-C.; Software, M.M.; Supervision, J.S.-C.; Validation, M.M. and J.S.-C.; Visualization, M.M.; Writing—original draft, M.M. and J.S.-C.; Writing—review and editing, J.S.-C. All authors have read and agreed to the published version of the manuscript.

Funding: This research was completed with support from the National Academies of Science, Engineering, and Medicine (NASEM) Gulf Research Program, Early Career Fellowship—Human Health and Resilience Track, Grant #2000012306.

Data Availability Statement: Limited data is available online as referenced in this paper. some data is not publicly available due to privacy restrictions.

Acknowledgments: We would like to extend our sincere appreciation to Feonix—Mobility Rising for their invaluable support and cooperation in providing data and insights that greatly contributed to the success of this research. Feonix—Mobility Rising has explicitly consented to be acknowledged in this publication.

Conflicts of Interest: The authors declare no conflict of interest.

References

1. Who.int Social Determinants of Health. Available online: https://www.who.int/health-topics/social-determinants-of-health (accessed on 2 February 2023).
2. CDC.com about Social Determinants of Health. Available online: https://www.cdc.gov/socialdeterminants/about.html (accessed on 14 March 2022).
3. Ogunwole, S.M.; Golden, S.H. Social Determinants of Health and Structural Inequities—Root Causes of Diabetes Disparities. *Diabetes Care* **2021**, *44*, 11–13. [CrossRef] [PubMed]
4. Wolfe, M.K. Access to Health Care: Perspectives on Transportation as a Social Determinant of Health. Ph.D. Dissertation, The University of North Carolina at Chapel Hill University Libraries, Chapel Hill, NC, USA, 2020.
5. Butler, S.M. What Is the Outlook for Addressing Social Determinants of Health? *JAMA Health Forum* **2021**, *2*, e213639. [CrossRef] [PubMed]
6. Transportation and Social Determinants of Health Destinations. Available online: https://nationalcenterformobilitymanagement.org/transportation-and-social-determinants-of-health-destinations/ (accessed on 13 October 2023).
7. Weinstein, J.N.; Geller, A.; Negussie, Y.; Baciu, A. National Academies of Sciences, Engineering, and Medicine Committee on Community-Based Solutions to Promote Health Equity in the United States. In *Communities in Action: Pathways to Health Equity*; National Academies Press: Washington, DC, USA, 2017.
8. Bardaka, E.; Hajibabai, L.; Singh, M.P. Reimagining Ride Sharing: Efficient, Equitable, Sustainable Public Microtransit. *IEEE Internet Comput.* **2020**, *24*, 38–44. [CrossRef]

9. Macfarlane, G.S.; Hunter, C.; Martinez, A.; Smith, E. Rider Perceptions of an On-Demand Microtransit Service in Salt Lake County, Utah. *Smart Cities* **2021**, *4*, 717–727. [CrossRef]
10. Doyle, T. *American Public Transportation Association Releases New Mobility Innovation Report*; American Public Transportation Association: Washington, DC, USA, 2021.
11. Shared Mobility Definitions | FTA. Available online: https://www.transit.dot.gov/regulations-and-guidance/shared-mobility-definitions (accessed on 21 April 2022).
12. Rossetti, T.; Broaddus, A.; Ruhl, M.; Daziano, R. Commuter Preferences for a First-Mile/Last-Mile Microtransit Service in the United States. *Transp. Res. Part A Policy Pract.* **2023**, *167*, 103549. [CrossRef]
13. National Academies of Sciences, Engineering, and Medicine. *Microtransit or General Public Demand–Response Transit Services: State of the Practice*; The National Academies Press: Washington, DC, USA, 2019.
14. THE 17 GOALS | Sustainable Development. Available online: https://sdgs.un.org/goals (accessed on 6 October 2023).
15. How Microtransit Helps Reduce Emissions. Available online: https://ridewithvia.com/resources/articles/how-microtransit-helps-reduce-emissions/ (accessed on 6 October 2023).
16. Rides to Wellness | Genesee County | MTA Flint. Available online: https://www.mtaflint.org/rides-to-wellness/ (accessed on 12 March 2022).
17. Osman, M. *Pinellas Suncoast Transit Authority's TD Late Shift Program*; American Public Transportation Association: Washington, DC, USA, 2019.
18. Gosselin, V.; Boccanfuso, D.; Laberge, S. Social Return on Investment (SROI) Method to Evaluate Physical Activity and Sport Interventions: A Systematic Review. *Int. J. Behav. Nutr. Phys. Act.* **2020**, *17*, 26. [CrossRef]
19. Feigon, S.; Murphy, C. *Shared Mobility and the Transformation of Public Transit*; American Public Transportation Association: Washington, DC, USA, 2016; ISBN 0-309-44582-5.
20. GoLink. Available online: https://www.dart.org/guide/transit-and-use/golink (accessed on 11 October 2023).
21. Jeon, C.M.; Amekudzi, A.A.; Guensler, R.L. Sustainability Assessment at the Transportation Planning Level: Performance Measures and Indexes. *Transp. Policy* **2013**, *25*, 10–21. [CrossRef]
22. Bertini, R.L.; El-Geneidy, A. Generating Transit Performance Measures with Archived Data. *Transp. Res. Rec.* **2003**, *1841*, 109–119. [CrossRef]
23. FlexRIDE On-Demand Service. Available online: https://www.rtcwashoe.com/public-transportation/flexride/ (accessed on 30 March 2022).
24. Shaheen, S.; Stocker, A.; Lazarus, J.; Bhattacharyya, A. *RideKC: Bridj Pilot Evaluation: Impact, Operational, and Institutional Analysis*; UC Berkeley: Berkeley, CA, USA, 2016; p. 67.
25. Available online: https://dialabus.org.uk/about/ (accessed on 16 October 2023).
26. Flexible On-Demand Transport Made to Fit Your Needs. Available online: https://www.gvh.de/en/timetable/sprinti/ (accessed on 16 October 2023).
27. Flexible Local Transport. Available online: https://translink.com.au/travel-with-us/on-demand (accessed on 17 October 2023).
28. Kempton, O.; Warby, A. *Measuring the Social Return on Investment of Stage 3 Adaptations and Very Sheltered Housing in Scotland*; Envoy Partnership: London, UK, 2012.
29. Emerson, J.; Wachowicz, J.; Chun, S. Social Return on Investment: Exploring Aspects of Value Creation in the Nonprofit Sector. *Box Set Soc. Purp. Enterp. Ventur. Philanthr. New Millenn.* **2000**, *2*, 130–173.
30. Emerson, J.; Twersky, F. *New Social Entrepreneurs: The Success, Challenge and Lessons of Non-Profit Enterprise Creation*; Homeless Economic Fund, the Roberts Foundation: San Fransisco, CA, USA, 1996.
31. Aeron-Thomas, D.; Nicholls, J.; Forster, S.; Westall, A. *Social Return on Investment: Valuing What Matters*; New Economics Foundation: London, UK, 2004.
32. Hutchinson, C.; Berndt, A.; Cleland, J.; Gilbert-Hunt, S.; George, S.; Ratcliffe, J. Using Social Return on Investment Analysis to Calculate the Social Impact of Modified Vehicles for People with Disability. *Aust. Occup. Ther. J.* **2020**, *67*, 250–259. [CrossRef]
33. Millar, R.; Hall, K. Social Return on Investment (SROI) and Performance Measurement: The Opportunities and Barriers for Social Enterprises in Health and Social Care. *Public Manag. Rev.* **2013**, *15*, 923–941. [CrossRef]
34. Cordes, J.J. Using Cost-Benefit Analysis and Social Return on Investment to Evaluate the Impact of Social Enterprise: Promises, Implementation, and Limitations. *Eval. Program Plan.* **2017**, *64*, 98–104. [CrossRef] [PubMed]
35. Kousky, C.; Ritchie, L.; Tierney, K.; Lingle, B. Return on Investment Analysis and Its Applicability to Community Disaster Preparedness Activities: Calculating Costs and Returns. *Int. J. Disaster Risk Reduct.* **2019**, *41*, 101296. [CrossRef]
36. McGrath, R.; Stevens, K. Forecasting the Social Return on Investment Associated with Children's Participation in Circus-Arts Training on Their Mental Health and Well-Being. *Int. J. Sociol. Leis.* **2019**, *2*, 163–193. [CrossRef]
37. Nicholls, J.; Lawlor, E.; Neitzert, E.; Goodspeed, T. *A Guide to Social Return on Investment*; Office of the Third Sector, The Cabinet Office: London, UK, 2012.
38. Bellucci, M.; Nitti, C.; Franchi, S.; Testi, E.; Bagnoli, L. Accounting for Social Return on Investment (SROI): The Costs and Benefits of Family-Centred Care by the Ronald McDonald House Charities. *SEJ* **2019**, *15*, 46–75. [CrossRef]
39. Bottero, M.; Ambrosini, G.; Callegari, G. Valuing the Impact of Social Housing Renovation Programs: An Application of the Social Return on Investment (SROI). In *Appraisal: From Theory to Practice*; Stanghellini, S., Morano, P., Bottero, M., Oppio, A., Eds.; Green Energy and Technology; Springer International Publishing: Cham, Switzerland, 2017; pp. 291–302, ISBN 978-3-319-49675-7.

40. Drabo, E.F.; Eckel, G.; Ross, S.L.; Brozic, M.; Carlton, C.G.; Warren, T.Y.; Kleb, G.; Laird, A.; Pollack Porter, K.M.; Pollack, C.E. A Social-Return-On-Investment Analysis Of Bon Secours Hospital's 'Housing For Health' Affordable Housing Program: Study Evaluates the Broader Social, Environmental, and Economic Benefits of Bon Secours Hospital's Housing for Health Program. *Health Aff.* **2021**, *40*, 513–520. [CrossRef]
41. Miller, M.C.; Rueda, J.A.; Gransberg, D.D. Applying Social Return on Investment to Risk-Based Transportation Asset Management Plans in Low-Volume Bridges. *Transp. Res. Rec.* **2015**, *2473*, 75–82. [CrossRef]
42. Ventura, R.; Bonera, M.; Carra, M.; Barabino, B.; Maternini, G. Evaluating the Viability of a Tram-Train System. A Case Study from Salento (Italy). *Case Stud. Transp. Policy* **2022**, *10*, 1945–1963. [CrossRef]
43. Khazraeian, S.; Hadi, M. Monte Carlo Simulation-Based Benefit-Cost Analysis Combined with Analytical Hierarchy Process to Support ITS Investment with Consideration of Connected Vehicle Technology. *Transp. Res. Rec.* **2018**, *2672*, 1–12. [CrossRef]
44. Arafat, M.; Iqbal, S.; Hadi, M. Utilizing an Analytical Hierarchy Process with Stochastic Return On Investment to Justify Connected Vehicle-Based Deployment Decisions. *Transp. Res. Rec.* **2020**, *2674*, 462–472. [CrossRef]
45. Wright, S.; Nelson, J.D.; Cooper, J.M.; Murphy, S. An Evaluation of the Transport to Employment (T2E) Scheme in Highland Scotland Using Social Return on Investment (SROI). *J. Transp. Geogr.* **2009**, *17*, 457–467. [CrossRef]
46. Shepherd, S.P. A Review of System Dynamics Models Applied in Transportation. *Transp. B Transp. Dyn.* **2014**, *2*, 83–105. [CrossRef]
47. Forrester, J.W. Industrial Dynamics: A Major Breakthrough for Decision Makers. *Harv. Bus. Rev.* **1958**, *36*, 37–66.
48. Sterman, J.D. *Business Dynamics: Systems Thinking and Modeling for a Complex World*; Nachdr.; Irwin/McGraw-Hill: Boston, MA, USA, 2000; ISBN 978-0-07-238915-9.
49. Abbas, K.A.; Bell, M.G. System Dynamics Applicability to Transportation Modeling. *Transp. Res. Part A Policy Pract.* **1994**, *28*, 373–390. [CrossRef]
50. Harrison, G.; Grant-Muller, S.M.; Hodgson, F.C. A Review of Transport-Health System Dynamics Models. *J. Transp. Health* **2021**, *22*, 101138. [CrossRef]
51. Hovmand, P.S. *Community Based System Dynamics*; SpringerLink; Springer: New York, NY, USA, 2014; ISBN 978-1-4614-8763-0.
52. Musich, S.; Wang, S.; Hawkins, K.; Klemes, A. The Impact of Personalized Preventive Care on Health Care Quality, Utilization, and Expenditures. *Popul. Health Manag.* **2016**, *19*, 389–397. [CrossRef] [PubMed]
53. Naydeck, B.L.; Pearson, J.A.; Ozminkowski, R.J.; Day, B.T.; Goetzel, R.Z. The Impact of the Highmark Employee Wellness Programs on 4-Year Healthcare Costs. *J. Occup. Environ. Med.* **2008**, *50*, 146–156. [CrossRef] [PubMed]
54. Triemstra, J.D.; Lowery, L. Prevalence, Predictors, and the Financial Impact of Missed Appointments in an Academic Adolescent Clinic. *Cureus* **2018**, *10*, e3613. [CrossRef]
55. Tsai, M.-H.; Xirasagar, S.; Carroll, S.; Bryan, C.S.; Gallagher, P.J.; Davis, K.; Jauch, E.C. Reducing High-Users' Visits to the Emergency Department by a Primary Care Intervention for the Uninsured: A Retrospective Study. *Inq. J. Health Care Organ. Provis. Financ.* **2018**, *55*, 0046958018763917. [CrossRef]
56. He, S.Y.; Thøgersen, J.; Cheung, Y.H.Y.; Yu, A.H.Y. Ageing in a Transit-Oriented City: Satisfaction with Transport, Social Inclusion and Wellbeing. *Transp. Policy* **2020**, *97*, 85–94. [CrossRef]
57. Velho, R.; Holloway, C.; Symonds, A.; Balmer, B. The Effect of Transport Accessibility on the Social Inclusion of Wheelchair Users: A Mixed Method Analysis. *Soc. Incl.* **2016**, *4*, 24–35. [CrossRef]
58. Sterman, J. *Business Dynamics*; Irwin/McGraw-Hill c2000: Boston, MA, USA, 2010; ISBN 0-07-231135-5.
59. Ercan, T.; Onat, N.C.; Tatari, O. Investigating Carbon Footprint Reduction Potential of Public Transportation in United States: A System Dynamics Approach. *J. Clean. Prod.* **2016**, *133*, 1260–1276. [CrossRef]
60. Coleman, E.A.; Eilertsen, T.B.; Kramer, A.M.; Magid, D.J.; Beck, A.; Conner, D. Reducing Emergency Visits in Older Adults with Chronic Illness. A Randomized, Controlled Trial of Group Visits. *Eff. Clin. Pr.* **2001**, *4*, 49–57.
61. Chu, L.; Sood, N.; Tu, M.; Miller, K. Reduction of Emergency Department Use in People with Disabilities. *Am. J. Manag. Care* **2017**, *23*, e409–e415. [PubMed]
62. How Many Working Days Are in a Year? Available online: https://www.symmetry.com/payroll-tax-insights/how-many-working-days-are-in-a-year (accessed on 3 June 2023).
63. Powdthavee, N. Putting a Price Tag on Friends, Relatives, and Neighbours: Using Surveys of Life Satisfaction to Value Social Relationships. *J. Socio-Econ.* **2008**, *37*, 1459–1480. [CrossRef]
64. Chhabra, K.R.; McGuire, K.; Sheetz, K.H.; Scott, J.W.; Nuliyalu, U.; Ryan, A.M. Most Patients Undergoing Ground And Air Ambulance Transportation Receive Sizable Out-Of-Network Bills: An Analysis of the Prevalence and Financial Impact of out-of-Network Billing for Ground and Air Ambulance Transportation. *Health Aff.* **2020**, *39*, 777–782. [CrossRef]
65. CMS.Gov. Available online: https://data.cms.gov/provider-data/search?page-size=50&theme=Physician%20office%20visit%20costs (accessed on 13 November 2022).
66. Stacker.Com. Available online: https://stacker.com/mississippi/what-common-medical-visits-cost-mississippi-and-how-they-compare-nearby-states (accessed on 13 November 2022).
67. Moore, B.J.; Liang, L. Costs of Emergency Department Visits in the United States, 2017. In *Healthcare Cost and Utilization Project (HCUP) Statistical Briefs [Internet]*; Agency for Healthcare Research and Quality (US): Rockville, MD, USA, 2020.
68. Mississippi: Average Annual Pay 2019. Available online: https://www.statista.com/statistics/732696/mississippi-annual-pay/ (accessed on 13 November 2022).

69. Maleki, M.; Mohammadpour, S.; Azadeh, S.R. The Effect of Infrastructural Integration of Regional Transport on Tourism Promotion: The Case of Guilan Province, Iran. *JURA* **2020**, *12*, 217–231. [CrossRef]
70. Coverage of Therapy and Mental Health Benefits. Available online: https://www.https://www.medicareplans.com/outpatient-mental-health-coverage//outpatient-mental-health-coverage/ (accessed on 27 March 2023).
71. How Much Do Antidepressants Cost? With & without Insurance. Available online: https://khealth.com/learn/antidepressants/how-much-do-antidepressants-cost/ (accessed on 27 March 2023).
72. Lamanna, M.; Klinger, C.A.; Liu, A.; Mirza, R.M. The Association between Public Transportation and Social Isolation in Older Adults: A Scoping Review of the Literature. *Can. J. Aging* **2020**, *39*, 393–405. [CrossRef]
73. Does Medicare Cover Physical Therapy in 2023? Available online: https://www.theseniorlist.com/medicare/medicare-cover-physical-therapy/ (accessed on 27 March 2023).
74. Officer, L.H.; Williamson, S.H. Computing 'Real Value'Over Time with a Conversion between UK Pounds and US Dollars, 1791 to Present. *Meas. Worth Accessed Dec.* **2019**, *27*, 2019.
75. U.S. Census Bureau EMPLOYMENT STATUS. Available online: https://data.census.gov/table?q=Holmes+County,+Mississippi&t=Employment&tid=ACSST5Y2020.S2301 (accessed on 2 June 2023).
76. U.S. Census Bureau DISABILITY CHARACTERISTICS. Available online: https://data.census.gov/table?q=Holmes+County,+Mississippi&t=Disability&tid=ACSST5Y2020.S1810 (accessed on 3 June 2023).
77. U.S. Census Bureau SEX BY AGE BY AMBULATORY DIFFICULTY. Available online: https://data.census.gov/table?q=Holmes+County,+Mississippi&t=Disability&tid=ACSDT5Y2020.B18105 (accessed on 3 June 2023).
78. U.S. Census Bureau COMMUTING CHARACTERISTICS BY SEX. Available online: https://data.census.gov/table?q=Holmes+County,+Mississippi&t=Employment&tid=ACSST5Y2020.S0801 (accessed on 3 June 2023).
79. datausa.io Holmes County, MS | Data USA. Available online: https://datausa.io/profile/geo/holmes-county-ms (accessed on 3 June 2023).
80. McCormick, T.H.; Salganik, M.J.; Zheng, T. How Many People Do You Know?: Efficiently Estimating Personal Network Size. *J. Am. Stat. Assoc.* **2010**, *105*, 59–70. [CrossRef]
81. The Average American Knows How Many People?—The New York Times. Available online: https://www.nytimes.com/2013/02/19/science/the-average-american-knows-how-many-people.html (accessed on 18 September 2023).
82. Motor Vehicle Licensing FAQs | DOR. Available online: https://www.dor.ms.gov/tagstitles/motor-vehicle-licensing-faqs (accessed on 6 June 2023).
83. Best Cheap Car Insurance in Mississippi for 2023. Available online: //www.usnews.com/insurance/auto/cheap-car-insurance-mississippi (accessed on 10 April 2023).
84. Available online: CarRegistration.com/blog (accessed on 10 April 2023).
85. Car Repair Costs Ranked State-by-State... Where Does Yours Rank?—Autoblog. Available online: https://www.autoblog.com/2011/06/24/car-repair-costs-ranked-state-by-state-where-does-yours-rank/ (accessed on 6 June 2023).
86. Taxi Driver Salary in Mississippi. Available online: https://www.salary.com/research/salary/benchmark/taxi-driver-salary/ms (accessed on 24 January 2023).
87. Customer Service Representative I Salary in Mississippi. Available online: https://www.salary.com/research/salary/benchmark/customer-service-representative-i-salary/ms (accessed on 24 January 2023).
88. Ridgeland Office Price per Sqft and Office Market Trends. Available online: https://www.commercialcafe.com/office-market-trends/us/ms/ridgeland/ (accessed on 7 June 2023).
89. FOTW #1237, May 9, 2022: Fuel Economy for All Vehicle Classes Has Improved Substantially Over the Past Two Decades. Available online: https://www.energy.gov/eere/vehicles/articles/fotw-1237-may-9-2022-fuel-economy-all-vehicle-classes-has-improved (accessed on 7 June 2023).
90. Feonix—We are Working to Bring Life-Changing Mobility Solutions. Available online: https://feonix.org/ (accessed on 26 October 2023).
91. Alirezaei, M.; Onat, N.; Tatari, O.; Abdel-Aty, M. The Climate Change-Road Safety-Economy Nexus: A System Dynamics Approach to Understanding Complex Interdependencies. *Systems* **2017**, *5*, 6. [CrossRef]
92. Qudrat-Ullah, H.; Seong, B.S. How to Do Structural Validity of a System Dynamics Type Simulation Model: The Case of an Energy Policy Model. *Energy Policy* **2010**, *38*, 2216–2224. [CrossRef]

systems

MDPI

Article

Acceptance Analysis of Electric Heavy Trucks and Battery Swapping Stations in the German Market

Florian Noto [1] and Hamid Mostofi [2,*]

[1] Sustainable Mobility Management, EUREF Campus, Technische Universität Berlin, 10623 Berlin, Germany; noto@campus.tu-berlin.de
[2] Mobility Research Cluster, Department of Work, Technology and Participation, Technische Universität Berlin, 10587 Berlin, Germany
* Correspondence: mostofidarbani@tu-berlin.de

Abstract: Heavy-duty vehicles are a major contributor to CO_2 emissions in the transportation sector, and it is necessary to develop clean and green technologies to replace diesel trucks. Electric trucks have not reached a breakthrough in the German market. In addition to technology development, customer acceptance of new technologies is a critical factor in the success of sustainable transportation policies. This study aims to fill this knowledge gap by investigating the perceptions regarding electric trucks and providing insights into the acceptance of these technologies. Data and arguments on the expected risks and benefits of heavy-duty electric trucks, with a special focus on the battery swapping solution, were collected through a survey and expert interviews in the German commercial transport sector. The authors collected a sample of 146 qualitative responses and 61 individual statements on the expected risks and benefits of electric trucks and battery swapping. While the responses to the classified questions are overwhelmingly positive, the individual statements show that there are still many open questions.

Keywords: electric heavy trucks; acceptance model; battery swapping station; sustainable transportation

Citation: Noto, F.; Mostofi, H. Acceptance Analysis of Electric Heavy Trucks and Battery Swapping Stations in the German Market. *Systems* **2023**, *11*, 441. https://doi.org/10.3390/systems11090441

Academic Editors: Mahyar Amirgholy and Jidong J. Yang

Received: 30 June 2023
Revised: 14 August 2023
Accepted: 22 August 2023
Published: 24 August 2023

1. Introduction

The number of electric vehicles in Germany and the EU is growing and has reached a significant share of passenger cars. For buses and some types of commercial vehicles, electric drives are also the leading alternative to internal combustion engines (ICE), as discussed in Section 2.2. The typical use of heavy trucks is still a challenge for electric drives; they carry heavy loads, travel long distances, and have changing destinations.

TU Berlin is involved in projects for applied science to research and develop electric trucks and battery swapping systems. The purpose of this paper is to provide insight into the subjective perception of electric heavy-duty trucks and their battery charging methods. To this end, we have defined two main research questions: What is the perception of electric heavy trucks in terms of technical, economic, and environmental aspects? And what is the perception of battery swapping stations in terms of perceived risks and benefits? These questions were analyzed through a survey (N = 146) and interviews (N = 4) with experts who are working with e-trucks, from logistics companies, research and development, and mechanical engineering.

In the EU, heavy-duty vehicles (HDV) including trucks, city buses, and long-distance buses generate over 25% of greenhouse gas (GHG) emissions from road transportation, and they make up more than 6% of the total GHG emissions in the region. These emissions have been steadily rising, particularly in the field of freight transport [1]. The European Union aims to reduce transport-related GHG emissions, air pollution, and energy dependence on imported fossil fuels. To achieve this goal, alternative drives should replace fossil fuels as the power source for HDV [2].

Customer acceptance is a key success factor for new technology and the implementation of policy objectives. Currently, heavy trucks highly depend on diesel fuel. It is an open race as to which alternative power source for HDV will be used in the future. The major truck manufacturers have developed electric versions of even the largest articulated semi-trucks, and new companies, which are specialized in e-trucks, are entering the market. However, the number of registrations of heavy electric trucks is still very low. Many studies have been published on the technical and economic aspects of electric trucks [3–6], but there is a lack of information on user acceptance. This paper examines the perception of e-trucks in the German market to provide insight into the acceptance of this technology. Previous research on the acceptance of technology has indicated that the way individuals perceive concerns and risks plays a significant role in shaping their attitudes toward new technologies [7–10]. Lee defined attitude as an individual's positive or negative thoughts about performing a particular behavior [9]. Classical technology acceptance models such as the Technology Acceptance Model (TAM) [11] and the Theory of Planned Behavior [12] suggest that a person's attitude towards an object determines, among other factors, their intention to use it. In addition to concerns and risks, studies on technology acceptance have shown that attitudes towards emerging technologies that people have not experienced before are influenced by perceptions of their advantages and potential benefits [8,9,11]. These studies indicate how customers' perceptions of the different dimensions of the benefits and usefulness affect their willingness to use them in the future.

User Acceptance of Electric Vehicles

Electric trucks are still new and rare (see Section 2.5); therefore, there is a lack of knowledge about user experiences and expectations. Few scientific studies deal with this topic. Existing studies about user experiences and preferences mainly focus on passenger cars [13,14] and refer to a lack of knowledge [15].

The German National Platform Future of Mobility (NPM) published a report in 2021 arguing that customer acceptance is the key to the market ramp-up of electric passenger cars [16]. The argument was based on the results of a representative survey (N > 1000) and interviews with experts. They gave a few, clear reasons for buying an electric vehicle (EV); these are:

- financial reasons (subsidies, lower total cost of ownership (TCO)),
- test-driving and driving pleasure,
- and environmental considerations.

According to this report, the reasons against EVs are diffuse and not clearly shaped. Apart from subsidies, the keys to growing acceptance of EV passenger cars are spreading knowledge, providing testing opportunities, and expanding the charging infrastructure.

In contrast to this, a study from Switzerland (N = 4149) with drivers of conventional cars questions the substantial effects of better information and test drives on EV adoption [17]. Most of the studies focusing on electric trucks deal with calculated economic or technical aspects [3,18–20] rather than user acceptance and expectations. Only one recent study, which was published in Hungary in 2021 [21], deals with major barriers to the adoption of electric trucks in the logistics system. This research is based on a survey among 60 professionals from the logistics industry in Budapest. The authors argue that the difference in the total costs of ownership (TCO) between e-trucks and diesel trucks is already marginal. Barriers are related to operational and infrastructural challenges such as a shorter range, a lack of charging infrastructure, and long charging times. Technological constraints are seen as a major barrier for e-trucks, including battery capacity and longevity, charging speed, and accessibility of charging points. A total of 67% of the respondents of the Hungarian survey were concerned about the short range, while 33% were concerned about a lack of charging infrastructure. Furthermore, high investment costs and a lack of incentives are reasons to avoid buying e-trucks. Investment costs for e-trucks are higher than for diesel trucks, but there is disinformation about the TCO.

An important global political initiative is called 'Global Drive to Zero'. Governments from Europe, the Americas, and New Zealand launched this in 2021 by signing a Memorandum of Understanding to accelerate the growth of global zero-emission commercial vehicles. The program is supported by companies and regional bodies (such as Berlin Partner, a local public–private partnership for economic and technological development in the German capital). The goal is a full transition to ZE-MHDV (zero-emission medium- and heavy-duty vehicles) in new fleets by 2040 and net-zero carbon emissions by 2050 [22].

2. Regulations and Market Analysis

The legal and economic framework conditions are briefly presented below to introduce the topic.

2.1. Legal Requirements for Clean Transport Vehicles in the EU

The term "clean vehicles" has different definitions for different categories of vehicles in the European Union legislation. The definition in Art. 4 (4) (b) of Directive (EU) 2019/1161 is: "clean vehicles" of categories N2 and N3 are those that use alternative fuels, as defined in Art. 2 (1) and (2) of the Directive 2014/94/EU. These alternative fuels include electric vehicles but also liquid and gaseous fuels that have the potential to contribute to the decarbonization and improvement of the environmental performance of the transport sector. Such fuels include hydrogen, biofuels, synthetic and paraffinic fuels, natural gas (compressed natural gas (CNG) and liquefied natural gas (LNG)), and liquefied petroleum gas (LPG). This means that clean trucks under Directive (EU) 2019/1161 do not have to be zero-emission vehicles. Instead, gaseous fossil fuels will still cause CO_2 emissions and air pollutants from combustion.

- Zero-emission HDV are those with a power source—electricity, hydrogen fuel cell, or ICE—that emits <1 g CO_2/kWh.
- Low-emission HDV are those with an ICE powered by hydrogen combustion, biofuels, synthetic and paraffinic fuels, compressed natural gas (CNG), liquefied natural gas (LNG), or liquefied petroleum gas (LPG).

The relevant legislation is presented in the following two subsections and in Table 1.

Table 1. Legal acts in the EU for clean transport vehicles.

Goal	Legal Act	Applies to	Vehicle Types	Until 2025	Until 2030	After 2030
Minimum procurement target	Directive (EU) 2019/1161	EU member states	Classes N2, N3 (commercial vehicles from 3.5 tonnes)	10% clean vehicles	15% clean vehicles	
CO_2 emission performance standards for new heavy-duty vehicles	Regulation (EU) 2019/1242	Manufacturers	Exceeding 16 tonnes technically permissible maximum laden mass		15% CO_2 emission reduction	30% CO_2 emission reduction

2.2. EU Directive for Clean and Energy-Efficient Road Transport Vehicles

To promote the use of clean and energy-efficient road transport vehicles, the EU has set minimum procurement targets for such vehicles in the Directive (EU) 2019/1161. The targets are different for each member state and for each vehicle class. Germany is one of the EU member states with the highest targets. For Germany, the minimum procurement target for the share of clean vehicles for heavy trucks (vehicle category N2 and N3) is 10% from 2021 to 2025 and 15% from 2026 to 2030 (Directive (EU) 2019/1161). The vehicle categories are defined in Article 4 of Regulation (EU) 2018/858. Category N vehicles are used for the carriage of goods. Category N2 are those with a maximum mass between 3.5 and 12 tonnes; category N3 are those above 12 tonnes including trailers.

The minimum procurement target of 15% for clean trucks by 2030 applies to 12 EU member states: Luxembourg, Sweden, Denmark, Finland, Germany, France, the Netherlands, Austria, Belgium, Italy, Ireland, and Malta.

2.3. EU Regulation Setting CO_2-Emission Performance Standards for New Heavy-Duty Vehicles

Regulation (EU) 2019/1242 applies to the manufacturers of heavy-duty vehicles of categories N2 and N3 with a maximum permissible mass exceeding 16 tonnes, 6 × 2 axle configuration, and tractor units. Manufacturers must reduce the specific emissions of their fleets by 15% for the reporting periods starting in 2025 and by 30% for the reporting periods starting in 2030 (Art. 1 lit. a and b). The reference period is from 1 July 2019 to 30 June 2020.

2.4. Market for Commercial Vehicles and Trucks—Data Analysis

By 1 January 2023, a total number of 3,641,755 vehicles used for the carriage of goods and a further 227,938 semi-truck tractor units were officially registered in Germany with the Federal Motor Transport Authority (Kraftfahrtbundesamt, KBA) [23]. Of this number, 85% had a technically permissible maximum laden mass below 3.5 t. Table 2 shows the number of vehicles for each segment and new registrations in 2022.

Table 2. Stock and new registrations of transport vehicles in Germany (2022) [23].

Vehicle Category	N1 (<3.5 t)	N2 (3.5–12 t)	N3		
			12–20 t	>20 t (rigid)	semi-trucks
Stock 1 January 2023	3,110,652	291,329	84,370	154,176	227,938
New registrations 2022		253,894			32,608
Renewal rate		7.0%			14.3%

For long hauling, mainly articulated semi-trucks are used. The renewal rate is significantly higher than for smaller commercial vehicles. The annual figures published by KBA on the stock and new registrations make it possible to calculate the renewal rate and average period of use. Table 3 shows the number of semi-trucks in recent years.

Table 3. Renewal rate and average usage time for semi-trucks in Germany over time (2018–2023 [23]).

Year (Y)	Stock of Semi-Trucks on 1st January (S)	Increase	New Registrations (N)	Renewal Rate (R)	Average Usage Time in Years (T)
2018	210,941	7513 (3.6%)	38,727	17.7%	5.6
2019	218,454	695 (0.3%)	38,620	17.6%	5.7
2020	219,149	−680 (−0.3%)	25,946	11.9%	8.4
2021	218,469	3810 (1.7%)	29,698	13.4%	7.5
2022	222,279	5659 (2.5%)	32,608	14.3%	7.0
2023	227,938				

The renewal rate (R) is calculated by dividing new registrations in a given year (N_Y) by the stock in the following year (S_{Y+1}), to take account of the annual increase (decrease in 2020). The average usage time (T) in years is the inversion of this result. The renewal rate and the average usage time are calculated using the following equations:

$$R = \frac{N_Y}{S_{Y+1}} \text{ and } T = R^{-1}$$

The renewal rate dropped significantly during the COVID-19 pandemic in 2020. The pandemic and its economic consequences (closures, disruption of delivery chains, and shortages of inputs) affected both the transport industry on the demand side, as buyers and users of trucks, and the automotive industry, as manufacturers and suppliers. In 2021 and 2022, the renewal rate increased slightly but has not reached the status quo ante yet. In 2018 and 2019, the average time in use of semi-trucks was less than 6 years. In 2022, it was 7.0 years. This does not necessarily mean that the trucks are scrapped; some may be sold to other countries. The data show that most of the heavy trucks that will be in use in 2030 have not yet been produced.

2.5. Alternative Drives for Heavy Trucks

Alternative powertrains for semi-trucks are still very rare. In 2022, only 52 battery-electric, 7 hybrid, 775 fossil gas, and 0 hydrogen-powered semi-trucks were newly registered in Germany. The share of alternative powertrains in new registrations was 2.6%. The situation is different for rigid trucks and vans. Of the 253,894 new registrations of trucks and vans in 2022, 8.5% (21,633) had an alternative drive. Most of these were battery-electric vehicles (a total of 18,322 = 7.2%). Smaller proportions used H2 fuel cells (30), fossil gas (2245), or were plug-in hybrids (77) or non-plug-in hybrids (959). Gas supply disruptions and rising prices in 2022 could give electric trucks a better position in the coming years [23]. Detailed figures for the different vehicle classes can be found in Table 4, while Figures 1 and 2 illustrate the large differences between semi-trucks and other vehicles.

Table 4. Alternative drives for different vehicle classes in Germany, 2022 [23].

Vehicle Class	New Registrations	Alternative Drives (% of New Registrations)	BEV (% of Alternative Drives)	Fossil Gas ICE	Other Alternatives (Hybrid EV, FCEV)
Passenger cars	2,651,357	1,315,567 (49.6%)	470,559 (35.8%)	16,852 (1.3%)	828,156 (63.0%)
Buses	4883	1750 (35.8%)	631 (36.1%)	26 (1.5%)	1093 (62.5%)
Commercial vehicles for the transport of goods	253,894	21,633 (8.5%)	18,322 (84.7%)	2245 (10.4%)	1066 (4.9%)
Semi trucks	32,608	834 (2.6%)	52 (6.2%)	775 (92.9%)	7 (0.8%)

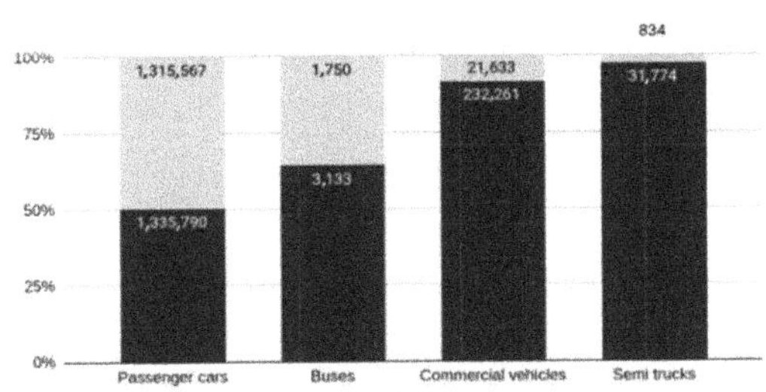

Figure 1. Share of alternative drives for new registrations in Germany by vehicle classes in 2022 [23].

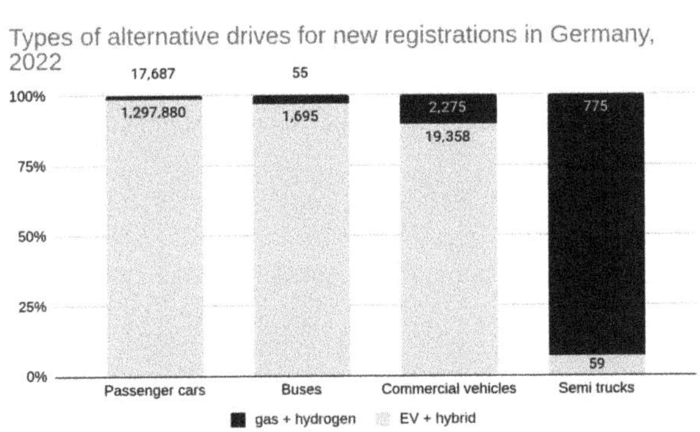

Figure 2. Types of alternative drives (gas + hydrogen vs. EV + hybrid) for new registrations in Germany by vehicle classes in 2022 [23].

Although the number of registrations of electric semi-trucks is low, all the major companies already offer this type of vehicle, and there are new companies specializing in this segment. MAN and Scania both belong to the Traton Group, which is part of Volkswagen AG. Daimler and VW together cover 70% of the market for semi-trucks in Germany; with DAF and Volvo, the share rises to 94%. More details can be found in Table 5.

Table 5. Semi-truck market shares in Germany, 2022 [23].

Producer	Stock of Semi-Trucks in Germany (1 January 2022)	Market Share
Daimler	66,639	30.0%
MAN	59,384	26.7%
Scania	29,678	13.4%
DAF	29,206	13.1%
Volvo	23,278	10.7%
Iveco, Renault, and other	14,094	6.3%
total	222,279	100%

3. Materials and Methods

Electric passenger cars are an emotive topic in Germany, and many people have reservations about them. In March 2023, the European Commission and member states agreed to ban the sale of polluting cars and vans in 2035. Before the decision was made, the German government blocked a compromise and demanded an exemption for ICE cars using synthetic e-fuels [24,25]. The question is whether there are similar reservations about electric trucks. A survey was conducted to find out the motives and fears for the use of heavy e-trucks.

3.1. STEEP Analysis

In order to find relevant topics for a survey on user perceptions of heavy e-trucks, the authors carried out a STEEP analysis. STEEP is an acronym for Social, Technological, Economic, Environmental, and Political factors. It is used to scan the business environment on a certain topic, such as e-trucks, and to find relevant aspects related to each factor. STEEP is a variant and extension of the classic PEST analysis, developed by Francis J. Aguilar [26], by adding the factor of environmental issues. Relevant STEEP factors and related aspects are shown in Table 6. The results are presented in Table 6 below.

Table 6. STEEP factors and related aspects for e-trucks.

Factor	Related Aspects
Social	Safety Labor shortage and attractiveness for young people to become truck drivers Working conditions General benefits for truck drivers
Technological	Effort for maintenance and repair Reliability Comparison to alternative fuels Comparison of battery swapping to direct charging Feasibility Wear and tear Time for charging or swapping

Table 6. *Cont.*

Factor	Related Aspects
Economic	Investment costs (for the procurement of the trucks and/or batteries) Operating costs (opex) Total costs of ownership (TCO) Risk of rising fuel prices General benefits for transport companies Flexibility Ownership of swappable batteries
Environmental	General benefits for the environment Ability to enter low-emission zones (restricted areas) Climate protection (lower CO_2 emissions) Noise Air pollution
Political	Taxes and tolls Infrastructure for charging/swapping/refueling Standardization of swappable batteries

3.2. *Questionnaire*

The questionnaire consisted of four demographic questions (see Table 7), 33 questions with a 1-to-5 scale (see Tables 8–10 and Figures 3–5), and one open text field. To keep it simple and user-friendly, the questions related to the five STEEP categories were grouped into three main segments: economic, technological, and social potential advantages. Environmental and political issues are integrated within these. The first page was in three languages: German, English, and Polish. This was followed by three pages of questions for each language. The questions were identical for each language. The questionnaire was created on Google Forms and was available online from November 2022 to March 2023. Participation was voluntary and no personal data was collected.

Table 7. Demographic results.

	Answer Options	Count	Percent
Language	German	71	49%
	English	72	49%
	Polish	3	2%
Are you a truck driver?	yes (12–44 t)	12	8%
	yes (3.5–12 t)	3	2%
	no	131	90%
Age	16–24	28	19%
	25–34	64	44%
	35–54	36	25%
	55–99	17	12%
Gender	Female	41	28%
	Male	102	70%
	Divers	3	2%

Table 8. Results for opinions on e-trucks (mean score and standard deviation). A lower score means more agreement (1 is strongly agree); a higher score means more disagreement with a statement (5 is strongly disagree).

Lead Questions	Sub-Questions	Mean Score	St. Dev.
What economic advantages do you expect from e-trucks (compared to diesel)?	Lower investment costs	3.83	1.13
	Lower operating costs	2.21	1.13
	Lower taxes and tolls	1.99	0.98
	Lower total costs of ownership	2.26	1.15
	Lower risk of rising fuel prices	2.09	1.17
What technical advantages do you expect from electric trucks (compared to diesel)?	Less effort for maintenance and repairs	2.16	1.11
	Better for the environment	1.68	1.01
	Just as reliable	2.12	1.1
	Safer for all road users	2.75	1.18
	Better than alternative fuels	2.41	1.19
What social benefits do you expect from electric trucks (compared to diesel)?	Attract young people to become truck drivers	3.16	1.18
	Better working conditions for truck drivers	2.75	1.14
	Good to enter low-emission zones	1.55	0.86
	A contribution to global climate protection	1.81	1.07
	Less noise and local air pollution	1.43	0.88

Table 9. Results for rollout and usage of e-trucks (mean score and standard deviation).

Proposed Development Rate	Mean Score	Standard Deviation
Your company gets one e-truck to gain experience with this technology	1.66	0.97
You should drive this truck (consider you're a truck driver)	1.82	1.02
From now on, 10% of the new vehicles (above 3.5 t) are e-trucks	2.08	1.18
From 2030, all new trucks should be e-trucks	2.34	1.29
From 2040, all new trucks should be e-trucks	2.15	1.31

Table 10. Results for statements related to battery swapping (mean score and standard deviation).

Lead Questions	Sub-Questions	Mean Score	St. Dev.
What do you think of battery swapping (compared to direct charging)?	It will be better for truck drivers	2.23	1.18
	It will be better for transport companies	2.17	1.16
	Battery swapping is necessary for e-trucks to establish	2.44	1.28
	In daily practice, battery swapping would be more feasible than direct charging	2.59	1.30
What are the biggest benefits of battery swapping compared to direct charging?	Time saving	1.77	1.04
	Higher flexibility	2.31	1.17
	Lower prices (charging when electricity is cheap)	2.50	1.13
	Allow driving long distances	2.37	1.16
	Lower investment costs for transport companies (batteries are rented instead of purchased)	2.63	1.13
What are the largest risks of battery swapping (compared to direct charging)?	There will be too few swapping stations	2.10	0.93
	Higher prices	2.42	0.95
	Higher wear and tear on the trucks	2.90	1.09
	Technical problems	2.52	1.15
Do you expect other benefits or risks of battery swapping?	Open text field to write comments and remarks		

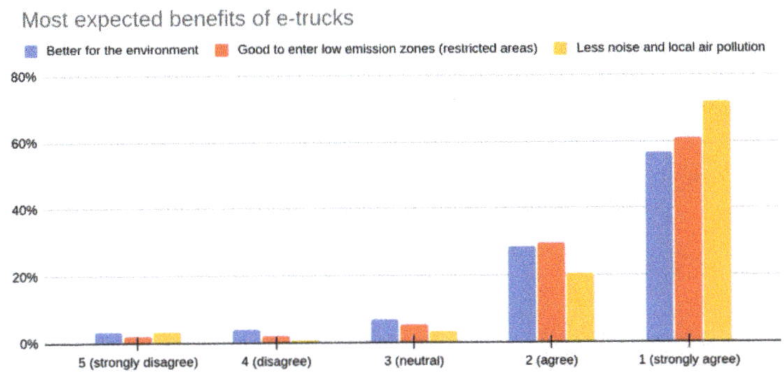

Figure 3. Histogram of most-expected benefits of e-trucks.

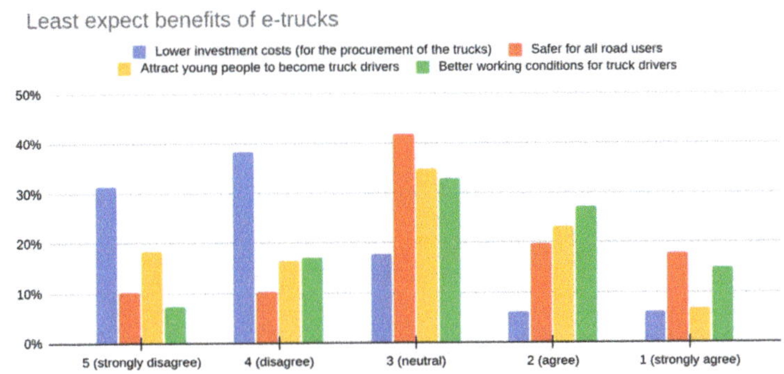

Figure 4. Histogram of least-expected benefits of e-trucks.

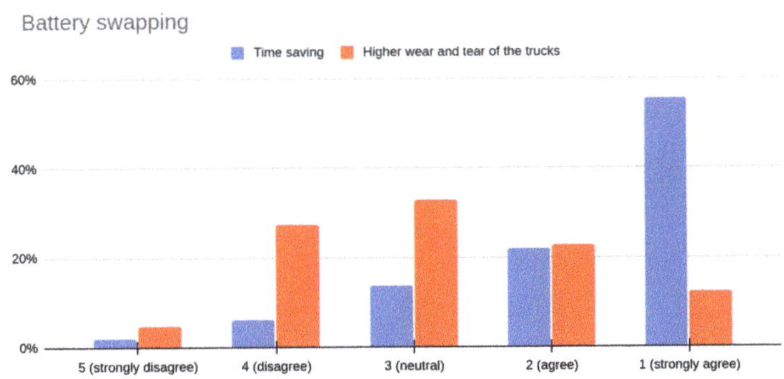

Figure 5. Histogram of most-expected benefit (time saving) and least-expected risk (higher wear and tear) for battery swapping.

The survey was announced personally at several events (seminars for students of TU Berlin, workshops for public consultations on sustainability issues in Berlin, working group meetings for applied science for e-trucks), and it was distributed in truck driver online forums on Facebook and other websites. The questionnaire was filled out 149 times, and 3 sets of answers were deleted because fewer than half of the questions were answered.

4. Results

The response option for the questionnaire was a 1-to-5 scale. A lower number indicates agreement with the statement, and a higher number indicates disagreement.

4.1. Opinions on Electric Trucks

The first 15 questions were to solicit opinions on electric trucks in general. The mean scores and standard deviations for all the questions are shown in Table 8.

The most-expected advantages of e-trucks compared to diesel trucks are all related to ecological factors: less noise and air pollution (1.43); good to enter low emission zones (1.55); better for the environment (1.68); a contribution to global climate protection (1.81).

The least-expected advantages of e-trucks compared to diesel are related to economic and social factors: lower investment costs (3.83); attract young people to become truck drivers (3.16); safer for all road users (2.75); better working conditions for truck drivers (2.75).

The histograms, Figures 3 and 4, show the distribution of responses for the most- and least-expected benefits. The histograms show a clear disagreement with the statement that e-trucks could have lower investment costs. Other ideas with a high average score are rated neutral by a majority of respondents.

4.2. Roll-Out and Usage of Electric Trucks

The following five questions were related to the rollout and usage of e-trucks. The participants were asked to suppose they work in a transport company. How do they rate the following developments? The results are shown in Table 9.

The agreement for having one e-truck to gain experience is very high (1.66), slightly lower when someone is asked personally to be the first user of the new vehicle. The agreement is still high for an immediate low share of e-trucks (10% of new purchases) and for the long-term goals (100% e-trucks from 2040). The lowest agreement is for 100% e-trucks from 2030 (2.34).

4.3. Battery Swapping Compared to Direct Charging

The last 13 questions were about battery swapping. An explanation and a short video from a swapping station in China [27] were given as an introduction to the topic. The explanation was:

"Battery swapping for e-trucks is successfully used in some countries. Empty batteries can be charged directly as usual. Or they are exchanged for full ones and charged outside the vehicle, and the truck can continue driving immediately. Please watch the short video (24 s). It shows a time-lapse example of how a mobile exchange station is set up and the batteries are exchanged".

The mean scores are shown in Table 10. It is expected that transport companies (2.17) and truck drivers (2.23) will have advantages from battery swapping. The most expected benefit is time saving (1.77). There is also hope for higher flexibility (2.31) and that it will allow for driving long distances (2.37). The most-expected risk is that there will be too few swapping stations (2.10). The least-expected risk is higher wear and tear.

The histogram, Figure 5, shows the distribution of responses for the two most outstanding results.

As a conclusion, the results of the survey for the STEEP factors (see Table 6 above) are shown in Table 11.

Table 11. Results of the survey related to STEEP factors.

Factor	Results
Social	No changes are expected for social aspects. All social consequences are rated neutral (2.7 to 3.2).
Technological	The biggest technological benefit of battery swapping is time saving (1.7). It is expected that e-trucks will be just as reliable as diesel and need less effort for maintenance and repair (2.1). The other technological aspects are rated slightly positive to neutral (2.4 to 2.9).
Economic	Lower investment costs (capex) are not expected (3.8). Economic advantages can be lower operating costs (opex), lower total costs of ownership (TCO), and lower risk of rising fuel prices (2.1 to 2.3). Battery swapping is expected to bring more flexibility and general benefits for transport companies (2.3). The other economic aspects are rated neutral (2.5 to 2.7).
Environmental	The most-relevant advantages are seen in ecological aspects. The average rating for all questions is 1.4 to 1.8.
Political	The highest risk is seen in a lack of swapping stations (2.0). Lower taxes and tolls for e-trucks are an expected benefit (2.0).

4.4. Analysis of Individual Comments and Remarks

The last question was: "Do you expect other benefits or risks of battery swapping?". The participants could write an individual response in an open text field. A total of 61 participants in the survey (42%) made individual statements. Nine comments such as "I don't know" were not included. Some statements had more than one argument, so 67 arguments were identified to be useful for further consideration: 20 expected benefits and 47 expected risks. Some risks or benefits were mentioned by more than one participant.

SWOT is an acronym for Strengths, Weaknesses, Opportunities, and Threats. A SWOT analysis is a planning tool to identify these aspects for companies or organizations. It will be used here to order the individual remarks and comments.

For a SWOT analysis, it is important to choose a point of view. One side's opportunity could be the other side's threat; rising energy prices could be an opportunity for energy providers and a threat to transport companies. For the purpose of this work—technology acceptance of heavy e-trucks—it is the user's perspective, i.e., trucking companies and truck drivers.

- Aspects that are helpful for freight transport companies are strengths and opportunities, whereas harmful aspects are weaknesses and threats.
- Aspects with an internal origin belong to strengths and weaknesses, while opportunities and threats have an external origin.

The basic scheme of a SWOT analysis is shown in Table 12. The numbers indicate how many arguments are mentioned in each category. It is noteworthy that 20 benefits and 46 risks are mentioned. This could be an indicator that, despite the positive results of the quantitative survey, there is a higher awareness of risks, more concerns, and underlying reservations than positive expectations of the new technology.

Table 12. SWOT analysis, number of comments for each section.

SWOT Analysis (66)	Helpful (20)	Harmful (46)
Internal origin (34)	Strengths (13)	Weaknesses (21)
External origin (32)	Opportunities (7)	Threats (25)

For further analysis, some arguments are clustered under a distinctive keyword. For qualitative analysis, it is not decisive how often an argument is mentioned, but the number is still given in brackets. For each keyword, one or a selection of statements are quoted.

It is important to remember that these are participants' views and not necessarily scientifically accepted benefits or risks. For example, five people mention an increased risk of fire. The German Insurance Association (GDV) disagrees with this. According to the data, there is no evidence that EV passenger cars have a higher risk of catching fire than ICE vehicles [28]. The Swedish Civil Contingencies Agency (MSB) published data proving that EVs are less likely to catch fire than ICE cars, and the number of fires per car in EVs has decreased over the last three years [29,30].

4.4.1. Expected Strengths

- Participation in technical progress and risk reduction (5): "You don't have to worry about the condition of the batteries, and you always have new models of batteries".
- Cheaper energy (3): "Battery replacement makes fast charging unnecessary. This eliminates very high costs for these charging stations; fast charging electricity is usually much more expensive per kWh and generates much higher charging losses in the battery and charging system".
- Faster (3): "Changing batteries could possibly be quicker than refueling. So, time saving".
- Lighter (2): "Weight savings, as it is not always necessary to drive with the largest battery".

4.4.2. Expected Weaknesses

- Mechanical problems (6): "I think that a crushing accident can occur while swapping battery". "Wear and tear due to constant installation and removal of batteries".
- Safety concerns, fire threat (5): "There are greater risks of fires with trucks. It is more difficult to fight fires with batteries".
- Shortage of swapping stations (4): "Congestion and energy bottlenecks at such stations"; "Imagine a traffic jam that lasted about 8 or 12 h which is a common occurrence, unfortunately".
- Higher total costs (4): "In total, there must be significantly more battery systems than e-trucks, and someone has to pay for that. Battery rental increases overall maintenance costs", "Higher prices, because the rental company has to earn money".
- Driver's problems (2): "More time pressure, less break time for drivers"; "Not really necessary as it can be charged during the breaks that are necessary anyway".

4.4.3. Expected Opportunities

- Easier to reuse/recycle (3): "Better second-life use of standardized exchangeable batteries".
- Electricity grid integration (2): "A great many batteries have to stand around unused in many places as a stockpile—but they can also be used as grid storage".
- Easier to maintain: "Large number of batteries should be maintained at one time in case of battery swapping".
- Political goals: "Battery replacement can be a good complement to the 100% target for commercial vehicles".

4.4.4. Expected Threats

- Threat of no common standard (8): "Disadvantage: Most of the OEMs have to join in order to have a uniform interface/standardization"; "Through standardization, a concerted, cross-brand development strategy is conceivable. But there is also a danger of monopolization and price dictation".
- Problems with the market introduction (6): "Introduction scenario difficult to imagine: standardization (monopoly?) and (international?) area-wide infrastructure prerequisite for use by early adopters, who however generate too little demand for the necessary investments".

- "I see a risk in the comprehensive market penetration and the long road to standardization on the part of the OEMs. Since there is already an agreement on MCS standardization, and all OEMs and infrastructure operators are already stepping on the gas, swapping may come very late". [MCS—Megawatt Charging System]
- Limited raw materials and environmental problems (4): "Limited materials for manufacturing batteries will be a risk, which will affect other sectors that use the same materials in their production processes".
- "More batteries needed than the number of vehicles, which might cause more waste".
- Space problems and grid dependency (3): "It takes up a lot of space".
- Limitations in the vehicle design (3): "Exchangeable batteries prevent space-optimized installations, e.g., under the driver's cab; the battery position on a 3-axle tractor shown in the video is impractical in Germany; 3-axle tractors are practically not used".
- Neglect of alternative modes of transport (1): "I hope that long overland journeys with trucks and lorries will soon be a thing of the past, and more traffic will be shifted to rail and inland waterways".

5. Discussion

The results of the survey in the section on the subjective perception of the economic aspects of e-trucks show a relatively high level of disagreement with the idea that e-trucks have lower investment costs compared to diesel trucks. A score of 3.83 suggests that there is a moderate level of skepticism or disagreement about the cost benefits associated with purchasing e-trucks. On the other hand, a score of 2.21 suggests that there is a notable level of acceptance or agreement with the notion that e-trucks can provide cost savings in terms of operational expenses. Overall, the results of the perception of the technical aspects of e-trucks show a generally positive perception of e-trucks across several factors. There is agreement that e-trucks require less maintenance effort, are better for the environment, are just as reliable as traditional trucks, and offer safety benefits for road users. However, there is relatively less agreement that e-trucks are better than those powered by alternative fuels.

While there is a relatively low level of agreement or acceptance for attracting young people to become truck drivers and improving working conditions for truck drivers, there is a higher level of agreement for e-trucks being seen as good vehicles for entering low-emission zones, making a significant contribution to global climate protection, and effectively reducing noise and local air pollution.

The level of agreement is significantly high for the decision to purchase one e-truck to gain experience (1.66), although it decreases slightly when individuals are asked to personally become the first user of the new vehicle. However, there is still a notable level of agreement for the immediate adoption of a small proportion of e-trucks (10% of new purchases) and for the long-term objective of moving to 100% e-trucks by 2040. The lowest level of agreement is observed for the proposal to reach 100% e-trucks by 2030.

The results of the perceived benefits and risks of a battery swapping solution indicate that both transport companies (2.17) and truck drivers (2.23) will benefit from battery swapping. The most anticipated advantage is time saving (1.77), with additional hopes for increased flexibility (2.31), and the ability to drive long distances (2.37). However, the biggest concern is the potential scarcity of swapping stations (2.10), while the least-expected risk is increased wear and tear on vehicles.

For the European trucking industry, different paths to the future are possible. Infrastructure is a major challenge for the introduction of alternative power sources in the trucking industry, as there is limited space for refueling or recharging facilities. When it comes to battery swapping, the availability of standardized batteries becomes a critical factor in determining success or failure. The establishment of a common standard can be driven by either OEMs or EU policy.

6. Conclusions

In conclusion, the pressing issue of CO_2 emissions from heavy-duty vehicles in the transportation sector necessitates the development of clean technologies to replace diesel trucks. This work aimed to study the critical role of customer acceptance in sustainable transportation policies. By investigating perceptions and acceptance of electric trucks, the study sheds light on potential pathways for adoption.

Through a survey and expert interviews within the German commercial transport sector, the study collected insights on heavy-duty electric trucks' anticipated benefits and risks and on battery swapping technology. While the classified responses generally exhibit positivity, individual statements reveal lingering uncertainties.

The perceptions regarding the economic aspects of e-trucks are mixed, with skepticism around investment costs but notable agreement on operational expense savings. The technical aspects are generally favorably perceived, including maintenance, environmental benefits, reliability, and road safety. However, alignment is relatively lower regarding e-truck superiority over alternative fuel vehicles.

Notably, e-trucks find strong support as vehicles for low-emission zones, climate protection, noise reduction, and local air pollution mitigation. The gradual transition to e-trucks enjoys significant agreement, while specific timelines, such as achieving 100% e-truck adoption by 2030, show less consensus.

Exploring the battery swapping solution highlights potential advantages for time saving, flexibility, and long-distance driving. Concerns mainly revolve around the scarcity of swapping stations, while wear and tear risks are less anticipated.

Ultimately, the study underscores the crucial role of standardized batteries and infrastructure availability in shaping the future of the European trucking industry. Successful implementation of alternative power sources hinges on collaboration between OEMs and EU policies. As the commercial transport sector seeks to reduce its carbon footprint, understanding perceptions and addressing acceptance barriers will be pivotal in shaping a cleaner and more sustainable future for heavy-duty vehicles.

Author Contributions: Conceptualization, F.N. and H.M.; methodology, F.N. and H.M.; formal analysis, F.N.; data curation, F.N. and H.M.; writing—original draft preparation, F.N. and H.M.; writing—review and editing, F.N. and H.M.; visualization, F.N. supervision, H.M. All authors have read and agreed to the published version of the manuscript.

Funding: This research received no external funding, and the APC was funded by the Technical University of Berlin.

Institutional Review Board Statement: Not applicable.

Informed Consent Statement: Not applicable.

Data Availability Statement: Not applicable.

Acknowledgments: We acknowledge support from the German Research Foundation and the Open Access Publication Fund of the Technical University of Berlin.

Conflicts of Interest: The authors declare no conflict of interest.

References

1. European Commission Q&A: CO_2 Emission Standards for Heavy-Duty Vehicles. Available online: https://ec.europa.eu/commission/presscorner/detail/en/qanda_23_763 (accessed on 30 June 2023).
2. European Commission Reducing CO_2 Emissions from Heavy-Duty Vehicles. Available online: https://climate.ec.europa.eu/eu-action/transport-emissions/road-transport-reducing-co2-emissions-vehicles/reducing-co2-emissions-heavy-duty-vehicles_en (accessed on 31 July 2023).
3. Bhardwaj, S.; Mostofi, H. Technical and Business Aspects of Battery Electric Trucks—A Systematic Review. *Future Transp.* **2022**, *2*, 382–401. [CrossRef]
4. Karlsson, J.; Grauers, A. Energy Distribution Diagram Used for Cost-Effective Battery Sizing of Electric Trucks. *Energies* **2023**, *16*, 779. [CrossRef]

5. Jahangir Samet, M.; Liimatainen, H.; van Vliet, O.P.R.; Pöllänen, M. Road Freight Transport Electrification Potential by Using Battery Electric Trucks in Finland and Switzerland. *Energies* **2021**, *14*, 823. [CrossRef]
6. Tol, D.; Frateur, T.; Verbeek, M.; Riemersma, I.; Mulder, H. *Techno-Economic Uptake Potential of Zero-Emission Trucks in Europe*; TNO: Den Haag, The Netherlands, 2022.
7. Bauer, R.A. Consumer Behavior as Risk Taking. In *Risk Taking and Information Handling in Consumer Behavior*; Harvard University Press: Cambridge, MA, USA, 1960; pp. 389–398.
8. Vijayasarathy, L.R. Predicting Consumer Intentions to Use On-Line Shopping: The Case for an Augmented Technology Acceptance Model. *Inf. Manag.* **2004**, *41*, 747–762. [CrossRef]
9. Lee, M.-C. Factors Influencing the Adoption of Internet Banking: An Integration of TAM and TPB with Perceived Risk and Perceived Benefit. *Electron. Commer. Res. Appl.* **2009**, *8*, 130–141. [CrossRef]
10. Im, I.; Kim, Y.; Han, H.-J. The Effects of Perceived Risk and Technology Type on Users' Acceptance of Technologies. *Inf. Manag.* **2008**, *45*, 1–9. [CrossRef]
11. Davis, F.D. Perceived Usefulness, Perceived Ease of Use, and User Acceptance of Information Technology. *MIS Q.* **1989**, *13*, 319. [CrossRef]
12. Ajzen, I. The Theory of Planned Behavior. *Organ. Behav. Hum. Decis. Process.* **1991**, *50*, 179–211. [CrossRef]
13. Daramy-Williams, E.; Anable, J.; Grant-Muller, S. A Systematic Review of the Evidence on Plug-in Electric Vehicle User Experience. *Transp. Res. Part D Transp. Environ.* **2019**, *71*, 22–36. [CrossRef]
14. Hardman, S.; Jenn, A.; Tal, G.; Axsen, J.; Beard, G.; Daina, N.; Figenbaum, E.; Jakobsson, N.; Jochem, P.; Kinnear, N.; et al. A Review of Consumer Preferences of and Interactions with Electric Vehicle Charging Infrastructure. *Transp. Res. Part D Transp. Environ.* **2018**, *62*, 508–523. [CrossRef]
15. Wicki, M.; Brückmann, G.; Quoss, F.; Bernauer, T. What Do We Really Know about the Acceptance of Battery Electric Vehicles?—Turns out, Not Much. *Transp. Rev.* **2023**, *43*, 62–87. [CrossRef]
16. Nationale Plattform Zukunft der Mobilität. *Kundenakzeptanz als Schlüssel für den Markthochlauf der Elektromobilität*; Arbeitsgruppe 2 "Alternative Antriebe und Kraftstoffe für nachhaltige Mobilität"; Nationale Plattform Zukunft der Mobilität: Berlin, Germany, 2021.
17. Brückmann, G. Test-Drives & Information Might Not Boost Actual Battery Electric Vehicle Uptake? *Transp. Res. Part A Policy Pract.* **2022**, *160*, 204–218. [CrossRef]
18. Çabukoglu, E.; Georges, G.; Küng, L.; Pareschi, G.; Boulouchos, K. Battery Electric Propulsion: An Option for Heavy-Duty Vehicles? Results from a Swiss Case-Study. *Transp. Res. Part C Emerg. Technol.* **2018**, *88*, 107–123. [CrossRef]
19. Shoman, W.; Yeh, S.; Sprei, F.; Plötz, P.; Speth, D. Battery Electric Long-Haul Trucks in Europe: Public Charging, Energy, and Power Requirements. *Transp. Res. Part D Transp. Environ.* **2023**, *121*, 103825. [CrossRef]
20. Schneider, J.; Teichert, O.; Zähringer, M.; Balke, G.; Lienkamp, M. The Novel Megawatt Charging System Standard: Impact on Battery Size and Cell Requirements for Battery-Electric Long-Haul Trucks. *eTransportation* **2023**, *17*, 100253. [CrossRef]
21. Qasim, M.; Csiszar, C. Major Barriers in Adoption of Electric Trucks in Logistics System. *Promet* **2021**, *33*, 833–846. [CrossRef]
22. Drive to Zero Global Memorandum of Understanding on Zero-Emission Medium- and Heavy-Duty Vehicles. Available online: https://globaldrivetozero.org/mou-nations/ (accessed on 31 March 2023).
23. Kraftfahrt-Bundesamt Statistics—Vehicles. Available online: https://www.kba.de/EN/Statistik_en/Fahrzeuge_Vehicles/vehicles_node.html (accessed on 31 March 2023).
24. Liboreiro, J. In Win for Germany, EU Agrees to Exempt e-Fuels from 2035 Ban on New Sales of Combustion-Engine Cars. *Euronews*. 2023. Available online: https://www.euronews.com/my-europe/2023/03/28/in-win-for-germany-eu-agrees-to-exempt-e-fuels-from-2035-ban-on-new-sales-of-combustion-en (accessed on 21 August 2023).
25. Posaner, J. Brussels and Berlin Strike Deal on 2035 Combustion-Engine Ban. Politico. Available online: https://www.politico.eu/article/brussels-and-berlin-strike-car-engine-combustion-zero-emissions-ban-deal/ (accessed on 21 August 2023).
26. Aguilar, F.J. *Scanning the Business Environment*; Macmillan: New York, NY, USA, 1967.
27. XCMGGroup. *XCMG Logistic Vehicle Battery Swap Station*; XCMGGroup: Xuzhou, China, 2020.
28. Gesamtverband der Deutschen Versicherungswirtschaft E-Autos in Tiefgaragen: Keine Erhöhte Brandgefahr Feststellbar. Available online: https://www.gdv.de/gdv/medien/medieninformationen/e-autos-in-tiefgaragen-keine-erhoehte-brandgefahr-feststellbar-66230 (accessed on 31 March 2023).
29. Bleakley, D. Petrol and Diesel Cars 20 Times More Likely to Catch Fire than EVs. Available online: https://thedriven.io/2023/05/16/petrol-and-diesel-cars-20-times-more-likely-to-catch-fire-than-evs/ (accessed on 31 July 2023).
30. MSB Bränder i Eltransportmedel under 2022. Available online: https://www.msb.se/sv/aktuellt/nyheter/2023/maj/brander-i-eltransportmedel-under-2022/ (accessed on 31 July 2023).

systems

MDPI

Article

Prediction of China Automobile Market Evolution Based on Univariate and Multivariate Perspectives

Debao Dai [1], Yu Fang [1,*], Shihao Wang [2] and Min Zhao [3]

[1] School of Management, Shanghai University, Shanghai 200444, China; ddb@shu.edu.cn
[2] Department of Digital Management, Shanghai MAHLE Thermal Systems Co., Ltd., Shanghai 201206, China; shihao.wang@smts-co.com
[3] SILC Business School, Shanghai University, Shanghai 201800, China; minzhao@163.shu.edu.cn
* Correspondence: fangyush@shu.edu.cn

Abstract: The automobile is an important part of transportation systems. Accurate prediction of sales prospects of different power vehicles can provide an important reference for national scientific decision making, flexible operation of enterprises and rational purchases of consumers. Considering that China has achieved the goal of 20% sales of new energy vehicles ahead of schedule in 2025, in order to accurately judge the competition pattern of new and old kinetic energy vehicles in the future, the automobile market is divided into three types according to power types: traditional fuel vehicles, new energy vehicles and plug-in hybrid vehicles. Based on the monthly sales data of automobiles from March 2016 to March 2023, the prediction effects of multiple models are compared from the perspective of univariate prediction. Secondly, based on the perspective of multivariate prediction, combined with the data of economic, social and technical factors, a multivariate prediction model with high prediction accuracy is selected. On this basis, the sales volume of various power vehicles from April 2023 to December 2025 is predicted. Univariate prediction results show that in 2025, the penetration rates of three types of vehicles will reach 43.8%, 44.4% and 11.8%, respectively, and multivariate prediction results show that the penetration rates will reach 51.0%, 37.9% and 11.1%, respectively.

Keywords: new energy vehicles; deep learning algorithm; Prophet; SVM; BP neural network; VAR; sales forecast; sustainable development

Citation: Dai, D.; Fang, Y.; Wang, S.; Zhao, M. Prediction of China Automobile Market Evolution Based on Univariate and Multivariate Perspectives. *Systems* **2023**, *11*, 431. https://doi.org/10.3390/systems11080431

Academic Editors: Mahyar Amirgholy and Jidong J. Yang

Received: 25 July 2023
Revised: 10 August 2023
Accepted: 15 August 2023
Published: 17 August 2023

1. Introduction

1.1. Background and Motivation

China's automotive industry is in a new stage of energy and technological innovation, and the era of multi-power with traditional fuel vehicles, battery electric vehicles and plug-in hybrid vehicles as the mainstay has begun. According to the data of the China Association of Automobile Manufacturers, in 2022, the operating income of China's automobile manufacturing industry reached 9289.99 billion CNY, up 6.8% year-on-year, accounting for 6.7% of the total operating income of industrial enterprises above the designated size. According to the statistics of the China Association of Automobile Manufacturers, in the first quarter of 2023, China exported 1.07 million automobiles, surpassing Japan for the first time and becoming the largest automobile exporter in the world. As a pillar industry, the automobile industry plays a key role in accelerating industrialization, promoting manufacturing innovation, stimulating domestic demand, increasing employment and promoting economic growth.

Under the dual pressure of environmental protection and energy shortages, developing new energy vehicles is a common strategic choice for all countries in the world. In 2022, the State Council, China, issued the Twelfth Five-Year National Strategic Emerging Industries Development Plan, which listed the new energy automobile industry as one of the seven strategic emerging industries in China. On 28 March 2023, the agreement of EU member states to ban the sale of new fossil fuel vehicles from 2035 was officially approved. In 2021, US President Biden announced an order that by 2030, the sales of new energy vehicles will account for half of the national car sales.

In 2022, Xu Changming, deputy director of the National Information Center, said that traditional fuel vehicles and new energy vehicles have their own advantages, and both have certain development space in the foreseeable future. In 2023, at the monthly meeting of the National Passenger Car Market Information Association, Cui Dongshu, Secretary-General of the Association, said that the development of new energy and the development of fuel vehicles are not simply antagonistic. While rationally arranging the new energy automobile industry, it is necessary to avoid the sudden drop in demand for traditional automobile products and ensure the steady transformation, upgrading and sustainable development of the industry.

The competition pattern of the automobile industry is influenced by many factors [1,2], such as the economic environment, technological progress, consumer demand, environmental laws and regulations, etc. The market demand and competition pattern of various power vehicles in the future are still uncertain. Under the condition of limited data, it is of great significance for the production planning of automobile enterprises and the scientific decision making of the government to effectively capture the historical sales laws of various types of automobiles in the automobile sales system, analyze the dynamic relationship among various influencing factors and various types of automobile sales, and then accurately predict the sales trend of various power types of automobiles.

1.2. Literature Review

1.2.1. Influencing Factors of Vehicle Sales

The sales of traditional cars and new energy vehicles are affected by many factors. Government policy is an important factor affecting the development of the automobile industry, which plays a key role in stabilizing consumption and releasing demand. In all stages of the development of new energy vehicles, government support has played an important role in its large-scale promotion [3]. Socioeconomically, the employment rate of residents and the level of the consumer price are usually considered as important factors affecting automobile sales. In addition, price is also one of the most widely studied socioeconomic factors affecting automobile sales, and most consumers are sensitive to price [4]. Subsidies can reduce the price of new energy vehicles, thus increasing consumers' willingness to buy [5]. In terms of technology, the progress and innovation of automobile technology are directly related to the use cost, cruising range, energy consumption and other issues of automobiles, and are the key factors affecting the promotion and popularization of automobiles [6]. There is a positive correlation between technological innovation and the adoption of electric vehicles [7]. Supporting infrastructure is directly related to whether consumers can enjoy convenient services after buying a car, so it is considered to have an impact on automobile promotion and sales. Studies have shown that perfect charging facilities for new energy vehicles will help to enhance consumers' willingness to buy [8]. However, Lin and Wu [3] think that in the case of short-distance travel, consumers mostly choose home charging, and the coverage of charging infrastructure has limited influence on the purchase intention. Some studies focus on the influence of consumer psychological factors on the sales of new energy vehicles. Yang, et al. [9] and other research shows that the comfort, handling, space and cost performance of electric vehicles have a significant impact on the sales of electric vehicles. Wang, et al. [10] and other studies show that perceived risk and environmental awareness have a significant impact on the acceptance of electric vehicles. Some studies think that the energy price is one of the driving forces of automobile promotion and sales [11,12]. For example, increase in the gasoline price is beneficial to the promotion of electric vehicles [13], but some studies show that the energy price has little influence on the promotion of electric vehicles [14].

1.2.2. Prediction Models of Vehicle Sales Volume

Statistical models are widely used in automobile sales forecasting. Li, et al. [15], based on the improved Bass model and grey theory, constructed the demand forecasting model of new energy vehicles, and its effectiveness is verified by using three data sets from Norway, France and Europe. Hsieh, et al. [16] developed a Monte Carlo model based on historical data. Kumar, et al. [17] used Gompertz, Logistic, Bass and Generalized Bass models to simulate the future demand for electric vehicles, and determined the best-fitting model for 20 major countries.

However, some studies believe that the traditional statistical model cannot effectively extract the nonlinear characteristics of automobile sales data [18]. Therefore, some scholars improved the traditional forecasting method. Zhou, et al. [19] put forward a new time-varying grey Bernoulli model, which better captured the nonlinear, complex and time-varying characteristics related to electric vehicle sales. Pei and Li [20] established a nonlinear grey Bernoulli model based on data grouping, and pointed out that in 2020, the sales of new energy vehicles will exceed 2 million. Liu, et al. [1] point out that by 2025, the sales of new energy vehicles in China will reach 8.84 million. Li, et al. [21], after improving the parameters of Bass model and LV model, showed it has a better forecasting effect on the sales of battery electric vehicles in China, and the Bass model is more accurate.

Some scholars also apply artificial intelligence algorithms to the problem of automobile sales forecasts. Zhang, et al. [22], using the LSTM algorithm, established a smart car sales forecasting model based on the KOL network public opinion and network search index, which improves the forecasting accuracy of smart car sales. Liu, et al. [23] proposed a multi-factor sales forecasting model combining discrete wavelet transform and BiLSTM, and obtained the MAE value, Maple value and RMSE value of the optimal DWT-BiLSTM model of 0.811, 5.671 and 1.001, respectively. Xia, et al. [24] used the XGBoost prediction algorithm for automobile sales prediction and achieved high prediction accuracy in a short runtime. Wu and Chen [25] combined a principal component analysis and neural network to predict the sales volume and growth rate of electric vehicles, and pointed out that the sales volume of electric vehicles in the world and China will continue to increase in the next 50 years, but the growth rate will continue to decline. With the development of machine learning technology, sentiment analysis technology has also been applied to the problem of automobile sales forecasts. Liu, et al. [26] proposed a combined forecasting model based on multi-angle feature extraction and sentiment analysis, which improved the forecasting accuracy of automobile sales. Ding, et al. [27] uses an online comment-driven combination forecasting model to improve the forecasting accuracy of new energy vehicle sales.

In addition to the innovation of methods, some studies focus on the competition and substitution between traditional fuel vehicles and new energy vehicles. Sun and Wang [28] predicted the market evolution of new energy vehicles and new energy vehicles in China based on the Lotka–Volterra model and system dynamics (SD) model. Guo, et al. [29] divided the passenger car market into four types—gasoline passenger car, natural gas passenger car, blade electric passenger car and plug-in hybrid passenger car—and predicted the evolution trend of the passenger car market based on the Lotka–Volterra model and historical data of automobile sales.

There is a lot of literature on the diffusion trend of automobile sales; however, few publications study the dynamic change trend of automobile sales with competition based on the perspective of prediction. On the basis of related research, this paper analyzes the diffusion of different power types of vehicles in 2025 from the perspectives of univariate prediction and multivariate prediction.

1.3. Contribution and Organization

The innovation and contribution of this paper are mainly reflected in two aspects. Firstly, the automobile industry is divided into three types, according to the power type: traditional fuel vehicles (ICEs), battery electric vehicles (BEVs) and plug-in hybrid vehicles (PHEVs). Based on the monthly sales data, from the perspective of univariate and multivariable, the prediction effects of various statistical models and machine learning models for automobile sales are compared, and high prediction accuracy is obtained. Secondly, in the multivariable forecasting part, the VAR model and BP neural network model are innovatively combined to analyze the lag effect of 17 influencing factors from economic, social and technical aspects on automobile sales, and a multivariable BP neural network forecasting model with lag characteristics is developed.

2. Materials and Methods

Figure 1 shows the framework of this study.

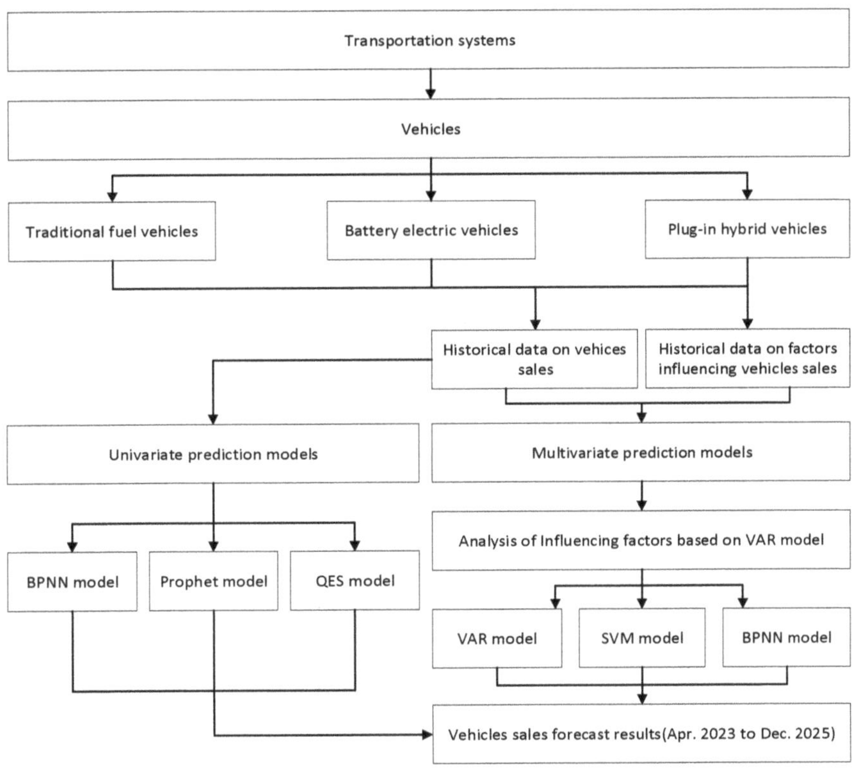

Figure 1. The research framework.

The vehicle is one of the basic components of transportation systems, and the reform of the automobile industry will profoundly affect the reconstruction of transportation systems. Considering that the monthly sales of traditional fuel vehicles and new energy vehicles have certain cyclical and seasonal characteristics, firstly, the automobile industry is divided into traditional fuel vehicles, battery electric vehicles and plug-in hybrid vehicles, according to power types. Secondly, based on the historical sales data of different power types from March 2016 to March 2023, the capturing effects of three univariate forecasting models, namely, the BP neural network (BPNN) model, quadratic exponential smoothing (QES) model and Prophet model, on the trends and laws of sales changes are compared. The univariate forecast results of automobile sales of three power types in 2025 are obtained by selecting the model with better overall performance. In addition, the multivariate prediction model can effectively identify the dynamic coupling relationship between the target variables and related influencing factors. Therefore, the influencing factors of automobile sales are selected from the aspects of economy, society and technology, and based on the automobile sales data and influencing factor data from March 2016 to March 2023, a vector autoregressive (VAR) model is established; the lagging order of influencing factors is analyzed; the input samples of the multivariate prediction model are constructed; and the prediction effects of the VAR model, support vector regression model (SVM) model and BP neural network model are compared.

2.1. Variable Selection and Data Description

Electric energy is one of the main power sources of new energy vehicles. The automobile market is divided into three categories: traditional fuel vehicles, battery electric vehicles and plug-in hybrid vehicles. Through the analysis of the influencing factors of automobile sales in Section 1.2.1, and considering the availability of data, 17 influencing factors of traditional fuel vehicles, battery electric vehicles and plug-in hybrid vehicles are selected from the aspects of economy, society and technology, as shown in Table 1.

Table 1. Influencing factors of vehicle sales.

Variable	Meaning	Variable	Meaning
Y1	Sales volume of traditional fuel vehicles (units)	X7	Average price of new energy vehicles (10,000 CNY)
Y2	Sales volume of battery electric vehicles (units)	X8	Patents granted for electric vehicles (units)
Y3	Sales volume of plug-in hybrid vehicles (units)	X9	Effective patents granted for power batteries (units)
X1	Consumer price index (%)	X10	Output of lithium-ion batteries (million units)
X2	Customs exports (100 million USD)	X11	Hydroelectric power generation (100 million kWh)
X3	Customs imports (100 million USD)	X12	Nuclear power generation (100 million kWh)
X4	Total value of imports (1000 USD)	X13	Employee index (%)
X5	Total export value (1000 USD)	X14	Highway passenger traffic (ten thousand people)
X6	Average price of traditional fuel vehicles (10,000 CNY)		

The statistical data cover the period from March 2016 to March 2023, with a total of 85 groups of data. The historical sales data of three types of cars are shown in Figure 1. Among them, automobile sales data come from the official website of the China Association of Automobile Manufactures, average automobile price data come from the car home website, patent data come from the China National Intellectual Property Administration website, customs import and export data come from the Oriental Fortune Network, and other data come from the National Bureau of Statistics and the China Statistical Yearbook. Some missing data are filled in by linear interpolation.

Car companies and governments pay more attention to the monthly data of car sales, while the annual data can better reflect the general development trend of different power types of cars. As can be seen from Figure 2, from 2011 to 2022, according to the growth of the penetration rate of new energy vehicles, the development of new energy vehicles in China can be roughly divided into three stages: (1) Before 2014: the initial stage; (2) From 2014 to 2020: the stage of rapid development; (3) From 2022 to present: the stage of transformation and upgrading. The red dotted line represents the dividing line of these three stages, which is used to distinguish them more clearly. The overall sales of traditional fuel vehicles are showing an obvious downward trend, while the overall sales of battery electric vehicles and plug-in hybrid vehicles are showing an upward trend.

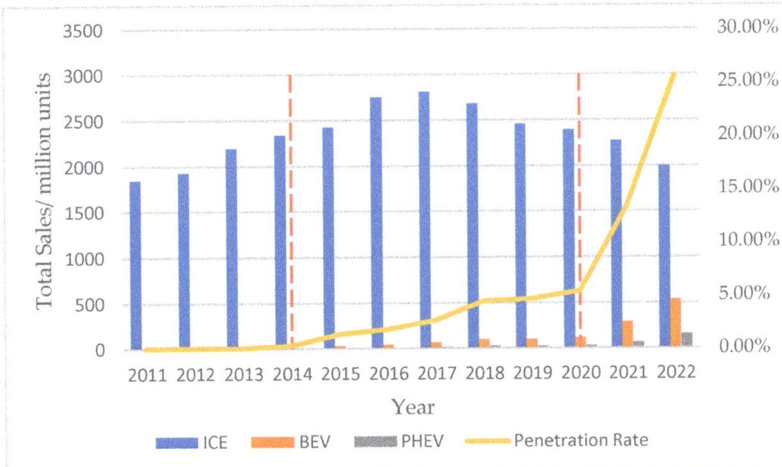

Figure 2. Annual sales of automobiles in China.

2.2. Methodology

2.2.1. Quadratic Exponential Smoothing Model

The exponential smoothing model is one of the classic statistical models in time series prediction [30], among which quadratic exponential smoothing model is suitable for the prediction of time series with a linear trend [31]. The formula of the quadratic exponential smoothing model is as follows:

$$S_t^{(1)} = \alpha Y_t + (1-\alpha) S_{t-1}^{(1)} \tag{1}$$

$$S_t^{(2)} = \alpha S_t^{(1)} + (1-\alpha) S_{t-1}^{(2)} \tag{2}$$

where $S_t^{(1)}$ and $S_t^{(2)}$ represent the first exponential smoothing value and the second exponential smoothing value in the t period, respectively; and $\alpha(0 < \alpha < 1)$ is the smoothing coefficient.

$$F_{t+T} = a_t + b_t * T \tag{3}$$

$$a_t = 2S_t^{(1)} - S_t^{(2)} \tag{4}$$

$$b_t = \frac{\alpha}{1-\alpha}\left(S_t^{(1)} - S_t^{(2)}\right) \tag{5}$$

According to Formula (5), the sales of traditional fuel vehicles, battery electric vehicles and plug-in hybrid vehicles from April 2023 to March 2024 are predicted respectively, where F_{t+T} represents the predicted value of the $t+T$ period, and T represents the number of predicted periods. a_t and b_t are the model coefficients of the t period.

2.2.2. Prophet Model

The Prophet model is a time series forecasting model put forward by FackBook Company in 2017, which can effectively fit the trend and seasonal characteristics of the series, and can also deal with holiday factors [32]. The expression is shown in formula (6):

$$y(t) = g(t) + s(t) + h(t) + \epsilon_t \tag{6}$$

where $g(t)$ represents the trend term for modeling the non-periodic change of sales of traditional fuel vehicles, battery electric vehicles and plug-in hybrid vehicles; $s(t)$ represents the seasonal term for modeling the periodic change of time series; $h(t)$ represents holidays; and $\in_t$ represents the error term.

There are two types of trend term $g(t)$: nonlinear saturated growth model and piecewise linear model. Because the automobile sales data did not show an obvious saturated growth trend, the piecewise linear model was chosen. The expression is as follows:

$$g(t) = \left(k + a(t)^T \delta\right) t + \left(b + a(t)^T \gamma\right) \tag{7}$$

where k represents the growth rate, b represents the offset, $a(t)$ represents the indicator function, and γ represents the offset of smoothing.

The Prophet model uses Fourier series to simulate the periodic change of time series, which is expressed as follows:

$$s(t) = \sum_{n=1}^{N} \left(a_n \cos\left(\frac{2\pi nt}{T}\right) + b_n \sin\left(\frac{2\pi nt}{T}\right)\right) \tag{8}$$

where N represents the total number of cycles, T represents cycles, and a_n and b_n are parameters to be estimated.

The expression of holiday item is as follows:

$$h(t) = Z(t)\gamma_i,\ \gamma \sim Normal\left(0, v^2\right) \tag{9}$$

where i represents various holidays, $Z(t) = [1(t \in D_1), \ldots, 1(t \in D_L)]$ represents the indicator function, D_1 represents holiday collection, and γ_i is the parameter of each holiday.

2.2.3. Vector Autoregressive Model

The vector autoregressive model (VAR) was originally proposed by Christopher Sims to study the dynamic relationship between variables [33,34]. The VAR model regards all variables as endogenous variables, which reduces the uncertainty in the simultaneous equations model caused by subjective judgment errors. The VAR model is used to determine the optimal lag order of the influencing factors of automobile sales, which is used as the basis for constructing the multivariate prediction model.

The form of the VAR (p) model is shown in Formula (10):

$$Y_t = A_0 + A_1 Y_{t-1} + A_2 Y_{t-2} + \ldots + A_n Y_{t-p} + \mu_t \tag{10}$$

where $Y_t = \begin{bmatrix} y_{1t} \\ y_{2t} \\ \cdots \\ y_{nt} \end{bmatrix}$, $A_0 = \begin{bmatrix} a_{10} \\ a_{20} \\ \cdots \\ a_{n0} \end{bmatrix}$, $e_t = \begin{bmatrix} e_{1t} \\ e_{2t} \\ \cdots \\ e_{nt} \end{bmatrix}$, $A_i = \begin{bmatrix} a_{11,i} & a_{12,i} & \cdots & a_{1n,i} \\ a_{21,i} & a_{22,i} & \cdots & a_{2n,i} \\ \cdots & \cdots & \cdots & \cdots \\ a_{n1,i} & a_{n2,i} & \cdots & a_{nn,i} \end{bmatrix}$, Y_t is the n-dimensional vector of endogenous variables, p is the lag order, A_0 and A_i are the matrices of coefficients to be estimated, and μ_t is the n-dimensional random perturbation term.

2.2.4. Support Vector Regression Model

The support vector regression model (SVM) is a classic machine learning model, and the support vector regression (SVR) is an important branch of SVM [35,36]. The expression of the SVM model is as follows:

$$y = \omega \varphi(x) + b \tag{11}$$

where ω is the weight vector, and b is the deviation.

According to the principle of structural risk minimization, SVM is transformed into the following optimization problem:

$$min \frac{1}{2} \|\omega\|^2 + C \sum_{i=1}^{N} (\delta_i + \delta_i^*) \tag{12}$$

$$\begin{cases} y_i - \omega\varphi(x) - b \leq \varepsilon + \delta_i \\ \omega\varphi(x) + b - y_i \leq \varepsilon + \delta_i^* \\ \delta_i \geq 0 \\ \delta_i^* \geq 0 \end{cases}, i = 1, 2, \ldots, N \tag{13}$$

where δ_i and δ_i^* are relaxation variables, c C is a penalty factor, and ε is a loss function.

2.2.5. BP Neural Network Model

The BP neural network, put forward by scholars such as Rinehart and MeClelland, is the most widely used artificial neural network [37], which has the characteristics of self-learning adaptation, parallel processing, strong learning ability and generalization [38], and generally consists of an input layer, hidden layer and output layer. Studies have shown that a three-layer BP neural network prediction model can approximate any nonlinear function [39,40]. Based on the Keras deep learning framework, this study constructs the univariate prediction model and multivariate prediction model of the BP neural network.

The general process of BP neural network prediction is as follows:

Step 1: Normalize the original data sequence, and input the normalized training samples into the network.

$$x_i' = \frac{x_i - x_{min}}{x_{max} - x_{min}} \tag{14}$$

Step 2: Initialize the network parameters. Set the number of neurons in each layer of the network, set the maximum number of iterations and learning rate, randomly assign initial values to the weights and deviations of the network, and determine the activation function.

Step 3: Calculate the input and output values of each layer, and compare the output value with the target value to determine the error.

Step 4: Based on the error, correct the weights and thresholds.

Step 5: Repeat the process from steps (3) and (4) until the model error drops to the preset value or the training times reach the preset value. After the training, the trained BP neural network can be used for data prediction.

2.3. Model Evaluation Index

The mean absolute percentage error (MAPE) and root-mean-square error (RMSE) are selected as evaluation indexes to compare the accuracy of the above models for automobile sales forecasting. The lower the index value, the better the forecasting effect of the model is proved.

$$MAPE = \frac{1}{n}\sum_{i=1}^{n}\left|\frac{\hat{y}_i - y_i}{y_i}\right| \tag{15}$$

$$RMSE = \sqrt{\frac{1}{n}\sum_{i=1}^{n}(\hat{y}_i - y_i)^2} \tag{16}$$

3. Results

3.1. Comparison of Univariate Prediction Models

The multivariate forecasting model can comprehensively consider the influence of various factors on automobile sales, but it cannot fully extract the trend information contained in the monthly automobile sales data. In some cases, the effect of univariate forecasting may be better than multivariate forecasting [41]. Therefore, based on the historical sales data of traditional fuel vehicles, battery electric vehicles and plug-in hybrid vehicles, univariate forecasting models are established individually. In order to remain consistent with the dimensions of subsequent multivariate prediction, the data series are processed by logarithm, and fitting and prediction are based on logarithm.

Taking the first 85% of data from March 2016 to February 2022 as the training set, and the remaining 15% of data from March 2022 to March 2023 as the test set, this paper compares the accuracy of the selected univariate forecasting model for the sales forecast of three power types. The univariate BP neural network model and Prophet model are realized using Python3.9 programming, and the quadratic exponential smoothing model is realized using Matlab2017b programming.

3.1.1. Prediction Results of Univariate BP Neural Network

Considering that the monthly sales data of automobiles is periodic with a step size of 12, a sliding window with a length of 12 is set; so, the input variable of the univariate BP neural network model is a 12-dimensional vector.

Taking 12 as a step, based on the historical sales data of traditional fuel vehicles, battery electric vehicles and plug-in hybrid vehicles from March 2016 to March 2023, 60 sets of training data from March 2017 to February 2022 and 13 sets of test data from March 2022 to March 2023 were constructed.

Based on the Keras deep learning framework, the univariate BP neural network models of traditional fuel vehicle sales, battery electric vehicle sales and plug-in hybrid vehicle sales are constructed individually, and the parameter settings are shown in Table 2.

Table 2. Parameter settings of univariate BP neural network model.

Target Sequence	Number of Plies	Activation Function of Each Layer	Number of Neurons	Training Times	Learning Rate	Loss Function
Traditional fuel vehicle	3	'relu', 'relu', 'linear'	$12 \times 6 \times 1$	500	0.0001	MSE
Battery electric vehicle	3	'relu', 'relu', 'linear'	$12 \times 7 \times 1$	600	0.0001	MSE
Plug-in hybrid vehicle	3	'relu', 'relu', 'linear'	$12 \times 7 \times 1$	500	0.0001	MSE

In total, 60 groups of pre-constructed training data of three types of power vehicles are individually input into the univariate BP neural network model, and they are trained according to the steps in Section 2.2.5, and the trained univariate BP neural network model and the fitting value of automobile sales from March 2017 to February 2022 are obtained. A total of 13 groups of data used for testing are input into the univariate BP neural network model based on the training set data, and the forecast value of automobile sales from March 2022 to March 2023 is obtained.

Comparing the fitted and predicted values with the actual values in the corresponding periods, the fitting and predicted results of the univariate BP neural network model for traditional fuel vehicles, battery electric vehicles and plug-in hybrid vehicles are shown in Figure 3.

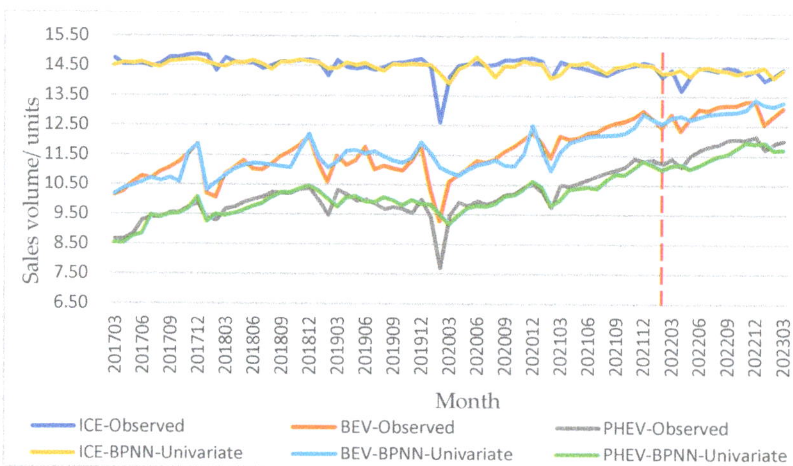

Figure 3. Fitting and prediction effect of univariate BP neural network model.

3.1.2. Prediction Results of Quadratic Exponential Smoothing

The input variable of the quadratic exponential smoothing model is a one-dimensional vector, that is, the historical sales data of the automobile. Based on the historical sales data of traditional fuel vehicles, battery electric vehicles and plug-in hybrid vehicles from March 2016 to February 2022, the quadratic exponential smoothing models of three types of power vehicles are constructed according to Formulas (1)–(5), and the quadratic exponential smoothing fitting values of the sales of three types of power vehicles from March 2016 to February 2022 are calculated. Based on the constructed quadratic exponential smoothing model and Formula (3), the predicted sales values of three types of power vehicles from March 2022 to March 2023 are calculated individually.

After many attempts, the quadratic exponential smoothing model has the best effect when the smoothing coefficient α is 0.3. Comparing the fitted and predicted values with the actual values in the corresponding periods, the fitting and predicted results of the quadratic exponential smoothing model for three types of power vehicles are shown in Figure 4.

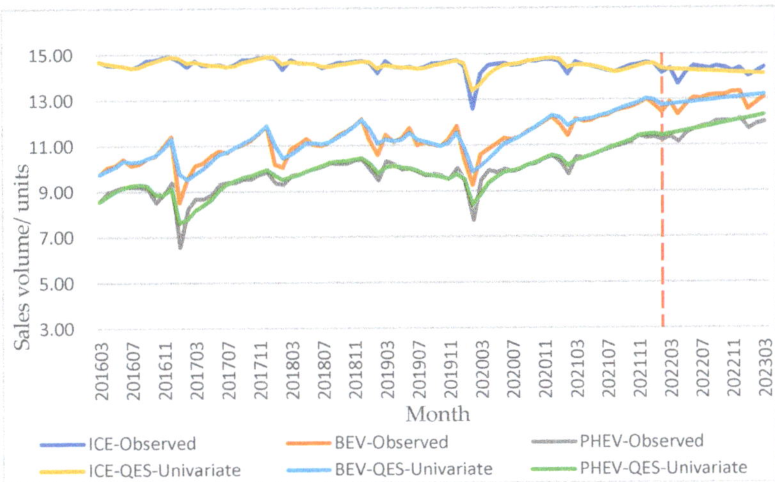

Figure 4. Fitting and prediction effect of quadratic exponential smoothing model.

3.1.3. Prediction Results of Prophet Model

The input variables of the Prophet model are two-dimensional vectors, namely, the date and corresponding historical sales data. The default parameters of the fbProphet library are used in this Prophet model. Based on the date series of traditional fuel vehicles, battery electric vehicles and plug-in hybrid vehicles from March 2016 to February 2022 and the corresponding vehicle sales data, according to Formulas (6)–(9), the prediction models of three power types of vehicles are constructed individually, and the fitting values of the vehicle sales of three power types are calculated. Based on the constructed Prophet model, the sales volume of three types of power vehicles from March 2022 to March 2023 are predicted by Formula (6).

Comparing the fitted and predicted values with the actual values in the corresponding periods, the fitted and predicted results of automobile sales of three power types are shown in Figure 5.

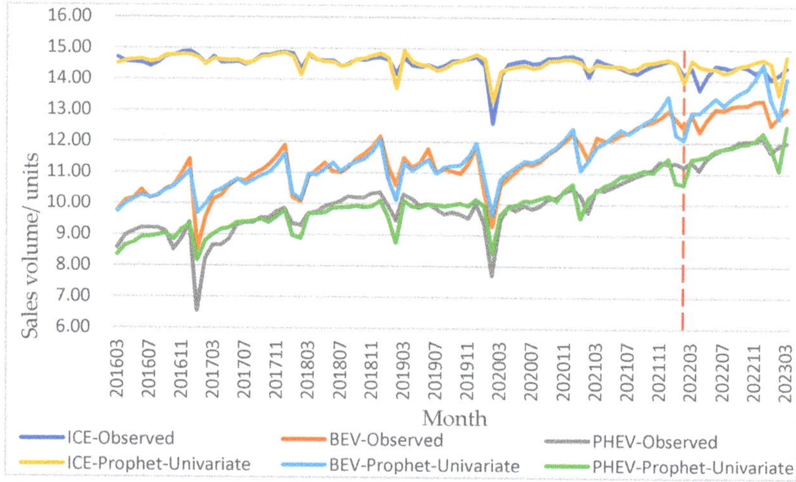

Figure 5. Fitting and prediction effect of Prophet model.

3.1.4. Error Comparison

Table 3 shows the fitting prediction errors of the BP neural network model, quadratic exponential smoothing model and Prophet model for the sales of three types of power vehicles. Among them, because the BP neural network model is based on the training data constructed by a sliding window, the fitting time range is from March 2017 to February 2022.

Table 3. Fitting and prediction errors of univariate prediction models.

Model	Error Type	Evaluating Indicator	ICE	BEV	PHEV
BP Neural Network	Fitting Error	MAPE	1.042%	2.488%	2.214%
		RMSE	0.262	0.405	0.329
	Predicting Error	MAPE	0.956%	1.921%	2.591%
		RMSE	0.240	0.298	0.335
Quadratic Exponential Smoothing	Fitting Error	MAPE	0.606%	1.621%	1.708%
		RMSE	0.150	0.276	0.241
	Predicting Error	MAPE	1.296%	1.816%	1.412%
		RMSE	0.234	0.274	0.231
Prophet	Fitting Error	MAPE	0.688%	1.715%	2.842%
		RMSE	0.142	0.267	0.357
	Predicting Error	MAPE	2.068%	3.382%	1.605%
		RMSE	0.363	0.596	0.296

Comparing the fitting and forecasting effects of each benchmark model on the sales volume of traditional fuel vehicles, the RMSE index shows that the fitting error of the Prophet model is the smallest, and the forecasting error of the exponential smoothing model is the smallest. The MAPE index shows that the fitting error of the quadratic exponential smoothing model is the smallest, and the prediction error of the BP neural network model is the smallest. For battery electric vehicles and plug-in hybrid vehicles, except RMSE index, the fitting error of Prophet model for pure electric vehicles is the smallest, and other RMSE indexes and MAPE indexes show that the fitting and prediction error of quadratic exponential smoothing model is the smallest.

Although the quadratic exponential smoothing model performs well on most indicators, it lacks the ability to identify the turning point of data, and the long-term prediction effect is poor, so it is difficult to meet the needs of out-of-sample prediction. Considering the fitting and forecasting performance of each benchmark model for three types of automobile sales, it is considered that the overall forecasting performance of the Prophet model is more in line with the requirements, and its average fitting MAPE and RMSE are only 1.748% and 0.255, respectively, and the average forecasting MAPE and RMSE are only 2.502% and 0.418, respectively. Therefore, the Prophet model is selected to forecast the sales volume of three types of power vehicles from April 2023 to December 2025.

3.2. Comparison of Multivariate Prediction Models

Taking the data from March 2016 to February 2022 as the training set, and the data from March 2022 to March 2023 as the test set, this paper compares the effectiveness of various multivariate prediction models, establishes the VAR prediction model based on Eviews9, and establishes the SVM model and the multivariate BP neural network prediction model based on Python3.9.

3.2.1. Analysis of Lag Effect Based on VAR Model

The VAR model is widely used to analyze the causal relationship between variables, so the VAR model is used to analyze the lag effect and dynamic mechanism of various factors on automobile sales, and to determine the input variables of the multivariate prediction model.

(1) Stationarity test. In order to avoid pseudo-regression, we test the stationarity of the original sales volume and influencing factor series based on the ADF test. In order to eliminate the influence of heteroscedasticity, this paper carries out logarithmic processing on automobile sales data and influencing factors data. As shown in Table 4, at the significance level of 1%, the 17 data series involved in the study are stationary.

Table 4. ADF test results.

Variable	Inspection Type	ADF	1% Critical Value	5% Critical Value	10% Critical Value	Prob	Stationarity
LNY1	(C, T, 0)	−6.3515	−4.0925	−3.4744	−3.1645	0.0000	stationary
LNY2	(C, T, 0)	−4.8055	−4.0925	−3.4744	−3.1645	0.0011	stationary
LNY3	(C, T, 0)	−4.6803	−4.0925	−3.4744	−3.1645	0.0017	stationary
LNX1	(C, T, 1)	−6.6209	−4.0946	−3.4753	−3.1650	0.0000	stationary
LNX2	(C, T, 0)	−5.8395	−4.0925	−3.4744	−3.1645	0.0000	stationary
LNX3	(C, T, 0)	−5.1623	−4.0925	−3.4744	−3.1645	0.0003	stationary
LNX4	(C, T, 0)	−5.2278	−4.0925	−3.4744	−3.1645	0.0003	stationary
LNX5	(C, T, 0)	−5.3067	−4.0925	−3.4744	−3.1645	0.0002	stationary
LNX6	(C, T, 0)	−7.7677	−4.0925	−3.4744	−3.1645	0.0000	stationary
LNX7	(C, T, 0)	−8.2087	−4.0925	−3.4744	−3.1645	0.0000	stationary
LNX8	(C, T, 0)	−8.9109	−4.0925	−3.4744	−3.1645	0.0000	stationary
LNX9	(C, T, 0)	−4.2731	−4.0925	−3.4744	−3.1645	0.0059	stationary
LNX10	(C, T, 5)	−7.5827	−4.1032	−3.4794	−3.1674	0.0000	stationary
LNX11	(C, T, 1)	−5.4392	−4.0946	−3.4753	−3.1650	0.0001	stationary
LNX12	(C, T, 0)	−7.9333	−4.0925	−3.4744	−3.1645	0.0000	stationary
LNX13	(C, T, 0)	−4.1356	−4.0925	−3.4744	−3.1645	0.0088	stationary
LNX14	(C, T, 0)	−6.3515	−4.0925	−3.4744	−3.1645	0.0000	stationary

(2) Determination of the optimal lag order. The establishment of the VAR model needs to choose the appropriate lag order. In order to fully reflect the dynamic characteristics of the established VAR model, many factors need to be considered when choosing the lag order. Based on LR, FPE, AIC, SC and HQ criteria, the optimal lag order is determined, and the maximum number of * is the optimal lag period. As shown in Table 5, the optimal lag order is 3.

Table 5. Optimal lag order.

Lag	LogL	LR	FPE	AIC	SC	HQ
0	1109.297	NA	1.99×10^{-35}	−31.66079	−31.11036	−31.44242
1	1783.608	996.8069	3.46×10^{-40}	−42.82922	−32.92144	−38.89847
2	2195.351	405.7754	4.64×10^{-41}	−46.38698	−27.12185	−38.74386
3	3304.615	546.5942 *	8.68×10^{-49} *	−70.16276 *	−41.54030 *	−58.80728 *

* indicates lag order selected by the criterion.

(3) Stability test of the VAR model. When a pulsating impact is applied to the process of an equation in the VAR model, the system is considered to be stable if the pulse disappears with the passage of time. When the modulus of the reciprocal of the characteristic root is less than 1, it means that the VAR model is stable. As shown in Figure 6, the feature roots are all located in the unit circle, which proves that the VAR (3) model is stable.

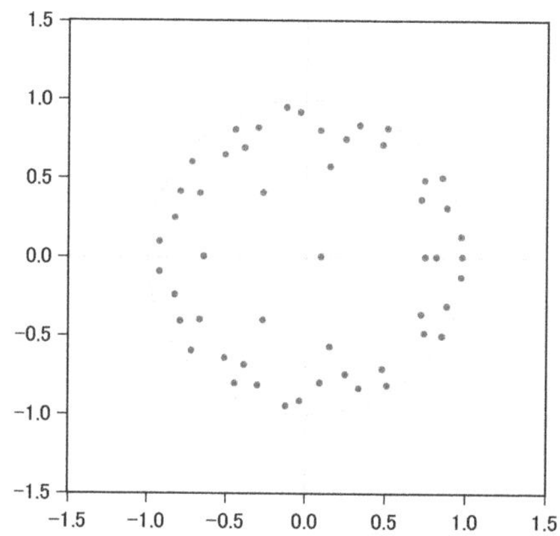

Figure 6. Stability test of VAR (3) model.

(4) The parameters of the unconstrained VAR (3) model constructed in this paper are shown in Table 6:

Table 6. Parameter estimation results of VAR (3) model.

Influencing Factors	LNY1	LNY2	LNY3	Influencing Factors	LNY1	LNY2	LNY3	Influencing Factors	LNY1	LNY2	LNY3
LNX1 (−1)	−0.631	12.767	18.913	LNX7 (−1)	−0.551	−2.106	−2.307	LNX13 (−1)	3.743	13.676	−1.277
LNX1 (−2)	3.979	−8.368	10.437	LNX7 (−2)	−0.342	−0.884	0.000	LNX13 (−2)	−4.586	2.206	11.620
LNX1 (−3)	7.134	52.144	54.138	LNX7 (−3)	0.183	1.522	0.118	LNX13 (−3)	3.248	5.035	1.930
LNX2 (−1)	−3.008	5.272	14.088	LNX8 (−1)	0.008	0.181	0.440	LNX14 (−1)	0.105	−2.728	−2.396
LNX2 (−2)	3.827	12.092	11.119	LNX8 (−2)	0.088	0.352	0.280	LNX14 (−2)	−0.253	1.793	0.745
LNX2 (−3)	1.151	3.442	0.740	LNX8 (−3)	0.037	0.289	0.043	LNX14 (−3)	−0.373	0.274	0.061
LNX3 (−1)	9.710	5.202	3.057	LNX9 (−1)	0.154	0.142	0.570	LNY1 (−1)	−0.502	−0.263	−0.054
LNX3 (−2)	0.049	−34.069	−60.848	LNX9 (−2)	0.176	0.533	1.072	LNY1 (−2)	0.259	−0.781	−1.590
LNX3 (−3)	−13.110	−29.194	−5.944	LNX9 (−3)	0.088	0.227	0.732	LNY1 (−3)	−0.006	−0.619	0.135
LNX4 (−1)	2.084	−6.401	−16.612	LNX10 (−1)	0.388	−0.436	0.449	LNY2 (−1)	0.183	0.198	−0.078
LNX4 (−2)	−5.447	−13.362	−13.635	LNX10 (−2)	−0.272	−1.021	−0.649	LNY2 (−2)	−0.025	−0.202	0.441
LNX4 (−3)	−1.815	−3.493	−1.794	LNX10 (−3)	0.044	1.302	0.321	LNY2 (−3)	0.118	0.461	0.089
LNX5 (−1)	−8.802	−4.380	−3.387	LNX11 (−1)	−0.105	1.280	0.911	LNY3 (−1)	−0.362	−0.846	−0.477
LNX5 (−2)	0.628	34.117	60.329	LNX11 (−2)	0.133	−1.581	−1.070	LNY3 (−2)	−0.054	0.326	0.276
LNX5 (−3)	12.731	26.898	6.157	LNX11 (−3)	0.794	0.948	0.718	LNY3 (−3)	0.163	0.933	0.736
LNX6 (−1)	−0.149	1.526	1.468	LNX12 (−1)	0.267	1.806	0.429	C	−12.037	−666.481	−697.181
LNX6 (−2)	−0.169	0.634	0.759	LNX12 (−2)	−0.763	−5.404	−2.570				
LNX6 (−3)	−0.245	−0.906	−0.932	LNX12 (−3)	−0.535	2.621	−1.327				

(5) Impulse response analysis. The impulse response function reflects the dynamic relationship between variables and the dynamic influence path of the impact of one variable on another [42].

Figures 7–9 show the trajectories of the impact of the sales of three types of power vehicles on 17 influencing factors. In Figures 7–9, the horizontal axis represents the number of periods, the vertical axis represents the magnitude of the impulse response function, the blue solid line represents the impulse response function, and the red dotted line represents the standard deviation band of plus or minus two times (±2S.E.).

Comparing the subgraphs, it can be found that in the long run, the sales of traditional fuel vehicles and plug-in hybrid vehicles and the average price of fuel vehicles tend to be stable, while the sales of battery electric vehicles have a negative impact on the three types of vehicles. In the short term, the consumer price index, the total export value, the number of patents granted for electric vehicles and the effective number of patents granted for power batteries all show the characteristics of fluctuation on the sales of the three types of vehicles, while in the long term, this influence tends to be positive. The impact trajectories of customs import, average price of new energy vehicles, nuclear power generation and employee index on the sales of three kinds of vehicles all show the characteristics of alternating ups and downs; in the long run, the average price of new energy vehicles has a positive impact on the sales of traditional fuel vehicles and a negative impact on the sales of battery electric vehicles and plug-in hybrid vehicles. The impact of customs imports, nuclear power generation and employee index on the sales of traditional fuel vehicles has gradually weakened, while it has a positive impact on the sales of battery electric vehicles and plug-in hybrid vehicles. The impact of customs export volume and road passenger volume on the sales of the three kinds of vehicles shows the characteristics of alternating ups and downs. In the long run, the impact on the sales of traditional fuel vehicles is gradually weakened, and it has a negative impact on the sales of battery electric vehicles and plug-in hybrid vehicles. In the short term, the effects of total import value, lithium-ion battery output and hydropower generation on the sales of the three types of vehicles all show the characteristics of alternating ups and downs, and in the long term, this effect tends to be negative.

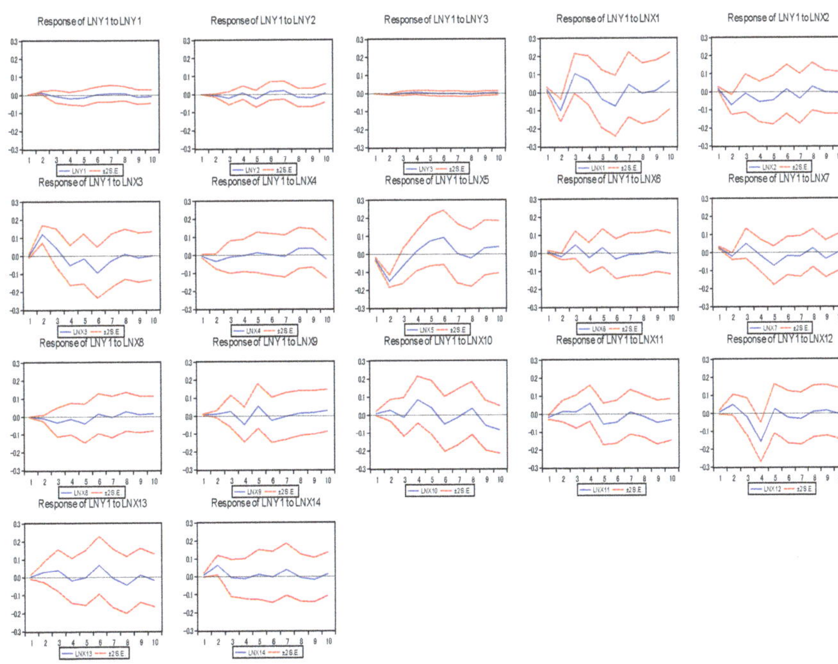

Figure 7. Impulse response of traditional fuel vehicle sales.

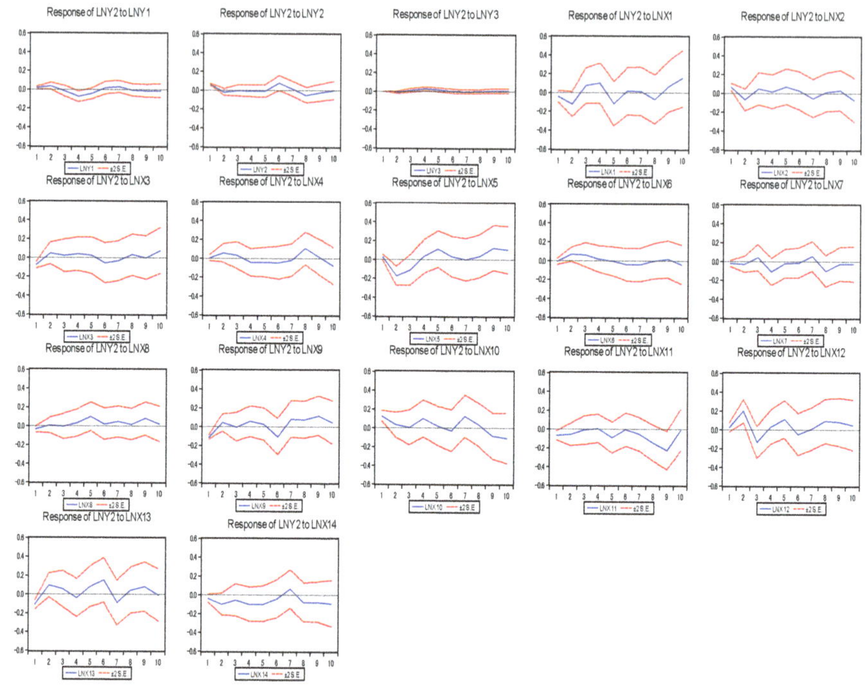

Figure 8. Impulse response of battery electric vehicle sales.

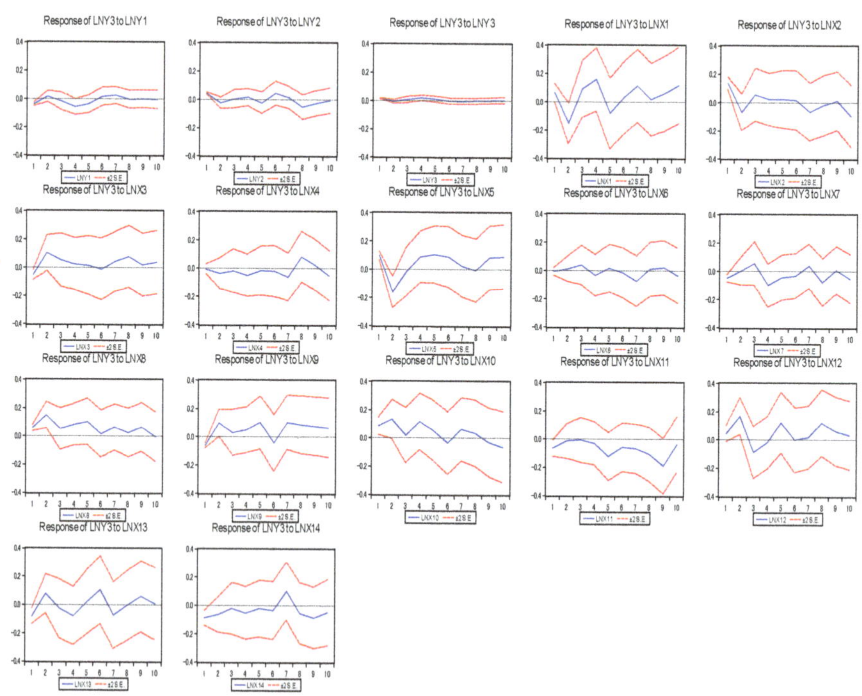

Figure 9. Impulse response of plug-in hybrid vehicle sales.

According to Figure 7, it can be found that the logarithm of 17 influencing factors, such as consumer price index, customs export volume and customs import volume, has a fluctuation characteristic of fluctuation in the impact trajectory of traditional fuel vehicle sales. The initial response value of traditional fuel vehicle sales to the unit impact of customs import volume, total export value, electric vehicle patent license amount and employee index is 0, and the subsequent response values alternate between positive and negative, indicating that these factors will not have an impact on traditional fuel vehicle sales at the initial stage. In the long run, the impact of customs import volume and electric vehicle patent license amount on traditional fuel vehicle sales tends to be stable, and the total export value and employee index will have a cyclical fluctuation impact on traditional fuel vehicle sales. The impact of consumer price index, customs export volume, total export value, average price of fuel vehicles, average price of new energy vehicles, effective patent authorization of power batteries, output of lithium-ion batteries, hydropower generation, nuclear power generation and road passenger traffic will make the sales of traditional fuel vehicles fluctuate to varying degrees. The sales of traditional fuel vehicles, battery electric vehicles and plug-in hybrid vehicles have a relatively stable influence on the sales of traditional fuel vehicles.

Figure 8 shows the effect of various influencing factors on the sales of battery electric vehicles. At the beginning, the response value of battery electric vehicle sales to the sales shock of traditional fuel vehicles fluctuated slightly, and finally stabilized. The response value of battery electric vehicle sales to its own impact fluctuated slightly in the middle and finally stabilized. The response value of battery electric vehicle sales to plug-in hybrid vehicle sales shock has been relatively stable. The impact from other influencing factors will make the sales of battery electric vehicles fluctuate to varying degrees.

Figure 9 shows the impact response of plug-in hybrid vehicle sales to various factors. The response value of plug-in hybrid vehicle sales to its own impact has been relatively stable. In the short term, the dynamic impact effect of traditional fuel vehicle sales and battery electric vehicle sales on plug-in hybrid vehicle sales presents the characteristics of ups and downs. In the long term, the influence of traditional fuel vehicle sales tends to be stable, but the influence of battery electric vehicle sales presents the characteristics of ups and downs. Consumer price index, customs import, electric vehicle patent authorization, power battery patent effective authorization and nuclear power generation have obvious positive effects on plug-in hybrid vehicle sales. The response value of plug-in hybrid vehicle sales to the impact of other influencing factors fluctuates up and down.

According to the analysis results of the VAR (3) model on influencing factors, based on 14 influencing factor data series and historical sales data of three types of power vehicles, the time window is set to 3, and the rolling time window is used to generate training samples as inputs of the VAR model, SVM model and multivariable BP neural network model.

3.2.2. Predictive Results of VAR Model

The input variables of the VAR (3) model are 17-dimensional vectors, namely, the historical sales data of three types of power vehicles and the data of 14 influencing factors. Based on the sales data of traditional fuel vehicles, battery electric vehicles and plug-in hybrid vehicles from March 2016 to February 2022 and the data of 14 influencing factors, according to the analysis of Formula (10) and Section 3.2.1, the VAR (3) prediction model is constructed, and the fitting values of the sales of three power types from June 2016 to February 2022 are calculated. Based on the constructed VAR (3) model and Formula (10), the predicted sales volume of three types of power vehicles from March 2022 to March 2023 is calculated.

Using the VAR (3) model, the fitting and forecasting results of the logarithm of the sales volume of traditional fuel vehicles, battery electric vehicles and plug-in hybrid vehicles are shown in Figure 10.

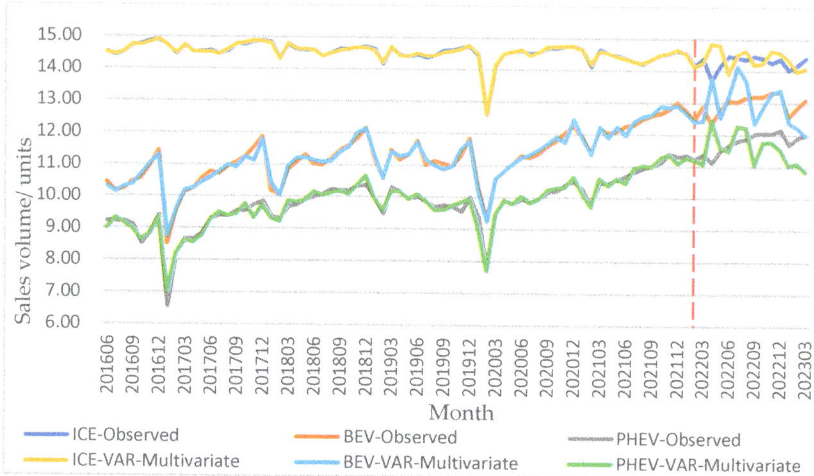

Figure 10. Fitting and prediction effect of VAR (3) model.

3.2.3. Predictive Results of SVM Model

Taking the construction process of the SVM forecasting model for the sales of traditional fuel vehicles as an example, based on the sales data of traditional fuel vehicles from March 2016 to March 2023 and the data of 14 influencing factors, 69 sets of training data from June 2016 to February 2022 and 13 sets of test data from March 2022 to March 2023 were constructed. In this SVM model, the input variables are 45-dimensional vectors.

Based on the principles shown in Equations (11)–(13), the SVM prediction model of traditional fuel vehicle sales is constructed by using the scikit-learn library of the Python platform. In total, 69 groups of pre-constructed training data are input into the SVM model and trained, and the trained SVM model and the fitting value of traditional fuel vehicle sales from June 2016 to February 2022 are obtained. A total of 13 groups of data used for testing are input into the SVM model based on the training set data, and the predicted sales value of traditional fuel vehicles from March 2022 to March 2023 is obtained.

The construction method of the SVM forecasting model for battery electric vehicle sales and plug-in hybrid vehicle sales is similar.

After many trainings, when forecasting the sales of traditional fuel vehicles, battery electric vehicles and plug-in hybrid vehicles, the kernel function is set to "Linear", the penalty coefficient is set to 10, and the default values of the scikit-learn library of the Python platform are used for other parameters. The fitting and forecasting results of the SVM model for automobile sales are shown in Figure 11.

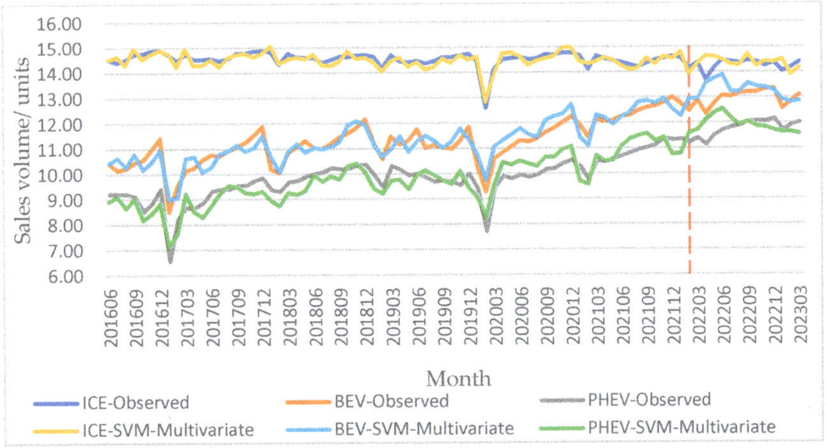

Figure 11. Fitting and prediction effect of SVM model.

3.2.4. Prediction Results of Multivariate BP Neural Network Model

Taking 3 as a step, based on the sales data of traditional fuel vehicles, battery electric vehicles and plug-in hybrid vehicles from March 2016 to March 2023 and the data of 14 influencing factors, 69 sets of training data from June 2016 to February 2022 and 13 sets of test data from March 2022 to March 2023 are constructed, so the input variable of the multivariate BP neural network model here is a 51-dimensional vector.

Based on the Keras deep learning framework, a multivariate BP neural network prediction model is constructed, and the parameter settings are shown in Table 7. We input 69 groups of pre-constructed training data into the multivariate BP neural network model, and trained them according to the steps in Section 2.2.5, and obtained the trained multivariate BP neural network model and the fitting value of automobile sales from June 2016 to February 2022. In total, 13 groups of data used for testing are input into the multivariate BP neural network model based on the training set data, and the forecast value of automobile sales from March 2022 to March 2023 is obtained.

Table 7. Parameter settings of univariate BP neural network model.

Parameter	Number of Plies	Activation Function of Each Layer	Number of Neurons	Training Times	Learning Rate	Loss Function
Value	3	'relu', 'relu', 'linear'	$51 \times 9 \times 3$	500	0.0001	MSE

The fitting results and prediction results of three kinds of vehicle sales trained by the multivariate BP neural network model are shown in Figure 12.

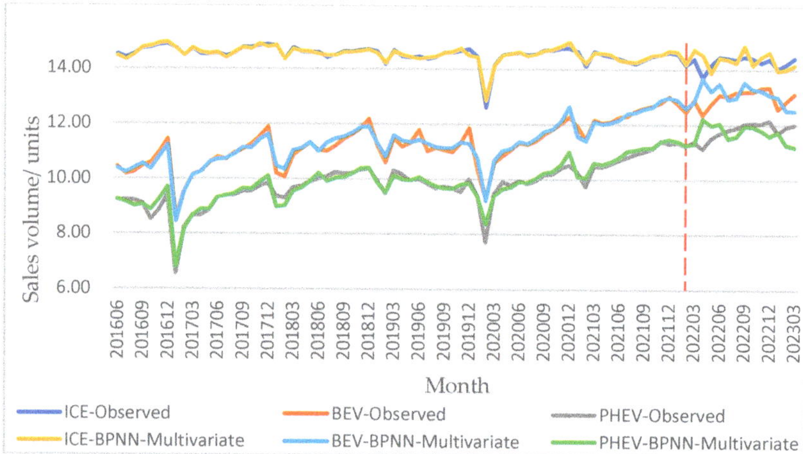

Figure 12. Fitting and prediction effect of multivariate BP neural network model.

3.2.5. Error Comparison

Table 8 shows the fitting and prediction errors of the VAR model, SVM model and multivariate BP neural network model for the sales of traditional fuel vehicles, battery electric vehicles and plug-in hybrid vehicles.

Table 8. Fitting and prediction errors of multivariate prediction models.

Model	Error Type	Evaluating Indicator	ICE	BEV	PHEV
VAR	Fitting Error	MAPE	0.145%	0.894%	1.014%
		RMSE	0.027	0.127	0.135
	Predicting Error	MAPE	2.565%	4.190%	4.863%
		RMSE	0.458	0.693	0.682
SVM	Fitting Error	MAPE	0.533%	1.488%	1.807%
		RMSE	0.147	0.304	0.345
	Predicting Error	MAPE	1.280%	2.389%	2.168%
		RMSE	0.381	0.598	0.512
BP Neural Network	Fitting Error	MAPE	0.321%	1.088%	1.357%
		RMSE	0.068	0.169	0.179
	Predicting Error	MAPE	1.810%	2.810%	3.384%
		RMSE	0.324	0.482	0.496

From the RMSE index, the multivariate BP neural network model has the best prediction effect on the automobile sales of three power types, and the VAR model has the best fitting effect on the automobile sales of three power types. From the MAPE index, the fitting error of the VAR model for three types of power vehicles is small, while the BP neural network model and SVM model have good forecasting effects for three types of power vehicles.

Based on the above analysis, it is considered that the multivariate BP neural network model has the best prediction performance for the sales of three power types of vehicles, and the fitting MAPE and RMSE are only 0.922% and 0.138, respectively, and the prediction MAPE and RMSE are only 2.668% and 0.434, respectively. Therefore, the multivariate BP neural network model is selected to predict the sales of three power types of vehicles outside the sample.

3.3. Comparison and Analysis of Prediction Accuracy

3.3.1. Comparison with Existing Literature

In Table 9, some of the existing literature about automobile sales forecast is sorted out. Comparing Tables 3 and 8, we can find that the prediction errors of the univariate Forecast Model and the multivariate BP neural network prediction model we selected are smaller than those in the literature in the table.

Table 9. Research and comparison on prediction of vehicles.

Authors (Year)	Models	Cases	Performance
Liu et al. (2023) [43]	GRA-DWT-BiLSTM	The electric vehicle sales in China	MAPE: 9.411%
Zhang, et al. (2022) [22]	LSTM	The vehicle sales of NIO	MAPE: 9.7718%
		The vehicle sales of XPeng	MAPE: 5.899%
Liu et al. (2022) [1]	DWT-BiLSTM	The electric vehicle sales in China	MAPE: 6.04%
Ding and Li (2021) [2]	ESOGM (1, 1)	Global electric vehicle sales	MAPE: 6.92%
Pei and Li (2022) [20]	DGA-based NGBM (1, 1)	The quarterly sales of new energy vehicles in China	RMSE: 23,907.59 MAPE: 7.26%

3.3.2. Comparison between Univariate Prediction Models and Multivariate Prediction Models

Comparing the errors in Tables 3 and 8, it can be found that the maximum fitting MAPE and RMSE of the three univariate forecasting models for three types of automobile sales are 2.842% and 0.405, respectively. However, the maximum MAPE and RMSE of the multivariable forecasting model for three types of power vehicles are 1.807% and 0.345, respectively. The maximum prediction MAPE and RMSE of the univariate prediction model are 3.382% and 0.596, respectively. The maximum prediction MAPE and RMSE of the univariate prediction model are 4.863% and 0.693, respectively. On the whole, the univariate and multivariate prediction models selected in this paper perform well, and the univariate prediction model is better in synthesis.

Table 10 summarizes the main advantages and disadvantages of the univariate forecasting model and multivariate forecasting model. The reason why this paper puts forward two kinds of models is not to compare the forecasting effects of the univariate forecasting model and multivariate forecasting model, but to consider that most of the current literature only forecasts automobile sales from one perspective. This paper hopes to combine the advantages of the two models to determine univariate forecasting and multivariate forecasting individually, so as to enhance the scientific and rigorous research and provide a more comprehensive reference for automobile sales forecasting.

Table 10. Advantages and disadvantages of univariate prediction models and multivariate prediction models.

Advantages and Disadvantages	Univariate Prediction Models	Multivariate Prediction Models
Advantages	Effectively capture the trends and changing rules contained in the historical sales data of automobiles.	Make full use of the relationship between automobile sales and various factors to improve the generalization ability of the automobile sales forecasting model.
Disadvantages	Only one characteristic parameter is considered, and the information extracted is limited, which cannot reflect the influence of other variables in the system on automobile sales.	It is difficult to define the system boundary, and too many influencing factors may affect the extraction of automobile sales trend information.

3.4. Extrasample Prediction Result

According to the above analysis, firstly, the sales of traditional fuel vehicles (ICEs), battery electric vehicles (BEVs) and plug-in hybrid vehicles (PHEVs) from April 2023 to December 2025 are predicted by the Prophet model. Among them, the default parameters of the Python platform fbprophet library are used for the out-of-sample prediction of traditional fuel vehicle sales, and the changepoint_range is set to 0.1 for the out-of-sample prediction of battery electric vehicle sales and plug-in hybrid vehicle sales, and the default values are used for other parameters.

Then, we predict the values of 14 influencing factors from April 2023 to December 2025 by using the Prophet model, and input the predicted values into the trained multivariate BP neural network prediction model. The univariate and multivariate predicted and restored values of traditional fuel vehicle sales logarithm, battery electric vehicle sales logarithm and plug-in hybrid vehicle sales logarithm from April 2023 to December 2025 are shown in Figure 13.

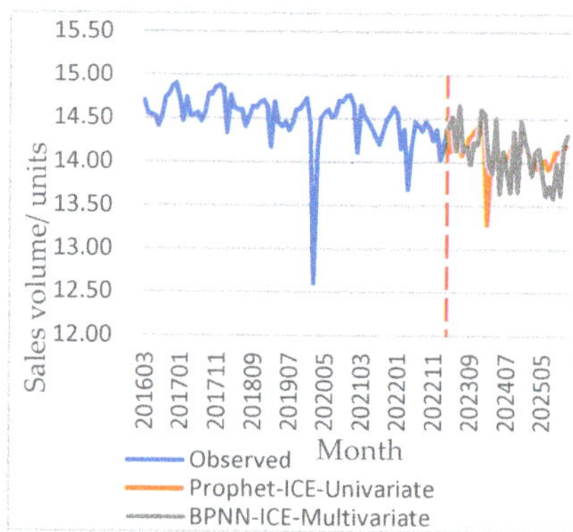

(a)

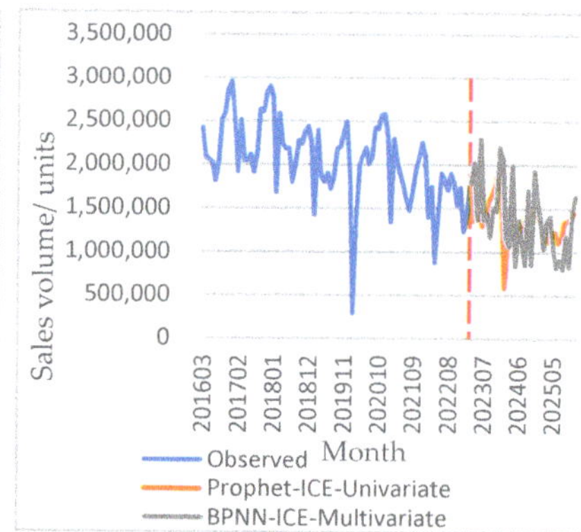

(b)

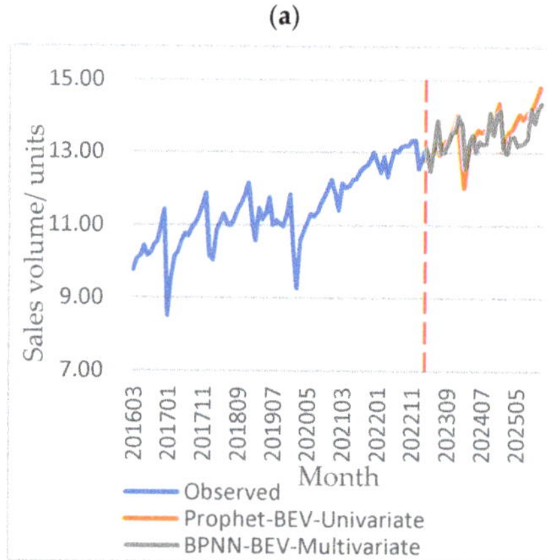

(c)

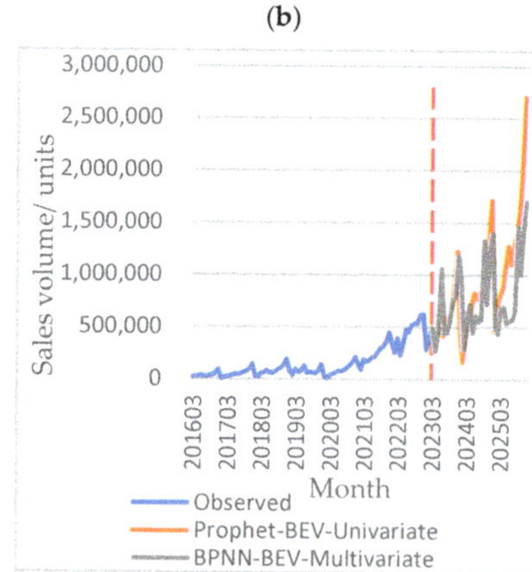

(d)

Figure 13. *Cont.*

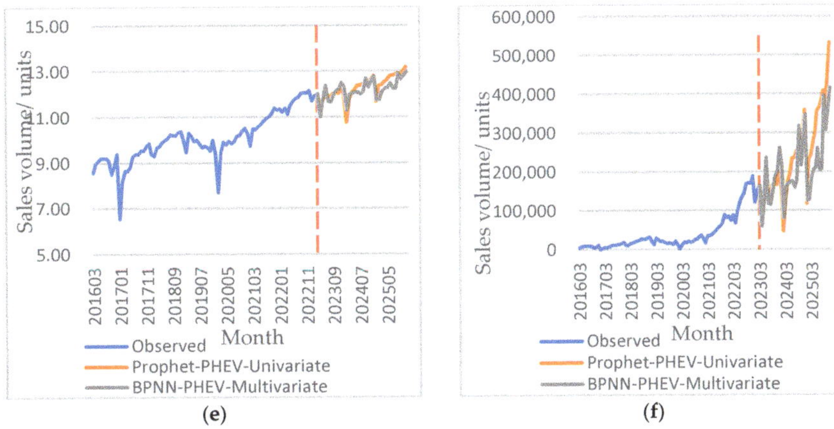

Figure 13. Logarithmic prediction results and reduction values of vehicle sales with different power types. (**a**) Logarithmic prediction of traditional fuel vehicle sales; (**b**) Reduction value of logarithmic prediction of traditional fuel vehicle sales; (**c**) Logarithmic prediction of battery electric vehicle sales; (**d**) Reduction value of logarithmic prediction of battery electric vehicle sales; (**e**) Logarithmic prediction of plug-in hybrid vehicle sales; (**f**) Reduction value of logarithmic prediction of plug-in hybrid vehicle sales.

According to the forecast results in Figure 13, it can be found that the Forecast Model and the Multivariate BP Neural Network forecast model have the same forecast trend for the sales of three types of power vehicles.

In 2022, the National Development and Reform Commission and the National Energy Administration issued the 14th Five-Year Plan for Modern Energy System, and the 14th Five-Year Comprehensive Work Plan for Energy Conservation and Emission Reduction issued by the State Council pointed out that by 2025, the sales of new energy vehicles should account for about 20% of the total sales of new vehicles in that year, and China has achieved this goal more than three years ahead of schedule. According to the univariate forecast, by the end of 2025 and the beginning of 2026, the sales volume of new energy vehicles in China will reach 56.2%, while the multivariate forecast shows that this proportion will reach 49.0%. The International Energy Agency (IEA) and Bloomberg New Energy Finance (BNEF) have also published relevant outlooks [44], and compared our forecast results with them. The Global EV Outlook 2023 released by IEA points out that under the guidance of China, the sales of electric vehicles, including battery electric vehicles and plug-in hybrid vehicles, continue to grow. According to the Electric Vehicle Outlook 2023 published by BNEF, by 2026, the sales of electric vehicles in China will reach 52.0%. It can be seen that the prediction results of this paper are very close to the prospects of IEV and BNEF.

At present, the Great Wall and other traditional car giants have accelerated the layout of new energy sources, and BYD announced that it would stop producing fuel vehicles in 2022. The intensification of competition in the new energy field has also promoted the research and development and application of new energy vehicles. In the short term, traditional fuel vehicles will not be completely eliminated, but their market share is gradually declining. According to univariate and multivariate forecast results, by 2025, the sales share of traditional fuel vehicles will drop to 43.8% and 51.0%, respectively. In 2020, the State Council released "New Energy Vehicle Development Plan (2021–2035)", pointing out that by 2035, battery electric vehicles will become the mainstream of sales. The forecast results of this paper are consistent with the national planning direction, and the sales volume of battery electric vehicles will maintain an upward trend. According to univariate and multivariate forecasts, by 2025, the market share of battery electric vehicles will reach 44.4% and 37.9%, respectively.

Compared with traditional fuel vehicles and battery electric vehicles, plug-in hybrid vehicles have certain advantages in fuel economy and cruising range. BYD, Great Wall and other car giants are shifting the main battlefield of new energy vehicle competition to the hybrid market, further promoting the maturity of hybrid technology. Univariate and multivariate forecasting results show that by 2025, the sales share of plug-in hybrid vehicles will rise to 11.8% and 11.1%, respectively, in the new car market.

It is not difficult to see from the forecast results that the new energy vehicles mainly based on electric energy will be the inevitable trend of the development of China's automobile industry.

From a global perspective, traditional car giants in various countries are accelerating the transformation of electrification. For example, BMW in Germany has set the goal that 50% of the group's sales will be battery electric vehicles by 2030. BYD plans to achieve the goal of global sales of 5 million vehicles by 2025. In addition, with the support of national policies, many new forces have emerged to build cars. Tesla in the United States has become the world's largest battery electric vehicle company with battery electric vehicles. China's Xpeng Motors, Leading Ideal and Nio Automobile have also taken advantage of the historic opportunity of the automobile industry reform to develop into the three giants of the new force of making cars. The global competition of new energy vehicles has started.

4. Conclusions

China's automobile industry is characterized by unprecedented diversification, and its development is accompanied by strong uncertainty. The new consumption reform of automobiles brings certain challenges to the country, enterprises and consumers. In order to accurately judge the development potential of several mainstream power vehicles on the market at present, the forecasting effects of various statistical models and machine learning on automobile sales are compared from the perspective of univariate and multivariate. Among them, the Prophet model and BP neural network model have achieved high prediction accuracy, and the developed VAR multivariable BP neural network model fully considers the lag effect of various factors on automobile sales. Therefore, we forecast the sales volume of traditional fuel vehicles, battery electric vehicles and plug-in hybrid vehicles in China from April 2023–December 2025 with the Prophet model as a univariate forecasting model and the BP neural network as a multivariate forecasting model. According to the univariate forecast, by 2025, the proportion of traditional fuel vehicles, battery electric vehicles and plug-in hybrid vehicles will be 43.8%, 44.4% and 11.8%, respectively. Multivariate prediction shows that by 2025, the sales of these three types of cars will account for 51.0%, 37.9% and 11.1%, respectively.

5. Suggestions

According to the above conclusions and analysis, the following suggestions are put forward: (1) Local governments and relevant departments should strengthen the construction of supporting infrastructure for new energy vehicles to better meet the demands of consumers. (2) In order to maintain China's competitive advantage in the global market, we should further enhance our independent innovation capability and increase investment in research and development of key core technologies, such as power batteries, operating systems and independent chips.

Although good results have been achieved for various types of cars, it is considered that there are some limitations in the same lag period of each influencing factor on car sales, and the heterogeneity of the lag effect of different factors on car sales can be considered in the future.

Author Contributions: Conceptualization, S.W. and Y.F.; methodology, M.Z. and Y.F.; software, D.D., Y.F. and M.Z.; project administration, D.D.; writing—original draft preparation, D.D., S.W., Y.F. and M.Z.; writing—review and editing, D.D., S.W., Y.F. and M.Z. All authors have read and agreed to the published version of the manuscript.

Funding: This research was funded by the project "Process Optimization and Data Governance Solution Research" of Shanghai MAHLE Thermal Systems Co., Ltd., (23H00706).

Data Availability Statement: Not applicable.

Conflicts of Interest: The authors declare no conflict of interest.

References

1. Liu, L.Y.; Liu, S.F.; Wu, L.F.; Zhu, J.S.; Shang, G. Forecasting the development trend of new energy vehicles in China by an optimized fractional discrete grey power model. *J. Clean Prod.* **2022**, *372*, 13. [CrossRef]
2. Ding, S.; Li, R.J. Forecasting the sales and stock of electric vehicles using a novel self-adaptive optimized grey model. *Eng. Appl. Artif. Intell.* **2021**, *100*, 13. [CrossRef]
3. Lin, B.Q.; Wu, W. Why people want to buy electric vehicle: An empirical study in first-tier cities of China. *Energy Policy* **2018**, *112*, 233–241. [CrossRef]
4. Huang, Y.L.; Qian, L.X. Consumer preferences for electric vehicles in lower tier cities of China: Evidences from south Jiangsu region. *Transport. Res. Part D-Transport. Environ.* **2018**, *63*, 482–497. [CrossRef]
5. Chen, K.; Ren, C.R.; Gu, R.; Zhang, P.D. Exploring purchase intentions of new energy vehicles: From the perspective of frugality and the concept of "mianzi". *J. Clean Prod.* **2019**, *230*, 700–708. [CrossRef]
6. Zhao, M.; Sun, T.; Feng, Q. Capital allocation efficiency, technological innovation and vehicle carbon emissions: Evidence from a panel threshold model of Chinese new energy vehicles enterprises. *Sci. Total Environ.* **2021**, *784*, 11. [CrossRef] [PubMed]
7. Egbue, O.; Long, S.; Samaranayake, V.A. Mass deployment of sustainable transportation: Evaluation of factors that influence electric vehicle adoption. *Clean Technol. Environ. Policy* **2017**, *19*, 1927–1939. [CrossRef]
8. Wang, Z.H.; Zhao, C.Y.; Yin, J.H.; Zhang, B. Purchasing intentions of Chinese citizens on new energy vehicles: How should one respond to current preferential policy? *J. Clean Prod.* **2017**, *161*, 1000–1010. [CrossRef]
9. Yang, Z.L.; Li, Q.; Yan, Y.M.; Shang, W.L.; Ochieng, W. Examining influence factors of Chinese electric vehicle market demand based on online reviews under moderating effect of subsidy policy. *Appl. Energy* **2022**, *326*, 120019. [CrossRef]
10. Wang, N.; Tang, L.H.; Pan, H.Z. Analysis of public acceptance of electric vehicles: An empirical study in Shanghai. *Technol. Forecast. Soc. Chang.* **2018**, *126*, 284–291. [CrossRef]
11. Baur, D.G.; Todorova, N. Automobile manufacturers, electric vehicles and the price of oil. *Energy Econ.* **2018**, *74*, 252–262. [CrossRef]
12. Austmann, L.M.; Vigne, S.A. Does environmental awareness fuel the electric vehicle market? A twitter keyword analysis. *Energy Econ.* **2021**, *101*, 105337. [CrossRef]
13. Ou, S.Q.; Lin, Z.H.; Xu, G.Q.; Hao, X.; Li, H.W.; Gao, Z.M.; He, X.; Przesmitzki, S.; Bouchard, J. The retailed gasoline price in china: Time-series analysis and future trend projection. *Energy* **2020**, *191*, 116544. [CrossRef]
14. Zhuge, C.X.; Wei, B.R.; Shao, C.F.; Dong, C.J.; Meng, M.; Zhang, J. The potential influence of cost-related factors on the adoption of electric vehicle: An integrated micro-simulation approach. *J. Clean Prod.* **2020**, *250*, 119479. [CrossRef]
15. Li, X.; Xiao, X.P.; Guo, H. A novel grey bass extended model considering price factors for the demand forecasting of European new energy vehicles. *Neural Comput. Appl.* **2022**, *34*, 11521–11537. [CrossRef]
16. Hsieh, I.Y.L.; Kishimoto, P.N.; Green, W.H. Incorporating multiple uncertainties into projections of Chinese private car sales and stock. *Transp. Res. Rec.* **2018**, *2672*, 182–193. [CrossRef]
17. Kumar, R.R.; Guha, P.; Chakraborty, A. Comparative assessment and selection of electric vehicle diffusion models: A global outlook. *Energy* **2022**, *238*, 16. [CrossRef]
18. Rietmann, N.; Hugler, B.; Lieven, T. Forecasting the trajectory of electric vehicle sales and the consequences for worldwide CO_2 emissions. *J. Clean Prod.* **2020**, *261*, 16. [CrossRef]
19. Zhou, H.M.; Dang, Y.G.; Yang, Y.J.; Wang, J.J.; Yang, S.W. An optimized nonlinear time-varying grey Bernoulli model and its application in forecasting the stock and sales of electric vehicles. *Energy* **2023**, *263*, 14. [CrossRef]
20. Pei, L.L.; Li, Q. Forecasting quarterly sales volume of the new energy vehicles industry in China using a data grouping approach-based nonlinear grey Bernoulli model. *Sustainability* **2019**, *11*, 1247. [CrossRef]
21. Li, S.X.; Chen, H.; Zhang, G.F. Comparison of the short-term forecasting accuracy on battery electric vehicle between modified bass and Lotka-Volterra model: A case study of China. *J. Adv. Transp.* **2017**, *2017*, 7801837. [CrossRef]
22. Zhang, M.Y.; Xu, H.Y.; Ma, N.; Pan, X.L. Intelligent vehicle sales prediction based on online public opinion and online search index. *Sustainability* **2022**, *14*, 10344. [CrossRef]
23. Liu, B.C.; Song, C.Y.; Wang, Q.S.; Zhang, X.M.; Chen, J.L. Research on regional differences of China's new energy vehicles promotion policies: A perspective of sales volume forecasting. *Energy* **2022**, *248*, 13. [CrossRef]
24. Xia, Z.C.; Xue, S.; Wu, L.B.; Sun, J.X.; Chen, Y.J.; Zhang, R. Forexgboost: Passenger car sales prediction based on XGBoost. *Distrib. Parallel Databases* **2020**, *38*, 713–738. [CrossRef]
25. Wu, M.F.; Chen, W. Forecast of electric vehicle sales in the world and China based on PCA-GRNN. *Sustainability* **2022**, *14*, 2206. [CrossRef]
26. Liu, J.P.; Chen, L.J.; Luo, R.; Zhu, J.M. A combination model based on multi-angle feature extraction and sentiment analysis: Application to EVS sales forecasting. *Expert Syst. Appl.* **2023**, *224*, 13. [CrossRef]

27. Ding, Y.; Wu, P.; Zhao, J.; Zhou, L.G. Forecasting product sales using text mining: A case study in new energy vehicle. *Electron. Commer. Res.* **2023**, 33. [CrossRef]
28. Sun, S.H.; Wang, W.C. Analysis on the market evolution of new energy vehicle based on population competition model. *Transport. Res. Part D-Transport. Environ.* **2018**, *65*, 36–50. [CrossRef]
29. Guo, D.; Yan, W.; Gao, X.B.; Hao, Y.J.; Xu, Y.; Wenjuan, E.; Tan, X.C.; Zhang, T.Q. Forecast of passenger car market structure and environmental impact analysis in China. *Sci. Total Environ.* **2021**, *772*, 17. [CrossRef]
30. Liu, Z.; Wang, X.H.; Zhang, Q.; Huang, C.Y. Empirical mode decomposition based hybrid ensemble model for electrical energy consumption forecasting of the cement grinding process. *Measurement* **2019**, *138*, 314–324. [CrossRef]
31. Zhang, C.Q.; Liu, M.; Wang, D.Y.; Ni, A.N.; Xiao, G.N.; Lu, W.T. Linkage mechanism of public transport subsidy: Considering passenger ridership, cost, fare and service quality. *Transp. Lett.* **2022**. [CrossRef]
32. Chaturvedi, S.; Rajasekar, E.; Natarajan, S.; McCullen, N. A comparative assessment of SARIMA, LSTM RNN and Fb prophet models to forecast total and peak monthly energy demand for India. *Energy Policy* **2022**, *168*, 113097. [CrossRef]
33. Guefano, S.; Tamba, J.G.; Azong, T.E.W.; Monkam, L. Forecast of electricity consumption in the Cameroonian residential sector by grey and vector autoregressive models. *Energy* **2021**, *214*, 118791. [CrossRef]
34. Alashari, M.; El-Rayes, K.; Attalla, M.; Al-Ghzawi, M. Multivariate time series and regression models for forecasting annual maintenance costs of EPDM roofing systems. *J. Build. Eng.* **2022**, *54*, 104618. [CrossRef]
35. Maqsood, H.; Maqsood, M.; Yasmin, S.; Mehmood, I.; Moon, J.; Rho, S. Analyzing the stock exchange markets of EU nations: A case study of brexit social media sentiment. *Systems* **2022**, *10*, 24. [CrossRef]
36. Ahmed, Q.I.; Attar, H.; Amer, A.; Deif, M.A.; Solyman, A.A.A. Development of a hybrid support vector machine with grey wolf optimization algorithm for detection of the solar power plants anomalies. *Systems* **2023**, *11*, 237. [CrossRef]
37. Guo, L.; Bai, L.F.; Liu, Y.X.; Yang, Y.Z.; Guo, X.H. Research on the impact of covid-19 on the spatiotemporal distribution of carbon dioxide emissions in China. *Heliyon* **2023**, *9*, e13963. [CrossRef]
38. Ran, Q.Y.; Bu, F.B.; Razzaq, A.; Ge, W.F.; Peng, J.; Yang, X.D.; Xu, Y. When will China's industrial carbon emissions peak? Evidence from machine learning. *Environ. Sci. Pollut. Res.* **2023**, *30*, 57960–57974. [CrossRef]
39. Liu, Q.; Liu, J.; Gao, J.X.; Wang, J.J.; Han, J. An empirical study of early warning model on the number of coal mine accidents in China. *Saf. Sci.* **2020**, *123*, 104559. [CrossRef]
40. Li, Y.; Huang, S.Y.; Miao, L.; Wu, Z. Simulation analysis of carbon peak path in China from a multi-scenario perspective: Evidence from random forest and back propagation neural network models. *Environ. Sci. Pollut. Res.* **2023**, *30*, 46711–46726. [CrossRef]
41. Zhang, Y.; Zhong, M.; Geng, N.N.; Jiang, Y.J. Forecasting electric vehicles sales with univariate and multivariate time series models: The case of China. *PLoS ONE* **2017**, *12*, 15. [CrossRef] [PubMed]
42. Ye, L.; Yang, D.L.; Dang, Y.G.; Wang, J.J. An enhanced multivariable dynamic time-delay discrete grey forecasting model for predicting China's carbon emissions. *Energy* **2022**, *249*, 123681. [CrossRef]
43. Liu, B.C.; Song, C.Y.; Liang, X.Q.; Lai, M.Z.; Yu, Z.C.; Ji, J. Regional differences in China's electric vehicle sales forecasting: Under supply-demand policy scenarios. *Energy Policy* **2023**, *177*, 13. [CrossRef]
44. Pelegov, D.V.; Chanaron, J.J. Electric car market analysis using open data: Sales, volatility assessment, and forecasting. *Sustainability* **2023**, *15*, 399. [CrossRef]

systems

Article

Gender Gaps in Mode Usage, Vehicle Ownership, and Spatial Mobility When Entering Parenthood: A Life Course Perspective

Hung-Chia Yang [1], Ling Jin [1,*], Alina Lazar [2], Annika Todd-Blick [1], Alex Sim [3], Kesheng Wu [3], Qianmiao Chen [1] and C. Anna Spurlock [1]

1 Lawrence Berkeley National Laboratory, Energy Analysis and Environmental Impacts Division, Berkeley, CA 94720, USA
2 Department of Computer Science and Information Systems, Youngstown State University, Youngstown, OH 44512, USA
3 Lawrence Berkeley National Laboratory, Computational Research Division, Berkeley, CA 94720, USA
* Correspondence: ljin@lbl.gov

Abstract: Entry into parenthood is a major disruptive event to travel behavior, and gender gaps in mobility choices are often widened during parenthood. The exact timing of gender gap formation and their long-term effects on different subpopulations are less studied in the literature. Leveraging a longitudinal dataset from the 2018 WholeTraveler Study, this paper examines the effects of parenthood on a diverse set of short- to long-term outcomes related to the three hierarchical domains of mobility biography: mode choice, vehicle ownership, spatial mobility, and career decisions. The progress of the effects is evaluated over a sequential set of parenting stages and differentiated across three subpopulations. We find that individuals classified as "Have-it-alls", who start their careers, partner up, and have children concurrently and early, significantly increase their car uses two years prior to childbirth ("nesting period"), and they then relocate to less transit-accessible areas and consequently reduce their reliance on public transportation while they have children in the household. In contrast, individuals categorized as "Couples", who start careers and partnerships early but delay parenthood, and "Singles", who postpone partnership and parenthood, have less pronounced changes in travel behavior throughout the parenting stages. The cohort-level effects are found to be driven primarily by women, whose career development is on average more negatively impacted by parenting events than men, regardless of their life course trajectory. Early career decisions made by women upon entering parenthood contribute to gender gaps in mid- to longer-term mobility decisions, signifying the importance of early intervention.

Keywords: life course; mobility biography; parenthood; mode use; gender gap; car ownership; spatial mobility

Citation: Yang, H.-C.; Jin, L.; Lazar, A.; Todd-Blick, A.; Sim, A.; Wu, K.; Chen, Q.; Spurlock, C.A. Gender Gaps in Mode Usage, Vehicle Ownership, and Spatial Mobility When Entering Parenthood: A Life Course Perspective. *Systems* **2023**, *11*, 314. https://doi.org/10.3390/systems11060314

Academic Editor: Oz Sahin

Received: 24 March 2023
Revised: 10 June 2023
Accepted: 14 June 2023
Published: 20 June 2023

1. Introduction

Our daily transportation mode choice (driving, biking, walking, etc.) bears substantial consequences for both the transportation system and the environment in the long term. The process of making mode choice decisions on a daily basis requires considerable effort, leading individuals to develop a repeated routine around their travel behavior, often without much conscious consideration of alternative mode options [1]. Clark et al. and Chatterjee et al. further support this notion and demonstrate that these habitual travel behaviors are more likely to be interrupted by a contextual change to an individual's life situation [2,3]. Hence, gaining insight into the key factors associated with the formation and alteration of these habits is therefore crucial for policymakers in designing effective strategies to encourage sustainable transportation choices, as well as for planners to better anticipate enduring trends and transitions in travel mode usage.

Travel behavior studies have utilized the mobility biography approach to capture the dynamics of travel choices over the life course with a focus on the influence of life events on individuals [1,4]. With the concept of mobility biographies, life events are linked to an individual's mobility-related decisions from a life course perspective. Among key events that are transport-relevant, entry into parenthood has been identified as having a significant impact on travel behavior. A large body of research provides strong evidence that the parents with young child(ren) tend to exhibit an increase in both car acquisition and car use [5–8]. However, recent literature suggests that this parenting effect on travel behavior can be influenced by a number of factors, including the parents' age, income, education level, as well as the location of their home and workplace, and that increased car-dependency may just be one of many patterns that can emerge [5,9–11]. When looking at gender differences in travel mode choices and car accessibility during parenthood, women are more likely to experience changes in travel behavior than men [7,12–15].

While most transportation literature that studied the effect of having a child focuses on the aggregated change in travel behavior either across a broad population or within a specific subpopulation based on static sociodemographics, limited research has investigated differences that may exist between subpopulations from a long-term perspective. Even less research exists on gender gaps in travel during parenthood among different subpopulations.

To improve the understanding of subpopulation differences in the role of parenthood from a long-term context, longitudinal data are needed for studying factors associated with the formation and persistence of gender gaps in mobility-related decisions over time among heterogeneous populations. In this paper, we rely on the life course cohorts previously derived based on social sequence clustering [16] of the life history calendar portion of the WholeTraveler Transportation Behavior study survey conducted in the San Francisco Bay Area in 2018 [17]. In previous work, we identified that career and family-related milestones, particularly those occurring at an earlier life stage, exhibit a stronger association with changes in mode-use behavior. Moreover, our findings reveal that the relative order of those events can also impact individuals' mode choices [16].

Following the same mobility biography approach, this study aims to take a dynamic perspective on travel choices over an individual's life course with a specific focus on the influence of parenthood on travel behaviors. As the family forming timeline for most people progresses from "no child" to "nesting", then to "having children at home", we differentiate the effect of being in one parenting stage relative to its previous stage. This analytical approach enables us to understand the progression of travel behavior and distinguish the turning points in mobility and career decisions and gender gap formation.

To achieve this, we employ linear regression models to estimate the marginal effects of parenting stages on a comprehensive set of short-term to mid- and long-term outcome variables associated with the three hierarchical domains of mobility biography. These outcome variables encompass diverse aspects of mobility decision making, such as choosing different travel modes (driving, public transportation, and walking or biking), car ownership, spatial mobility as indicated by public transit availability, and career decisions. Through the analysis, we aim to address the following research questions: (1) To what extent does parenting alter the short- (mode use), mid- (vehicle ownership), and long-term (residential location) mobility decisions across subpopulations with different life course trajectories? (2) To what extent does parenting alter career decisions across subpopulations with different life course trajectories? (3) Are these differences further elucidated by differences across gender? By addressing these research questions, we seek to contribute to the existing body of knowledge, advancing our understanding of the complex interplay between parenthood, life course trajectories, mobility decisions, career choices, and gender gaps.

The paper commences with a brief review of relevant studies examining the parenting effect on travel behaviors and gender differences in mobility decision making using the mobility biography approach. Details about the WholeTraveler Transportation Behavior Study survey, the design of the life history calendar, the derivation of life course cohorts, and the model specifications to estimate the marginal parenting effects are presented in Section 3. The results from the model estimations and interpretations are discussed in Section 4. The Section 5 that links the main takeaways with the literature and discusses the limitations to our study are shown in Section 5. Finally, the paper concludes with a Section 6 that highlights the policy implications and future research.

2. Relevant Literature

The theoretical foundation of mobility biographies is that travel behaviors are generally considered habitual, and people do not tend to change their travel patterns on a daily basis [18–20]. However, habitual travel behavior may be disrupted when major life events occur, providing opportunities to influence mobility decisions [21–27]. The extent and manner in which individuals change their travel mode in response to a specific life event depend on the long-term decision contexts within which these decisions are made. This long-term decision context consists of an individual's current situation, past experiences, and future plans [21], and this context varies across the population.

Traditional travel behavior studies have predominantly relied on the use of cross-sectional survey data to examine mobility decisions and their correlation with concurrent individual attributes, attitudes, and the built environment at a point in time (as discussed in [26,28,29]). However, the dynamic and multi-dimensional decision context can only be revealed in a long-term life history rather than via a life cycle stage lens in which the life cycle stages are defined statically at the moment an event occurs. Schoenduwe et al. conducted an extensive review of transportation studies employing the "mobility biography" approach to analyze the dynamics of travel choices over the life course, providing a longer-term decision context [28]. In a key study within this domain, Beige and Axhausen investigated the interdependence between various life course dimensions, such as changes in education, employment, and mobility behavior [30]. Rau and Manton introduced the concept of "mobility milestone", highlighting the important role of transport-related structural conditions in shaping travel behaviors through major life events [31]. Haas et al. applied a latent transition analysis to examine the effect of life events and other exogenous variables on transitions between different travel patterns [32]. Their findings suggest that individuals who strictly adhere to unimodal travel modes, such as car or bike usage, are less likely to alter their travel patterns in response to life events than those who engage in multimodal travel.

A mobility biography typically consists of three hierarchical domains: travel mode usage (driving, public transportation, and walking or biking), car ownership, spatial mobility (in our case, indicated by public transit availability), and career decisions. Spatial mobility and career decisions at the top level of the hierarchy represent the longest-term decisions that may further condition the short-term travel behaviors and/or serve as fundamental triggers of other related events in the accessibility and mobility domains [33–35]. Dieleman and Mulder and Adhikari et al. have provided evidence showing that residential relocation frequently coincides with significant life events related to both professional and family domains, such as marriage, childbirth, and job change [36,37]. Furthermore, changes of residence may impact the availability and choice of travel mode. Career decisions may also reinforce the labor divisions in the household, which was previously found as an underlying driver of gender gaps in travel behavior [12]. Even though it is beyond our scope, a growing body of literature has started to look at the effect of workplace relocation on commuting behavior, as well as commuting satisfaction and subjective well-being [38,39].

Among key life milestones covered in mobility biography studies, the transition to parenthood is found to have an important impact on all aspects of an individual's life, and gender gaps in mobility and career decisions are evident as the traditional gendered division of labor remains operative in households [8,14,40,41]. While some studies found that being a parent is associated with a decrease in public transit usage and cycling and an increase in car use and ownership [8,15,26,32,42], other studies showed that some parents maintained a similar travel behavior to before having children, or even increased walking and biking [9]. A number of factors can influence the impact of parenthood on travel behavior, including parents' age, income, education, as well as the location of their home and workplace. For example, McCarthy et al. found that individuals living in a city, having a lower income, having a partner who works part-time or not at all, and not having a second child have a lower likelihood of shifting toward a car-centric lifestyle after childbirth [10]. In a recent California study by Chakrabarti and Joh, childcare infrastructure and greener mode availability are shown to play a part in the decision to change travel behavior in response to the childbirth event [7].

After entering parenthood, care-related journeys become a significant part of everyday mobility in the household [13]. Women are more likely to be responsible for childcare and household tasks, which can lead to an increase in car use by these women [1,12,15]. It is found that the birth of a first child has a significant effect on women's decision to shift car accessibility from shared access to full access [15]. With the need to juggle multiple duties in their daily lives, mothers tend to increase their car dependence more than men [6]. Scheiner summarized underlying drivers of gender differences in travel behavior: spatial ties imposed by household work and caregiving duties make women travel shorter distances; women tend to have more complex trip chaining in order to gain efficiency; more variable activities with more anchor points; and all above reasons encourage women to use flexible modes of transport such as the car to juggle all their duties [1,12].

It is worth noting the reliability concern of conducting mobility biography studies, as the data collection typically relies on retrospective surveys, where individuals are asked to recall changes in their lifestyle domain, accessibility domain, and mobility domain in the past [43]. In most cases, major life events, such as marriage, childbirth, employment, and relocation, are remembered reliably. The acquisition or disposal of means of transport does not happen on a daily basis; thus, recall is less likely to be biased. However, recollection of daily travel patterns may not be as accurate, but any changes in travel behavior triggered by disruptive events might be remembered better [44].

3. Data and Method

3.1. Survey Data Description

We use data collected through the WholeTraveler Transportation Behavior Study, which is part of the U.S. Department of Energy's Systems and Modeling for Accelerated Research in Transportation (SMART) Mobility Consortium. The survey was administered in the nine core counties in the San Francisco Bay Area, California through a survey implementation firm with expertise in large-scale representative sampling and outreach. With a target sample size of 900 respondents, invitations to respond to an online survey were initially sent to 30,000 residential addresses via a mailed letter, given an expected response rate of around 3%. (We set our target sample size at 900 respondents based on two facts. First, we reviewed previous published surveys of technology adoption behavior and use and found their sample sizes to range widely from about 700 to 9000 (as shown in [45,46]). Second, we considered the methodological work that shows that even for a large population, there is little to be gained in terms of margin of error by increasing the sample size beyond 1000 (see, e.g., the discussion in [47]). It quickly became apparent that the response rate was trending lower than anticipated. In response, a reminder postcard was sent to non-respondents of the initial 30,000 households, and the invitation letter was mailed to an additional 30,000 randomly selected residential addresses. Of the 60,000 households that were sent invitations, 997 completed the survey.

Data collected from this survey have been successfully used to study emerging vehicle technology and service adoption [48], travel mode choices [49,50], and the impact of e-commerce on travel demand differentiated by households with and without children [51]. While the portion of female and male respondents closely aligns with the Bay Area population, highly educated and high-income respondents are disproportionally represented in the sample. Overall, the sociodemographics of the WholeTraveler sample are consistent with those of other similarly scaled S.F. Bay Area transportation studies [52–54]. A more detailed discussion regarding the representation of the survey sample and variable definitions are included in [16]. A breakdown of the sociodemographic characteristics of the WholeTraveler sample is presented in Appendix A.

The retrospective life history calendar portion of the survey design gathered longitudinal data on mobility-related behavioral changes and major life events and circumstances occurring around those changes. Specifically, the respondents were asked to indicate their household composition (childbirth, relocation, cohabitation, and household size), education and employment status, public transit availability, transportation modes, and vehicle ownership on an annual basis starting at age 20 and up to age 50. Respondents less than 20 years of age were not asked to answer the life history calendar portion of the survey. Key variables concerning regularly used (used at least twice per week) transportation modes, car ownership, availability of public transportation, and employment status are used in this study.

Prior to analysis, we cleaned the data to account for missing observations and erroneous responses, which removed 172 respondents. We then restrict the analysis to the 17,777 annual observations from the 569 respondents who were aged 35 or older at the time they took the survey (in 2018) as the sample of the panel regressions, which accounts for 71% of the remaining data. This selection ensures that we observe complete responses between ages 20 and 35 and thus capture a life period that presents the greatest heterogeneity among the population [30,55]. Our sample size may seem small for a survey study; however, our sample size is consistent with previous life history calendar studies in transportation and mobility research reviewed by [4,28,31].

3.2. Description of Life Course Cohorts

In our previous paper [16], we used sequence clustering to construct cohorts of respondents who share similar life course trajectories. Around 85 percent of the respondents fall into three cohorts (listed below and visually described in Figure 1).

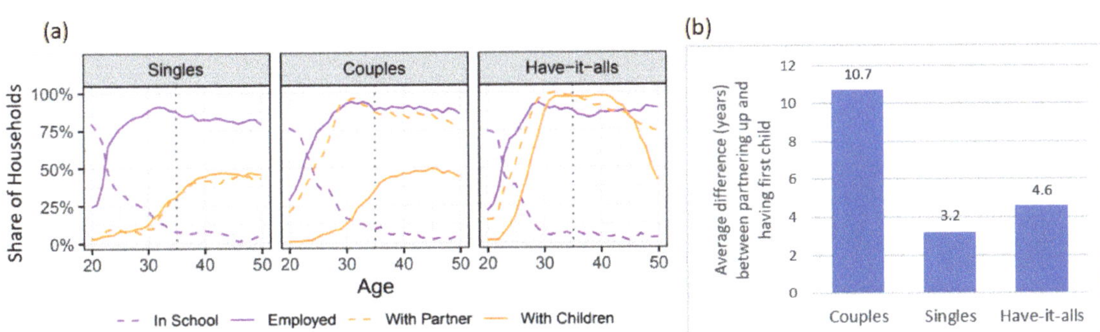

Figure 1. Life course patterns of family and career status in the three major life course cohorts analyzed in this study (**a**), and the average difference in years between partnering up and having a first child for each of the three cohorts (**b**).

- Singles (40% of the sample) tend to finish school and enter the workforce at an earlier age while opting to defer or forgo having a partner or children.
- Couples (27%) tend to finish school, work, and partner up at an earlier age while opting to defer having children.
- Have-it-alls (18%) finish school and enter the workforce early in life, and partner up and have children only very slightly thereafter.

Only these three cohorts are included in our analysis because they have a relatively large sample size and exhibit distinct family patterns, which is the focus of this paper. Specifically, while they all finish school early and become employed early, they differ in the timing of having a partner and children. Since decisions made in the past can affect the options for long-term decisions in the future, restricting the analysis to these three cohorts that share similar school and career trajectories could help minimize the influence associated with decisions made in the career domain on later decisions in the family domain. Thus, the analysis in this study is restricted to these three cohorts.

3.3. Empirical Analysis Specifications

The research questions we aim to address in this paper are: (1) To what extent does parenting alter the short- (mode use), mid- (vehicle ownership), and long-term (residential location) mobility decisions across subpopulations with different life course trajectories? (2) To what extent does parenting alter career decisions across subpopulations with different life course trajectories? (3) Are these differences further elucidated by differences across gender?

There are seven separate outcome variables of interest (Y_{igt}), which are defined for each person i of age g in year t:

- Three binary choices of transportation modes (more than twice per week on average): drive own car ("Drove"), use public transit ("Public Transit"), walk/bike ("Bike or Walk").
- Two vehicle ownership variables ("NumCar" and "Nor.NumCar"): total number of cars owned and the same quantity normalized by the number of household members above driving age.
- One location-characteristic variable ("PT available"): whether public transit (PT) was available at the residence location, regardless of usage.
- One employment variable ("Employ"): whether the person was employed (worked at least 35 h per week on average).

We run ordinary least squares panel data regressions separately for each outcome variable to estimate the parenting effect on that outcome variable. In the analysis, we separate the parenting event era into two stages: (1) "nesting", defined as the two years prior to the year in which the first child entered the household (To identify the mobility-relevant planning period around the child event, we examine the predictors to the event of "moved, or place of school or work changed" recorded in the life history calendar survey using logistic regression. It is found that two years prior to the first childbirth significantly predicts the move or changes to school or work location even after controlling for other events corresponding to school or work changes and relationship changes.), and (2) "having a child(ren)", defined as having at least one child less than 18 years old at home. We make the distinction between the "nesting" period and the "having a child(ren)" period because our data indicate that people tend to move two years prior to having their first childbirth.

Including this preparation period for parenting can help distinguish the turning points of mobility and career decisions. As the family-forming timeline for most people progresses from no child to nesting, then to having children at home, considering both "nesting" and "having a child(ren)" in the regression can differentiate the effect of being in one parenting stage relative to its previous stage (i.e., "nesting" relative to no child; "having a child(ren)" relative to nesting). In contrast to other studies that focus on the before and after changes of a given event (e.g., [9,56]), our approach estimates the marginal effects averaged over the whole period (or era) of the two parenting stages relative to its previous stage.

We estimate the marginal effects differentiated across the life course cohorts ($cohort_c$) defined by the clustering analysis (Equation (1)).

$$Y_{igt} = \alpha_i + \sum_c (\beta_{c,1} ChildNest_{it} \cdot cohort_c + \beta_{c,2} ChildiHome_{it} \cdot cohort_c) + \delta_g + \varepsilon_{it} \quad (1)$$

$ChildNest_{it}$ is a binary variable that equals one when the respondent (i) is in the "nesting" or having at least one child under the age of 18 at year t, and zero otherwise. $ChildiHome_{it}$ is a binary variable that equals one when the respondent (i) has at least one child under the age of 18 in the household at year t, and zero otherwise. We incorporate a person-fixed effect (α_i) to account for individual-specific characteristics that remain constant over time, effectively controlling for any inherent differences among individuals. Additionally, an age-fixed effect (δ_g) is included to capture age-specific factors that are experienced by all individuals. The error terms are assumed to follow an independently and identically distributed (IID) normal distribution. To account for serial correlation across time observations within individuals, we cluster the standard errors of the estimates at the individual level. This layered specification of parenting eras (i.e., "*ChildNest*" and "*ChlidinHome*") allows for differentiating the effect of one stage relative to its previous stage as one person progresses from no child (the omitted category) to nesting, to having children at home. Here, the parameter $\beta_{c,1}$ captures the marginal parenting effect during the nesting period relative to the "no child" period, and $\beta_{c,2}$ captures the marginal parenting effects during the child-in-home period relative to the nesting period.

To identify the potential gender effects of parenting within life course cohorts on outcome variables, we add female interaction terms with both parenting eras in the model specification, as shown in Equation (2).

$$
\begin{aligned}
Y_{igt} = \alpha_i + \textstyle\sum_c \quad & (\beta'_{c,1} ChildNest_{it} \cdot cohort_c + \beta'_{c,2} ChildiHome_{it} \cdot cohort_c) \\
& + \textstyle\sum_c (\gamma_{c,1} ChildNest_{it} \cdot cohort_c \cdot fem_i + \gamma_{c,2} ChildiHome_{it} \cdot cohort_c \cdot fem_i) + \delta_g + \varepsilon_{it}
\end{aligned} \quad (2)
$$

where fem_i is an indicator variable equal to one if respondent i identifies as female, and zero otherwise. Here, the parameters $\beta'_{c,1}$ and $\beta'_{c,2}$ capture the parenting effects of the male respondents, and $\gamma_{c,1}$ and $\gamma_{c,2}$ represent the marginal effects of the female respondents relative to their male counterparts in the same cohort. In order to avoid the confounding effects on car ownership resulting from child(ren) in the household reaching driving age (16 years old), time periods after children turned 16 years old are excluded from the analysis (supporting information available upon request).

3.4. Summary Statistics

Table 1 provides the summary statistics of the life history data used in our analysis, broken down by cohort and gender. Besides variables for "# vehicles in the household", "Normalized # vehicles in the household, "Number of modes used", and "Age in Lifecycle Calendar"), all other variables are binary variables. The mean value indicates the average portion of the life history calendar where the corresponding subpopulation experienced the given event. Equivalently, the mean value can also be interpreted as the average fraction of the subpopulation that experienced the given event between age 20 and age 50. The respondents, on average, were employed during the majority of the time period (70% to 85%) covered in the life history. However, we find a significant difference in employment status between women and men in all three cohorts. Men, on average, have a higher employment rate than women, and the difference is especially considerable among "Have-it-alls". We further investigate the employment status during 30 to 35 years of age of each respondent, which is the period that many key family and career events occur concurrently. An even greater discrepancy in employment status between men and women is found in "Have-it-alls" from 30 to 35 years of age, which suggests that women in "Have-it-alls" likely play the primary childrearing role in the household and may be more likely than their male counterparts to give up their job once children enter the home.

When it comes to car ownership, men on average have higher car ownership between the ages of 20 and 50 than women, except for "Have-it-alls" where women have higher car ownership between the ages of 20 and 50 than men. After normalizing the number of vehicles by the number of household members above driving age, a similar finding holds true. More than 80% of respondents within each cohort and gender group indicated that public transit was available for regular use, whether or not it was used. Most respondents used a personal vehicle as a regular transportation mode (two or more times per week), and relatively few respondents reported taking public transit and walking or biking regularly. In terms of the level of multimodality, it ranges between 1.06 and 1.23 modes in a given year across cohort and gender groups, with significant differences between women and men in "Couples" and "Have-it-alls", but not in "Singles".

Table 1. Life History Calendar Summary Statistics *.

Variables	Couples		Singles		Have-It-Alls	
	Female $N = 73$	Male $N = 76$	Female $N = 115$	Male $N = 111$	Female $N = 40$	Male $N = 60$
Nesting	0.03	0.04	0.03	0.03	0.07	0.07
Has Child (<18 y.o.) in house	0.14	0.28	0.22	0.22	0.60	0.61
Employed (≥35 h/wk avg) in the year	0.78	0.82	0.76	0.79	0.70	0.85
Employed between 30 and 35 y.o.	0.91	0.95	0.84	0.89	0.67	0.98
# vehicles in the household	1.53	1.64	1.33	1.39	1.80	1.61
Normalized # vehicles in the household	0.83	0.83	0.84	0.86	0.91	0.81
Drove regularly (2+ time/wk)	0.68	0.71	0.60	0.66	0.76	0.71
Took public transit regularly	0.26	0.28	0.34	0.29	0.15	0.29
Walked or biked regularly	0.23	0.22	0.23	0.21	0.14	0.18
Number of modes used	1.19	1.23	1.18	1.19	1.06	1.19
Public transit available	0.85	0.88	0.84	0.83	0.88	0.80
Age in Lifecycle Calendar	33.07	33.69	33.36	33.50	33.49	33.86

* The table provides the mean values of individual-level variables.

Figure 2 illustrates how different mode use and car ownership patterns evolve within each cohort in the life history calendar. Respondents reported increasing regular use of a personal vehicle as they grow older, while there was a decreasing trend in taking public transit and active modes, such as walking and biking, over time. When taking a deeper look at the driving pattern across cohorts, "Singles" started driving more after the age of 20, and the car usage remained constant until it started to drop at age of 45; "Couples" increased their car usage consistently throughout their life between 20 and 50 years old; "Have-it-alls" exhibited a substantial increase in car use between the ages of 20 and 35, and then their car use remained relatively steady. In plot (b), the number of vehicles in the household increased during the period covered under the life history calendar for all cohorts, with the exception that the car ownership of "Have-it-alls" actually decreased after turning 45 years old. As the number of vehicles owned in the household is likely correlated to the household size and household composition, we also plot the normalized number of vehicles trend (shown in light blue, with a corresponding y-axis on the right-hand side). Interestingly, the dark blue and light blue lines overlap with each other very closely for "Couples", which suggests that when Couples partner up, the partner tends to bring an additional vehicle to the household.

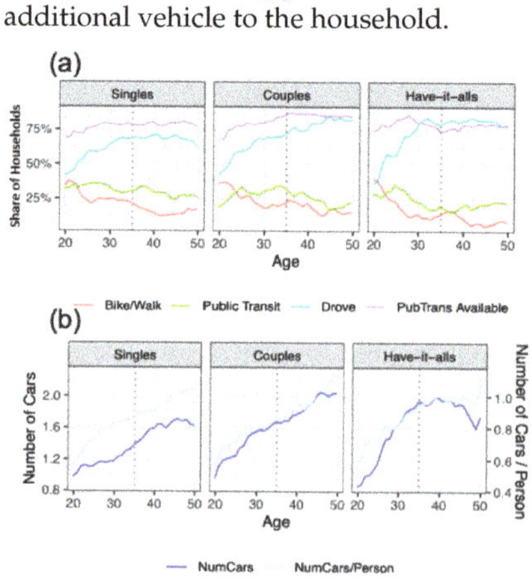

Figure 2. Life course patterns of regularly used modes and public transportation availability (**a**); the number of cars and per-person number of cars in the household (**b**), with dark blue line corresponding to the left y-axis and light blue lines corresponding to the right y-axis.

4. Results

4.1. *Effect of Parenting on Mobility and Career Decision by Cohort*

The marginal effects $\beta_{c,1}$ and $\beta_{c,2}$, estimated from Equation (1), are the change of the mobility outcomes averaged over a specific parenting era relative to the preceding era. For binary outcomes such as mode choice, public transit availability, or employment status, these coefficients can be interpreted as the percentage point change in the probability of the outcome occurring during this parenting era relative to the preceding era. For the number of cars, these coefficients represent a marginal change in number of cars during this parenting era relative to the preceding era. We explicitly account for the age effect by controlling for it in our analyses as a fixed effect, so our results are meant to represent the effects of parenting events above and beyond the underlying overall socioeconomic status evolution with aging.

Figure 3 summarizes the regression results for each mobility outcome variable (labeled in columns) by cohorts (labeled in rows), with the hollow bars indicating results that are not statistically significantly different from zero and solid bars indicating estimates that are significant at the 10% level or smaller. From left to right, the columns are arranged in the order of short-term, mid-term, then long-term mobility and career decisions.

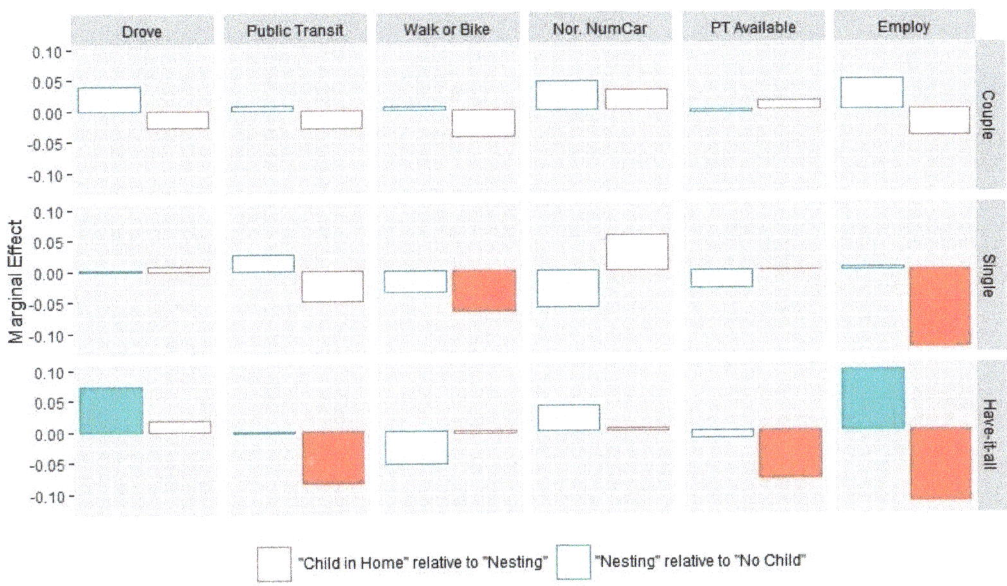

Figure 3. Marginal effects of "Nesting" and "Child in Home" by cohorts. Refer to the column variable definition in Section 3.3. Solid colors indicate significant effects at 10% level.

During the nesting period, no significant change of modes takes place relative to prior to the nesting period, except that driving significantly increases by 7.2 percentage points for the Have-it-alls. The total number of vehicles in the households all significantly increases by 0.27 to 0.33 cars. However, judging by per-person car ownership, the changes are not significant. All three cohorts do not show significant movement between transit-rich and transit-poor areas, as indicated by the changes in public transit availability. There is an increase in employment for Have-it-alls, but no change in employment status for Couples and Singles.

After the child enters the household, the parenting effects are more heterogeneous across the cohorts. Couples, on average, do not further change their mode use, car ownership, home locations, or employment status. Singles further reduce active mode usage (walk/bike) and employment. Have-it-alls, on average, tend to move to a transit-poor area (indicated by a 7.7 percentage point reduction in their reported transit availability) and simultaneously reduce their public transit use by 8.4 percentage points. Employment for Have-it-alls is reduced by 11.4 percentage points, which offsets the increase we observed during the nesting era and results in an overall decrease in employment over the two parenting events. Overall, car use and per-person car ownership do not change in this parenting era relative to the nesting era across all cohorts.

4.2. Gender Gaps on Mobility and Career Decision by Cohorts

The cohort-level average treatment effects may mask the different gender behaviors, which are further explored in this section. In Equation (2), the parameters $\beta_{c,1}$ and $\beta_{c,2}$ capture the parenting effects of the male respondents, and $\gamma_{c,1}$ and $\gamma_{c,2}$ represent the marginal effects of the female respondents relative to their male counterparts in the same cohort. Figures 4 and 5 summarize the regression results by cohorts for male and female respondents, respectively. Note that the interpretation of the bars in Figure 5 should be the marginal female effect relative to the effect of their male counterparts shown in Figure 4.

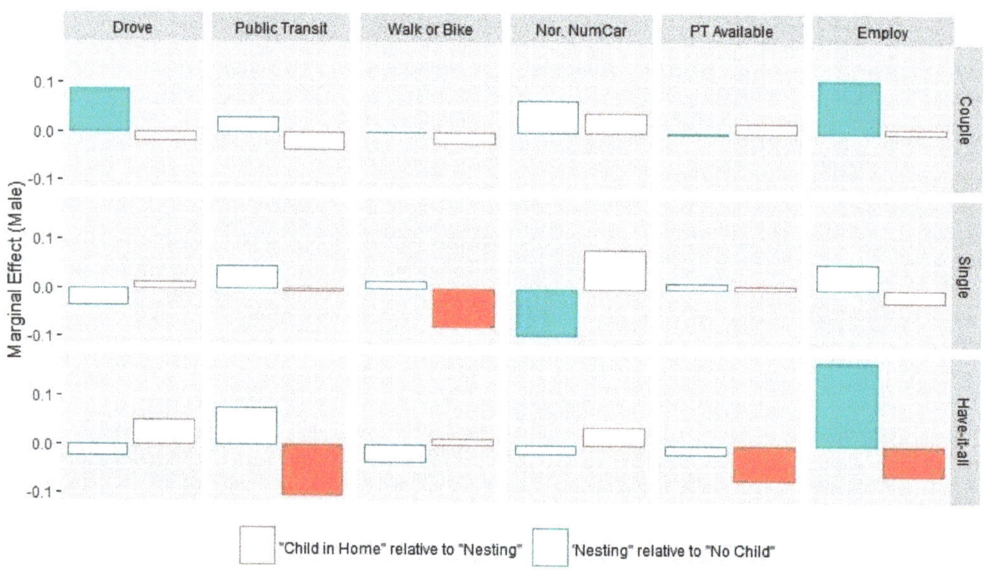

Figure 4. Marginal effects of "Nesting" and "Child in Home" for men by cohorts. Refer to the column variable definition in Section 3.3. Solid colors indicate significant effects at 10% level.

During the nesting phase, male Singles tend to reduce per-person car ownership (0.10 car/person less), which suggests that male Singles' households tend to shed a car. Female Have-it-alls, on the other hand, show a larger increase in per-person car ownership (0.15 car/person more) than their male counterparts during the nesting phase, implying that female Have-it-alls households often acquire additional cars while preparing for the birth of their first child. Unlike the cohort level average behavior in the Couples cohort, male respondents here significantly increase driving (9%) and employment (11%). On the contrary, female Couples drive significantly less and have significantly lower employment than their male counterparts during the nesting phase. Male and female Singles, on average, do not show much difference in terms of transportation mode choices, residential location, and career decisions when they are in the nesting phase.

From the gender analysis, we see that the previously identified cohort-level average behavior of increased car use for Have-it-alls is mainly driven by female Have-it-alls, whereas the increase in employment is the result of the offsetting effect from male Have-it-alls' increased employment and female Have-it-alls' decreased employment. Even though Have-it-alls at the cohort level do not change their public transit use during the nesting phase, female Have-it-alls significantly reduce their public transit use more than their male counterparts.

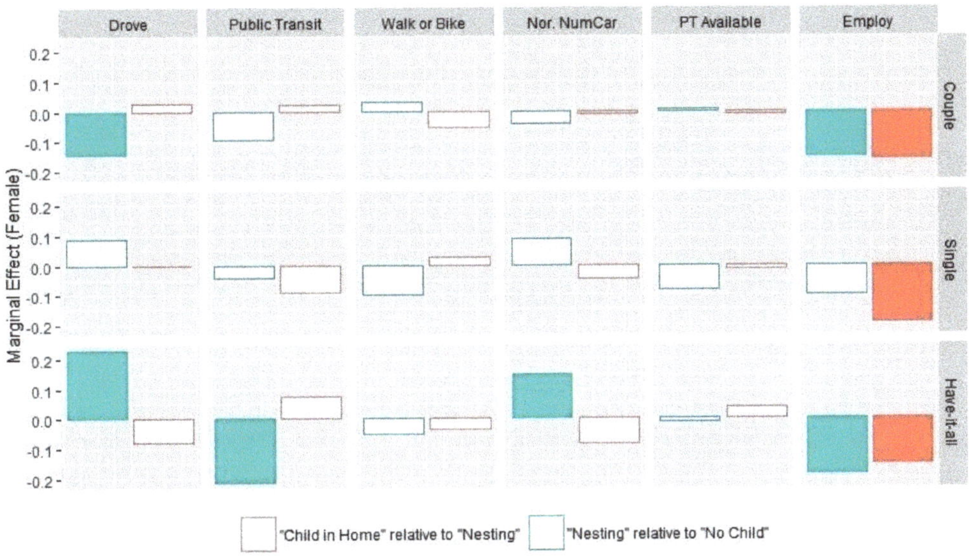

Figure 5. Marginal effects of "Nesting" and "Child in Home" for women relative to men by cohorts. Solid colors indicate significant effects at 10% level.

After the child enters the household, males in the Couples cohort do not change their mobility decisions, car ownership, and career as compared to the nesting phase. However, the results show that female Couples further reduce their employment when having child(ren) at home relative to their male counterparts after an already reduced employment during the nesting phase. In fact, we observe that women across all three cohorts have significantly lower employment than their male counterparts after having child(ren). When adding that to the negative impact of nesting on employment for women, women's career development is, on average, more negatively impacted by parenting events than men, regardless of their life course trajectory. Male Singles take less active modes after the child enters the household.

4.3. *Effect of Career Decision on Gender Gaps in the Have-It-Alls Cohorts*

When women give up careers upon entry to parenthood, labor division in the household is likely to get reinforced, which may lead to longer-term behavior lock-in. Given the most pronounced heterogeneity observed in Have-it-alls between men and women, we are interested in learning whether the career decisions that female Have-it-alls made between 30 and 35 years old, the period when they entered parenthood, could be an underlying driver of gender gaps in mid- to longer-term mobility.

According to our life history calendar data, Have-it-all women take two distinct career-related paths: (1) women who continue to work full time while forming a family, hereafter referred to as "employed women", and (2) women who gave up working full-time while forming a family, hereafter referred as "underemployed women". Approximately 65% of the Have-it-alls women fall into the "employed women" category, a much lower percentage than women in the other two cohorts (84% in Singles and 89% in Couples).

To explore how employment status interplays with gender differences in relation to parenting events in other mobility decisions, we performed a similar analysis using the model specification in Equation (2) but restricting the sample to only either employed women and employed men or underemployed women and employed men. The estimated marginal effects for employed and underemployed Have-it-all women are summarized in Figure 6. Note that the interpretation of the bars in Figure 6 should be the marginal effect relative to the effect of employed men.

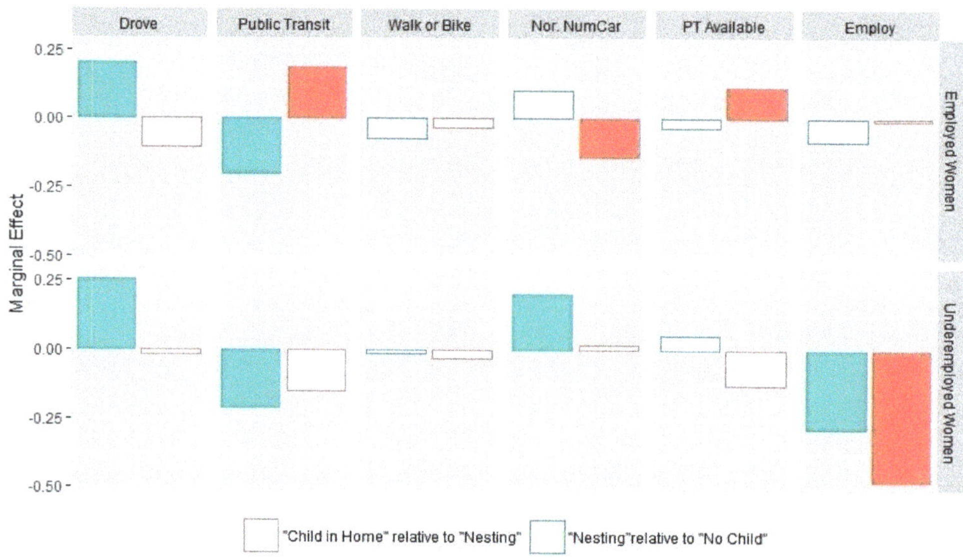

Figure 6. Gender gaps in parenting effects in Have-it-all cohort by early career decision: Have-it-alls women employed during age 30–35 relative to employed men (top); Have-it-alls women underemployed during age 30–35 relative to employed men (bottom). Solid colors indicate significant effects at 10% level.

In transitioning to the nesting period, both employed and underemployed women increase car use and reduce public transit usage relative to employed men. Interestingly, while we see no change in per-person car ownership during the nesting period for employed women, underemployed women have a significantly larger increase in per-person car ownership than employed men. This is likely due to the fact that the employed women households own more cars per person because of the need to commute to work, whereas fewer underemployed women may need a car until they are preparing for their first childbirth.

After children enter the household and over time, employed women tend to relocate to a more transit-friendly neighborhood, increase public transit usage, and reduce normalized car ownership relative to men (Figure 6, first row). On the other hand, there is no change in public transit availability, public transit usage, and car ownership for underemployed women, indicating that the initial behavior shifts toward a car-oriented lifestyle are persistent over the longer term. Regarding long-term career differences, women who remain employed at the beginning of parenthood remain employed at the same level as males over time. In contrast, the gender gap in employment widens over time among women who give up their careers at the beginning of parenthood.

5. Discussion

Many recent travel behavior studies use the mobility biography approach to study the dynamics of people's travel behavior by considering their life course trajectories. This study builds upon our previous work on exploring the heterogeneous effects of life events on habitual travel modes through a life course perspective by focusing on how the timing of parenting affects short- to long-term mobility decisions and how these impacts differ across three cohorts with distinct life history contexts, namely "Couples", "Singles", and "Have-it-alls". We further disaggregate these impacts by gender and employment status to understand if entering parenthood impacts women and men differently and whether early career decisions drive the trend of gender differences in the mobility domains over the longer term.

Our research framework provides two advancements beyond the traditional approach of mobility biography as applied in gender and travel behavior studies. First, the progression and turning points of mobility decisions are studied among different cohorts, constructed based on the similarity in the timing and order of the sociodemographic attributes rather than all individuals. It allows us to discover the underlying pathway leading to the heterogeneous effects of parenting on travel behavior by subpopulation. Second, we look at the parenting effects on three hierarchical domains, including different travel mode usage (driving, public transportation, and walking or biking), car ownership, spatial mobility as indicated by changes in public transit availability, and career decisions, to provide a comprehensive view on how parenting affects short-, mid-, and long-term mobility decisions.

As shown in [1–3,21], an individual's travel patterns are mainly habitual and developed over time; thus, entering parenthood should cause variable impacts on the mobility decisions of individuals who experience different life trajectories. Past studies have shown mixed results regarding the effect of parenting on patterns of mode use. Some studies found that being a parent is associated with a decrease in public transit usage and cycling and an increase in car use and ownership [8,15,26,32,42], and other studies show that some parents reported maintaining similar travel behavior or even increased walking and biking [9]. Based on our results, we also found differentiated parenting effects on travel behavior across cohorts. Have-it-alls significantly increase their car use during the nesting period; after that, they take less public transit when having children at home and tend to relocate to where public transit is less available. In contrast, we do not observe many changes in mode use for Couples and Singles during the nesting and having-children-at-home periods. As mentioned previously, the most distinct difference between Couples and Singles versus Have-it-alls is the timing of family formation; the first two cohorts have children much later than Have-it-alls. Hence, the life conditions and built environment for Couples and Singles when having children are likely to be at a more stable stage. Thus, their mobility decision is less affected by parenting events.

The gender gaps observed in travel behavior are also found to have different magnitudes of impact among cohorts. We observe the most pronounced gender differences in car use and car ownership in the Have-it-alls cohort. While male Have-it-alls experience no significant change in car use and car ownership during the parenting event era relatively to previously, female Have-it-alls show an increase in car usage, per-person car ownership, and a decrease in public transit use relative to their male counterparts in transitioning to the nesting period, and the gender gap remains after having children at home. Overall, when differentiating the parenting effects on mobility decisions and car ownership by gender, Singles have a minimal gender gap, while Have-it-alls have the widest gender gap in both car ownership and mode uses.

Employment is often tied to one's mobility decisions, as one may change a job or give up their job because of relocation or vice versa [36]; or one could change their travel behavior because of job changes or vice versa [57]. We find that women across all three cohorts have significantly lower employment than their male counterparts after having child(ren). If including the negative impact of parenting on employment during the nesting period for women in Couples and Have-it-alls, the overall impact of having children at home is even more significant in these cohorts. When further looking into the career decisions made by women between 30 and 35 years old, the period when they entered parenthood, a higher share of female Have-it-alls were not employed during that age period compared to the other two cohorts, and that may be attributed to the imbalanced division of labor between men and women in this particular cohort. We find evidence supporting that the early career decisions made by Have-it-alls women between 30 and 35 years old are associated with the gender difference in mobility observed over mid- to longer terms. Have-it-all women who remain employed at the beginning of parenthood remain employed at the same level as males over time. In contrast, the gender gap in employment widened over time among women who give up their careers at the beginning of parenthood. Although both employed and underemployed Have-it-alls women increased car use and reduced public transit usage during the nesting period relative to their male counterparts, only underemployed Have-it-alls women maintained this in the longer term. Employed women, on the other hand, tend to shift toward less car ownership, relocate to transit-friendly places, and use more public transit and thus gradually close the gender gap over time during parenthood.

These findings indicate that upon entering parenthood, although women who choose to hold on to their careers tend to temporarily become more car-reliant than their male counterparts at the beginning of parenthood, over time, these women are more likely to change mobility lifestyle. In contrast, we see that the gender gap remains the same or widens in the mobility outcome variables among the women who gave up their careers at the beginning of parenthood, which is accompanied by a further widening career gap from men over time. This exemplifies how women's career decisions at the beginning of parenthood could have differentiated impacts on their mid- to long-term mobility decisions.

While our study yields comprehensive insights into the dynamics of travel choices over the life course through the mobility biography approach, it is important to acknowledge certain constraints within the survey scope. First, our study focuses on individuals' travel behavior in a single region. While this approach aligns with many existing mobility biography studies, the generalizability of our results to a broader population is limited. To enhance the applicability of our results in a policy context, future research endeavors should consider deploying surveys with a similar life history calendar approach across a wider geographical region. Such an expansion would contribute to the improvement and robustness of the findings. Second, our survey was conducted prior to the onset of the COVID-19 pandemic. Consequently, certain findings from our study may have limited applicability in the present context due to the substantial life event changes experienced by individuals during the pandemic, which may have influenced their decision-making process around mobility. Nonetheless, the retrospective survey approach presented in this paper still offers significant value for researchers examining the impact of COVID-19 on individual travel behavior. Lastly, given the scope of our work, we adopted a more limited perspective regarding gender, specifically focusing on the binary categorization of men and women. The inclusion of gender into our analysis primarily illustrates the contrasting mobility patterns during parenting events between men and women. However, we did not delve into the examination of power dynamics within a gendered social context, which significantly influence the varied mobility patterns as discussed in [58].

6. Conclusions

Our results highlight the role of life course context on the dynamics of the short- (mode use), mid- (vehicle ownership), and long-term (residential location) mobility decisions, and career decisions in response to parenting events. The clear distinction in the mode use changes across the life trajectory cohorts signifies the importance of both past experiences and future expectations in mobility decision-making. Further differentiating the parenting effects by gender and employment status can help planners and policymakers better understand the tendencies and constraints faced by different individuals.

Our study identifies the Have-it-all cohort as the primary driver of gender gaps observed at the population level. Both short- and mid-/long-term travel behaviors of Have-it-all women are significantly influenced by parenting events, potentially because their family and career formation are intertwined at the same time. Therefore, implementing a transportation policy designed to address the unique needs of this particular subpopulation in the event of childbearing and caring for child(ren) may incentivize them to have a less car-centric travel pattern. Additionally, as women's career decisions in early parenthood are found to have lasting effects on mobility over time, more targeted policies could be tailored to help these vulnerable subpopulations during this particularly challenging life phase. For example, as supported by our empirical analysis, helping career moms remain employed (such as providing a flexible schedule and child care support) may help them recover from car dependency and become more accepting of multimodal lifestyles later on. Such policy measures are especially important given current circumstances, as career women have shown to be disproportionally affected by the caregiving responsibilities associated with children at home than their male counterparts since the onset of the COVID-19 pandemic in early 2020 [59–61].

While our research centers on the effect of parenting life events on mobility decisions, the data collection method and mobility biography approach used here can be applied to studying the impact of many other life events in the world reshaped by the COVID-19 pandemic. Although certain effects stemming from the stay-at-home measure may be temporary, such as limited mobility, increased e-commerce activities, and reduced ride-hailing usage, the pandemic has also triggered major life events that would have a long-lasting impact on individual travel patterns [62–64]. For instance, the growing demand for larger living spaces and access to natural amenities has led to the relocation from high-density city centers to suburban areas, potentially resulting in increased car dependency. Job loss or job switching due to the economic slowdown has the potential to disrupt one's commute routine. These long-term impacts of COVID-19 on mobility decisions through the triggering of major life events may also differ across heterogeneous populations, which highlights the need to understand an individual's mobility biography when studying their travel behavior.

Author Contributions: Conceptualization, L.J., A.T.-B., A.S., K.W. and C.A.S.; Methodology, L.J., A.L., A.T.-B., A.S., K.W. and C.A.S.; Software, L.J., A.L., A.S. and K.W.; Formal analysis, H.-C.Y., L.J., A.L. and A.T.-B.; Data curation, H.-C.Y., L.J., A.L., A.T.-B., A.S. and K.W.; Writing—original draft, H.-C.Y. and L.J.; Writing—review & editing, H.-C.Y., L.J. and Q.C.; Visualization, H.-C.Y. and L.J.; Supervision, L.J. and C.A.S.; Project administration, L.J. and C.A.S.; Funding acquisition, C.A.S. All authors have read and agreed to the published version of the manuscript.

Funding: This work was sponsored by the U.S. Department of Energy (DOE) Vehicle Technologies Office (VTO) under the Systems and Modeling for Accelerated Research in Transportation (SMART) Mobility Laboratory Consortium, an initiative of the Energy Efficient Mobility Systems (EEMS) Program, under Lawrence Berkeley National Laboratory Contract No. DE-AC02-05CH11231.

Data Availability Statement: Data collected from the WholeTraveler Transportation Behavior study survey are available at: https://livewire.energy.gov/ds/wholetraveler/phase1.00 (accessed on 3 October 2019).

Acknowledgments: The authors would like to thank T. Kwasnik, S. Carmichael, T. Chan, and M. Faulkner for their help with technical and administrative support and data collection; and A. Duvall, M. Taylor, S. Belal, G. Wong-Parodi, V. Walker, and A. Gopal for contributing to the discussion surrounding the survey design and analysis plan.

Conflicts of Interest: The authors declare no conflict of interest. The funders had no role in the design of the study; in the collection, analyses, or interpretation of data; in the writing of the manuscript; or in the decision to publish the results.

Appendix A

Table A1. Sociodemographic characteristics of the WholeTraveler sample.

	Percentage in Sample
Education	
12th grade or less, no diploma	0.4%
High school diploma/GED	1.5%
Some college	7.9%
Associate's degree	4.9%
Bachelor's degree	37.8%
Master's degree	30.6%
Professional degree (for example: M.D., DDS, DVM, J.D.)	5.6%
Doctoral degree (for example: PhD, EdD)	8.6%
N/A	2.8%
Income	
Less than USD 10,000	1.4%
USD 10,000 to USD 14,999	1.1%
USD 15,000 to USD 24,999	2.5%
USD 25,000 to USD 34,999	2.5%
USD 35,000 to USD 49,999	5.9%
USD 50,000 to USD 74,999	9.3%
USD 75,000 to USD 99,999	10.6%
USD 100,000 to USD 149,999	18.4%
USD 150,000 to USD 199,999	12.6%
USD 200,000 to USD 299,999	12.9%
USD 300,000 to USD 399,999	5.1%
USD 400,000 or more	3.1%
N/A	14.7%
Employment	
Self-employed	2.1%
Out of work and looking for work	0.3%
Out of work but not currently looking for work	0.1%
A homemaker	0.7%
A student	2.8%
Retired	1.8%
Unable to work	0.3%
Prefer not to answer	2.9%
Multiple	88.9%
N/A	0.2%
Ethnicity	
White	54.2%
Hispanic, Latino, or Spanish origin	6.0%
Black or African American	2.3%
Asian	25.5%
Middle Eastern or North African	1.2%
American Indian or Alaska Native	0.8%
Native Hawaiian or Other Pacific Islander	1.4%
Some other race or origin	1.9%
Prefer not to answer	6.6%
N/A	0.2%
Gender	
Male	48.9%
Female	47.4%
Other	0.1%
N/A	3.6%

Table A1. *Cont.*

	Percentage in Sample
Household has any children	
No	84.5%
Yes	15.5%
Speak another language at home	
No	69.4%
Yes	24.7%
N/A	6.0%

References

1. Scheiner, J. Why is there change in travel behaviour? In search of a theoretical framework for mobility biographies. *Erdkunde* **2018**, *72*, 41–62. [CrossRef]
2. Clark, B.; Chatterjee, K.; Melia, S. Changes to commute mode: The role of life events, spatial context and environmental attitude. *Transp. Res. Part A Policy Pract.* **2016**, *89*, 89–105. [CrossRef]
3. Chatterjee, K.; Clark, B.; Bartle, C. Commute mode choice dynamics: Accounting for day-to-day variability in longer term change. *Eur. J. Transp. Infrastruct. Res.* **2016**, 16. [CrossRef]
4. Müggenburg, H.; Busch-Geertsema, A.; Lanzendorf, M. Mobility biographies: A review of achievements and challenges of the mobility biographies approach and a framework for further research. *J. Transp. Geogr.* **2015**, *46*, 151–163. [CrossRef]
5. Toasin, O.M.A.; Manting, D.; Nijland, H. Dynamics in car ownership: The role of entry into parenthood. *Eur. J. Transp. Infrastruct. Res.* **2016**, 16. [CrossRef]
6. Schwanen, T. Car use and gender: The case of dual-earner families in Utrecht, The Netherlands. In *Auto Motives*; Emerald Group Publishing Limited: Bingley, UK, 2011; pp. 151–171.
7. Chakrabarti, S.; Joh, K. The effect of parenthood on travel behavior: Evidence from the California Household Travel Survey. *Transp. Res. Part A Policy Pract.* **2019**, *120*, 101–115. [CrossRef]
8. McCarthy, L.; Delbosc, A.; Currie, G.; Molloy, A. Trajectories and transitions: Mobility after parenthood. *Transportation* **2021**, *48*, 239–256. [CrossRef]
9. Lanzendorf, M. Key Events and Their Effect on Mobility Biographies: The Case of Childbirth. *Int. J. Sustain. Transp.* **2010**, *4*, 272–292. [CrossRef]
10. McCarthy, L.; Delbosc, A.; Currie, G.; Molloy, A. 'Transit Faithfuls' or 'Transit Leavers'? Understanding mobility trajectories of new parents. *Transp. Policy* **2019**, *78*, 105–112. [CrossRef]
11. McCarthy, L.; Delbosc, A.; Currie, G.; Molloy, A. Factors influencing travel mode choice among families with young children (aged 0–4): A review of the literature. *Transp. Rev.* **2017**, *37*, 767–781. [CrossRef]
12. Scheiner, J. Gendered key events in the life course: Effects on changes in travel mode choice over time. *J. Transp. Geogr.* **2014**, *37*, 47–60. [CrossRef]
13. Plyushteva, A.; Schwanen, T. Care-related journeys over the life course: Thinking mobility biographies with gender, care and the household. *Geoforum* **2018**, *97*, 131–141. [CrossRef]
14. Scheiner, J.; Holz-Rau, C. A comprehensive study of life course, cohort, and period effects on changes in travel mode use. *Transp. Res. Part A Policy Pract.* **2013**, *47*, 167–181. [CrossRef]
15. Oakil, A.T.M. Securing or sacrificing access to a car: Gender difference in the effects of life events. *Travel Behav. Soc.* **2016**, *3*, 1–7. [CrossRef]
16. Jin, L.; Lazar, A.; Sears, J.; Todd-Blick, A.; Sim, A.; Wu, K.; Yang, H.-C.; Spurlock, C.A. Clustering Life Course to Understand the Heterogeneous Effects of Life Events, Gender, and Generation on Habitual Travel Modes. *IEEE Access* **2020**, *8*, 190964–190980. [CrossRef]
17. SMART Mobility: Mobility Decision Science Capstone Report. Art. no. DOE/EE-2063, July 2020. Available online: https://trid.trb.org/view/1925927 (accessed on 19 March 2023).
18. Chorus, C.G.; Dellaert, B.G.C. Travel Choice Inertia: The Joint Role of Risk Aversion and Learning. *J. Transp. Econ. Policy* **2012**, *46*, 139–155.
19. Banister, D. The influence of habit formation on modal choice ?a Heuristic model. *Transportation* **1978**, *7*, 5–33. [CrossRef]
20. Gärling, T.; Axhausen, K.W. Introduction: Habitual travel choice. *Transportation* **2003**, *30*, 1–11. [CrossRef]
21. Zhang, J.; Van Acker, V. Life-oriented travel behavior research: An overview. *Transp. Res. Part A Policy Pract.* **2017**, *104*, 167–178. [CrossRef]
22. Larouche, R.; Rodriguez, U.C.; Nayakarathna, R.; Scott, D.R. Effect of Major Life Events on Travel Behaviours: A Scoping Review. *Sustainability* **2020**, *12*, 10392. [CrossRef]
23. Oakil, A.T.; Ettema, D.; Arentze, T.; Timmermans, H. Changing household car ownership level and life cycle events: An action in anticipation or an action on occurrence. *Transportation* **2014**, *41*, 889–904. [CrossRef]

24. Zhang, J.; Yu, B.; Chikaraishi, M. Interdependences between household residential and car ownership behavior: A life history analysis. *J. Transp. Geogr.* **2014**, *34*, 165–174. [CrossRef]
25. Clark, B.; Chatterjee, K.; Melia, S. Changes in level of household car ownership: The role of life events and spatial context. *Transportation* **2016**, *43*, 565–599. [CrossRef]
26. Klein, N.J.; Smart, M.J. Life events, poverty, and car ownership in the United States: A mobility biography approach. *J. Transp. Land Use* **2019**, *12*, 395–418. [CrossRef]
27. Schäfer, M.; Jaeger-Erben, M.; Bamberg, S. Life Events as Windows of Opportunity for Changing Towards Sustainable Consumption Patterns? *J. Consum. Policy* **2012**, *35*, 65–84. [CrossRef]
28. Schoenduwe, R.; Mueller, M.G.; Peters, A.; Lanzendorf, M. Analysing mobility biographies with the life course calendar: A retrospective survey methodology for longitudinal data collection. *J. Transp. Geogr.* **2015**, *42*, 98–109. [CrossRef]
29. Guo, J.; Feng, T.; Timmermans, H.J. Time-varying dependencies among mobility decisions and key life course events: An application of dynamic Bayesian decision networks. *Transp. Res. Part A Policy Pract.* **2019**, *130*, 82–92. [CrossRef]
30. Beige, S.; Axhausen, K.W. Interdependencies between turning points in life and long-term mobility decisions. *Transportation* **2012**, *39*, 857–872. [CrossRef]
31. Rau, H.; Manton, R. Life events and mobility milestones: Advances in mobility biography theory and research. *J. Transp. Geogr.* **2016**, *52*, 51–60. [CrossRef]
32. De Haas, M.; Scheepers, C.; Harms, L.; Kroesen, M. Travel pattern transitions: Applying latent transition analysis within the mobility biographies framework. *Transp. Res. Part A Policy Pract.* **2018**, *107*, 140–151. [CrossRef]
33. Salomon, I. Life styles—A broader perspective on travel behaviour. *Recent Adv. Travel Demand Anal.* **1983**, 290–310.
34. Lanzendorf, M. Mobility Styles and Travel Behavior: Application of a Lifestyle Approach to Leisure Travel. *Transp. Res. Rec. J. Transp. Res. Board* **2002**, *1807*, 163–173. [CrossRef]
35. Farinloye, T.; Mogaji, E.; Aririguzoh, S.; Kieu, T.A. Qualitatively exploring the effect of change in the residential environment on travel behaviour. *Travel Behav. Soc.* **2019**, *17*, 26–35. [CrossRef]
36. Dieleman, F.M.; Mulder, C.H. *The Geography of Residential Choice*; Bergin & Garvey: Westport, CT, USA, 2002. Available online: https://dare.uva.nl/search?arno.record.id=121039 (accessed on 22 May 2023).
37. Adhikari, B.; Hong, A.; Frank, L.D. Residential relocation, preferences, life events, and travel behavior: A pre-post study. *Res. Transp. Bus. Manag.* **2020**, *36*, 100483. [CrossRef]
38. Gerber, P.; El-Geneidy, A.; Manaugh, K.; Lord, S. From workplace attachment to commuter satisfaction before and after a workplace relocation. *Transp. Res. Part F Traffic Psychol. Behav.* **2020**, *71*, 168–181. [CrossRef]
39. Maheshwari, R.; Van Acker, V.; De Vos, J.; Witlox, F. A multi-perspective review of the impact of a workplace relocation on commuting behaviour, commuting satisfaction and subjective well-being. *Transp. Rev.* **2023**, *43*, 385–406. [CrossRef]
40. Havet, N.; Bayart, C.; Bonnel, P. Why do Gender Differences in Daily Mobility Behaviours persist among workers? *Transp. Res. Part A Policy Pract.* **2021**, *145*, 34–48. [CrossRef]
41. Craig, L.; van Tienoven, T.P. Gender, mobility and parental shares of daily travel with and for children: A cross-national time use comparison. *J. Transp. Geogr.* **2019**, *76*, 93–102. [CrossRef]
42. Whittle, C.; Whitmarsh, L.; Nash, N.; Poortinga, W. Life events and their association with changes in the frequency of transport use in a large UK sample. *Travel Behav. Soc.* **2022**, *28*, 273–287. [CrossRef]
43. Lanzendorf, M. Mobility Biographies. A New Perspective for Understanding Travel Behaviour. 2003. Available online: https://www.semanticscholar.org/paper/Mobility-biographies.-A-new-perspective-for-travel-Lanzendorf/46e666c293a342d9575f9b2a4e72f521204e0b64 (accessed on 19 March 2023).
44. Müggenburg, H. Beyond the limits of memory? The reliability of retrospective data in travel research. *Transp. Res. Part A Policy Pract.* **2021**, *145*, 302–318. [CrossRef]
45. Axsen, J.; TyreeHageman, J.; Lentz, A. Lifestyle practices and pro-environmental technology. *Ecol. Econ.* **2012**, *82*, 64–74. [CrossRef]
46. Krizek, K.J. Lifestyles, Residential Location Decisions, and Pedestrian and Transit Activity. *Transp. Res. Rec. J. Transp. Res. Board* **2006**, *1981*, 171–178. [CrossRef]
47. Forsyth, A.; Agrawal, A.W.; Krizek, K.J. Simple, Inexpensive Approach to Sampling for Pedestrian and Bicycle Surveys: Approach Developed in Pedestrian and Bicycling Survey. *Transp. Res. Rec. J. Transp. Res. Board* **2012**, *2299*, 22–30. [CrossRef]
48. Spurlock, C.A.; Sears, J.; Wong-Parodi, G.; Walker, V.; Jin, L.; Taylor, M.; Duvall, A.; Gopal, A.; Todd, A. Describing the users: Understanding adoption of and interest in shared, electrified, and automated transportation in the San Francisco Bay Area. *Transp. Res. Part D Transp. Environ.* **2019**, *71*, 283–301. [CrossRef]
49. Wells, E.M.; Small, M.; Spurlock, C.A.; Wong-Parodi, G. Factors associated with emerging multimodal transportation behavior in the San Francisco Bay Area. *Environ. Res. Infrastruct. Sustain.* **2021**, *1*, 031004. [CrossRef]
50. Lazar, A.; Ballow, A.; Jin, L.; Spurlock, C.A.; Sim, A.; Wu, K. Machine Learning for Prediction of Mid to Long Term Habitual Transportation Mode Use. In Proceedings of the 2019 IEEE International Conference on Big Data (Big Data), Los Angeles, CA, USA, 9–12 December 2019; pp. 4520–4524. [CrossRef]
51. Spurlock, C.A.; Todd-Blick, A.; Wong-Parodi, G.; Walker, V. Children, Income, and the Impact of Home Delivery on Household Shopping Trips. *Transp. Res. Rec. J. Transp. Res. Board* **2020**, *2674*, 335–350. [CrossRef]

52. Cervero, R.; Tsai, Y.-H. City CarShare in San Francisco, California: Second-Year Travel Demand and Car Ownership Impacts. *Transp. Res. Rec. J. Transp. Res. Board* **2004**, *1887*, 117–127. [CrossRef]
53. Circella, G.; Tiedeman, K.; Handy, S.; Alemi, F.; Mokhtarian, P. What Affects Millennials' Mobility? Part I: Investigating the Environmental Concerns, Lifestyles, Mobility-Related Attitudes and Adoption of Technology of Young Adults in California, May 2016. Available online: https://escholarship.org/uc/item/6wm51523 (accessed on 2 May 2023).
54. NuStats Research Solutions. 2010–2012 California Household Travel Survey. *Calif. Dep. Transp.* June 2013. Available online: https://trid.trb.org/view/1308918 (accessed on 2 May 2023).
55. Enam, A.; Konduri, K.C. Time Allocation Behavior of Twentieth-Century American Generations: GI Generation, Silent Generation, Baby Boomers, Generation X, and Millennials. *Transp. Res. Rec. J. Transp. Res. Board* **2018**, *2672*, 69–80. [CrossRef]
56. Beige, S.; Axhausen, K.W. The dynamics of commuting over the life course: Swiss experiences. *Transp. Res. Part A Policy Pract.* **2017**, *104*, 179–194. [CrossRef]
57. Scheiner, J. Mobility Biographies: Elements of a Biographical Theory of Travel Demand (Mobilitätsbiographien: Bausteine zu einer biographischen Theorie der Verkehrsnachfrage). *Erdkunde* **2007**, *61*, 161–173. [CrossRef]
58. Hanson, S. Gender and mobility: New approaches for informing sustainability. *Gender Place Cult.* **2010**, *17*, 5–23. [CrossRef]
59. Sevilla, A.; Smith, S. Baby steps: The gender division of childcare during the COVID-19 pandemic. *Oxf. Rev. Econ. Policy* **2020**, *36* (Suppl. S1), S169–S186. [CrossRef]
60. Carli, L.L. Women, Gender equality and COVID-19. *Gend. Manag. Int. J.* **2020**, *35*, 647–655. [CrossRef]
61. Cardel, M.I.; Dean, N.; Montoya-Williams, D. Preventing a Secondary Epidemic of Lost Early Career Scientists. Effects of COVID-19 Pandemic on Women with Children. *Ann. Am. Thorac. Soc.* **2020**, *17*, 1366–1370. [CrossRef]
62. De Palma, A.; Vosough, S.; Liao, F. An overview of effects of COVID-19 on mobility and lifestyle: 18 months since the outbreak. *Transp. Res. Part A Policy Pract.* **2022**, *159*, 372–397. [CrossRef] [PubMed]
63. Matson, G.; McElroy, S.; Lee, Y.; Circella, G. Longitudinal Analysis of COVID-19 Impacts on Mobility: An Early Snapshot of the Emerging Changes in Travel Behavior. *Transp. Res. Rec. J. Transp. Res. Board* **2022**, *2677*, 298–312. [CrossRef] [PubMed]
64. Engle, S.; Stromme, J.; Zhou, A. Staying at Home: Mobility Effects of COVID-19. *SSRN Electron. J.* **2020**, 86–102. [CrossRef]

systems

MDPI

Article

Ramp Spacing Evaluation of Expressway Based on Entropy-Weighted TOPSIS Estimation Method

Jie Ma, Yilei Zeng and Dawei Chen *

School of Transportation, Southeast University, Nanjing 211189, China
* Correspondence: dw_chen@seu.edu.cn

Abstract: The main objective of this study is to design a method for evaluating the reasonability of ramp spacing of the expressway in a specific district. The study proposes an entropy-weighted Technique for Order Preference by Similarity to an Ideal Solution (TOPSIS) estimation method, in which the entropy weight method determines the indicator weights, and TOPSIS is employed to compare different alternatives of ramp spacing. Four patterns of evaluation indicators are taken into account representing traffic efficiency, safety, traffic accessibility, and economy, respectively. Using the Beijing–Hong Kong–Macao Expressway in Henan Province as a case study, the validity of the method is verified, and the optimal ramp spacing is obtained as 14 km for the given scenario. The results of the study show: (1) extreme spacing values are not conducive to the overall benefits of the expressway; (2) ramp spacing settings that allow for coordinated sharing of traffic demand along the route (TDAR) are a prerequisite for an expressway to have great overall benefits; and (3) appropriately shortening ramp spacing will allow the expressway to effectively respond to increased TDAR. The estimation method proposed in this study provides a theoretical reference for the local authority to plan ramp spacing that can satisfy regional traffic demand and ensure the overall benefits of expressways in a sustainable urban context.

Keywords: expressway; ramp spacing; multi-criteria estimation method; TOPSIS; entropy

Citation: Ma, J.; Zeng, Y.; Chen, D. Ramp Spacing Evaluation of Expressway Based on Entropy-Weighted TOPSIS Estimation Method. *Systems* **2023**, *11*, 139. https://doi.org/10.3390/systems11030139

Academic Editors: Mahyar Amirgholy and Jidong J. Yang

Received: 12 February 2023
Revised: 28 February 2023
Accepted: 2 March 2023
Published: 4 March 2023

1. Introduction

As the national transport corridor connecting large centers of activity over long-haul transportation, expressways play a pivotal role in the entire highway transportation system, which is crucial to ensuring robust economic growth and sustainable development [1–3]. In the wake of continuously expanding urbanization, travelers per se have higher requirements for service quality and traffic efficiency [4]. Once travelers select to use an expressway service, they need to enter or exit from ramps along the expressways. That is to say, the placement of ramps and associated spacing has a sizeable influence on the convenience of travelers with exits and entrances regarding the expressway service [5]. When ramp spacing is set to be shorter, traffic accessibility will be improved as travelers can more easily enter/exit the expressway. On the other hands, shorter ramp spacing means more ramps being constructed per unit length of the expressway, which results in some negative impacts. For example, the increase in the number of weaving areas formed by vehicles entering and exiting the ramps causes more frequent disruptions to the traffic flow on the expressway and even brings about safety concerns [6,7]; further, the project cost of constructing more ramps increases concomitantly. Thus, a methodology is necessitated to evaluate the reasonability of different ramp spacing settings, which is addressed in this paper.

Currently, a number of previous studies have been conducted to analyze the reasonable setting of expressway ramp spacing from different perspectives. From the perspective of traffic safety, studies based on accident data near ramps quantified the direct relationship between ramp spacing and safety [8–10]. Le and Porter indicate that vehicle crashes increase with the decreasing ramp spacing [11]. For such traffic safety concerns, some existing standards specify the minimum ramp spacing pertaining to geometric design variables concerning weaving volume, ability to sign, and lengths of speed-change lanes. For instance, the American Association of State Highway and Transportation Officials (AASHTO) specifies a minimum ramp spacing of 3 km in rural areas [12]; the specification developed by the China Highway Engineering Consultants Corporation recommends that the spacing between adjacent ramps should be at least 4 km [13]. Further, some studies investigated ramp spacing from the perspective of traffic efficiency. The Highway Capacity Manual 2010 manifests that ramp spacing is related to vehicle speed [14]. Several scholars consider the setting of ramp spacing by analyzing vehicle operating characteristics along the expressway. For instance, Chen et al. calculated reasonable ramp spacing by applying a statistical package for the social sciences and multiple linear regressions to analyze the speed characteristics within an expressway section [15]. Wang et al. determined the ultimate expressway ramp spacing based on constructing the GPS floating car speed data-driven model and the gap acceptance model [16]. Using acceleration noise to represent the stability of traffic conditions, Jang reported the minimum ramp spacing under higher-speed traffic conditions (around 160 km/h) [17]. To compensate for the lack of acceleration noise's discrimination between index per minimum spacing, Kim et al. utilized the surrogate safety assessment model to analyze conflicts and estimated the minimum ramp spacing for expressways that allow travel speed above 140 km/h [18]. Moreover, the issue of economic factors is also taken into account in determining ramp spacing, as the construction cost of ramps is expensive in the expressway system [19]. According to surveys of ramp construction costs [20,21], a ramp's (a complete structure including at least one entrance and one exit) cost can range from USD 5 million to USD 15 million or more when the mainline of an expressway costs range from USD 7 million to USD 15 million per kilometer to build. Therefore, ramp spacing being set excessively short unnecessarily increases the overall project cost. Ramp spacing that considers the economic factors is conducive to guaranteeing the successful construction of the expressway project and enhancing social benefits [22].

Most previous studies about ramp spacing focused on the setting of minimum spacing considering one or two indicators (e.g., safety, economy). However, these studies did not compare the effects of different ramp spacing settings, while lacking consideration of more factors reflecting practical circumstances. For example, Winkler and Fan indicated that when ramp spacing is set longer, there are fewer weaving areas along the expressway, which contributes to increased expressway capacity, improved safety, and lower construction costs [23]. However, Chen et al. point out that excessively long ramp spacing poses an inconvenience as travelers have to travel long distances to the nearest ramp [24]; it can also reduce travelers' motivation to choose the expressway and wastes road resources, because travelers prefer to enter the expressway at the on-ramp closest to their origin and exit at the off-ramp closest to their destination. Hence, the setting of ramp spacing is a nontrivial task, which needs to simultaneously allow for several factors so that the ramps can be set to accommodate the multiple benefits of the expressway, especially in the context of urbanization expansion and rapid land use development around cities.

In practice, the evaluation of expressway ramp spacing involves multi-criteria, and various patterns of relevant indicators need to be taken into account. In recent years, several popular estimation methods of multi-criteria decision making have been developed, such as Analytic Hierarchy Process (AHP), Technique for Order Preference by Similarity to an Ideal Solution (TOPSIS), PROMETHEE, and Fuzzy AHP [25]. Among these, TOPSIS has a relatively simple calculation algorithm, which can analyze quantitative data and fully use data information [26,27]. Recently, a growing number of estimation studies in the field of transportation have applied TOPSIS [28–31], e.g., quality and safety of public transport, scenarios for developing public transport systems, choice of investment location, and road transport. However, none of the prior studies have particularly applied estimation methods to analyze the effects of ramp spacing settings on the level of expressway service.

To fill such a gap, the objective of this study is to propose an entropy-weighted TOPSIS estimation method that quantitatively estimates different settings of ramp spacing along the expressway within a specific district. The evaluation indicator system contains four patterns of relevant indicators spanning traffic efficiency, safety, traffic accessibility, and economy, respectively. Further, the Beijing–Hong Kong–Macao Expressway within Henan Province in China is applied as the case study to obtain estimation results based on the proposed estimation method. Finally, the evaluation results are analyzed to explore the potential principles for setting the ramp spacing of expressways within specific districts and to provide a theoretical basis for the practical applications. The method proposed in this study takes a long expressway section as a research object for setting ramp spacing from the regional level. Compared with studies that separately consider factors such as vehicle safety [32], traffic demand [33] or operating characteristics [15–18], respectively, the results of this research have critical theoretical support for setting ramp spacing that satisfies the whole district's traffic demand and gives the whole expressway system good comprehensive benefits. In addition, it can provide preliminary recommendations for setting ramp spacing for studies that consider multiple factors for ramp siting [24]. In the context of urbanization expansion, this study has implications for promoting the efficient use of road resources and sustainable transportation development.

The remainder of this paper is organized as follows. Section 2 below describes the problem and evaluation indicators. Section 3 introduces the estimation model. Then, a case study is conducted in Section 4, and discussions are presented in Section 5. Finally, conclusions are provided in the last section.

2. Problem Statement and Evaluation Indicators

2.1. Problem Statement

In this paper, we study the setting of ramp spacing along the expressway within a specific district. The expressway in the district attracts demand from enroute cities, counties, and towns. For the sake of simplicity, these demand regions are divided into U towns (denoted as the units of demand) and represented by a set of corresponding centroids $T = \{T_1, T_2, \ldots, T_U\}$. Note that all these towns constitute the service area dubbed "study district". This study proposes m possible alternatives for ramp spacing settings according to the characteristics along the expressway. It is assumed that under each spacing alternative, travelers always choose the closest ramps to their origins or destinations to employ the expressway service. Table 1 depicts the variables used in the evaluation indicator system.

2.2. Evaluation Indicator System

The setting of ramp spacing is of fundamental importance to the sustainability of expressway service, particularly regarding accessibility. Shortening ramp spacing can improve accessibility obviously. However, the shorter spacing requires more ramps to be built and increases the construction cost. Furthermore, excessively short spacing makes the weaving behaviors of vehicles more frequent, which will adversely affect traffic efficiency and operational safety for vehicles when using the expressway service [24]. To determine the appropriate setting of ramp spacing, an evaluation indicator system simultaneously consisting of traffic efficiency, safety, traffic accessibility, and economy needs to be constructed. Evaluation indicators are employed in the estimation method in Section 3.

2.2.1. Traffic Efficiency

Traffic efficiency is characterized by the average speed $\overline{v}$ and average delay $\overline{d}$ of the vehicles traveling on the expressway section (including mainline and ramps). The meanings of these two indicators are described by Equations (1) and (2), respectively:

$$\overline{v} = \frac{\sum_{i=1}^{N} v_i}{N}, \tag{1}$$

$$\overline{d} = \frac{\sum_{i=1}^{N} t_i}{N}, \tag{2}$$

where v_i is the speed of vehicle i on the expressway section (km/h); d_i denotes the delay of vehicle i on the expressway section (s); N represents the sample size of vehicles in the whole process of simulating traffic operation on the expressway section with VISSIM.

Table 1. List of notations in the evaluation indicator system.

Variable	Notation
Parameters	
$\overline{v}$	the average speed
$\overline{d}$	the average delay
v_i	the speed of vehicle i
d_i	the delay of vehicle i
N	the sample size of vehicles
ϵ	the accident rate of 100 million vehicle- kilometers
σ	the standard deviation of the speed of all vehicles
L_i	the comprehensive level of service of roads within the town i
α	the grade of roads
len_α	the length of the road with grade α
h_α	the evaluation value of the road with class α
Len_i	the total length of the roads passing through the town i
a_i	the accessibility of town i
l_{ij}	the distance from town i to ramp j
M_i	the comprehensive aggregation scale of town i
P	the number of evaluation indicators for the comprehensive aggregation scale
γ_k	the weight of the $k-$th evaluation indicator
u_{ik}	the value of the $k-$th evaluation indicator of town i
ϕ	the accessibility of the study district
Ω	the project cost of constructing all ramps along the expressway section
ρ	the density of ramps along the expressway section
l_c	
δ_c	the relevant parameter of the project cost
β_c	
Sets	
R	the set of ramps set according to spacing alternative
R_i	the set of ramps that can serve the town i, $R \in R_i$
T	the set of towns

2.2.2. Safety

Due to the normal situation of traffic flow being seriously affected once an accident occurs on the expressway, this study chose traffic accident rate as the safety indicator. As per the work of Pei and Cheng [34], the accident rate increases exponentially with the increase in vehicle speed standard deviation. The fitting relationship between accident rate, vehicle speed, and standard deviation of vehicle speed is shown as below:

$$\epsilon = 9.583\,\exp^{0.055\sigma}, \tag{3}$$

where ϵ is the accident rate per 100 million vehicle-kilometers (case/(km$\cdot 10^{-8} \cdot$veh^{-1})); Let $\sigma = \sqrt{\frac{\sum_{i=1}^{N}(v_i-\overline{v})^2}{N-1}}$,$\sigma$ denotes the standard deviation of the speed of all vehicles on the expressway section (km/h).

2.2.3. Traffic Accessibility

Traffic accessibility is utilized to estimate the convenience of vehicles accessing the expressway service. It is assumed that traffic accessibility is related to the comprehensive level of service (LOS) of roads within the town and the comprehensive aggregation size (CAS) of the town.

In terms of LOS of the road, different grades of roads are assigned according to the relevant specification for urban road design [35]: expressway = 4, trunk road = 3, secondary road = 2, and slip road = 1. The comprehensive LOS of roads within the town can be calculated as below:

$$L_i = \frac{len_\alpha h_\alpha}{Len_i}(i \in T), \tag{4}$$

where len_α denotes the length of the road with grade α; h_α is the evaluation value of the road with class α; Len_i signifies the total length of the roads passing through town i.

Then, using the reciprocal of the distance from town i to ramps, the accessibility of town i is calculated as:

$$a_i = \sum_{j \in R_i} L_i \frac{1}{l_{ij}}(i \in T), \tag{5}$$

where R_i represents the set of ramps that can serve town i, $card(R_i)$ is less than the number of ramps in the corresponding spacing alternative; l_{ij} represents the distance from town i to ramp j, $j \in R_i$.

CAS comprehensively represents the quality and quantity of economic activity aggregation scale within a town [36]. The CAS of a town is generally assessed based on the intensity of development, economic prosperity, scale of land development, and population. The comprehensive aggregation size M_i can be expressed by:

$$M_i = \sum_{k=1}^{P} \frac{\gamma_k u_{ik}}{u_k}(i \in T), \tag{6}$$

where P is the number of evaluation indicators about CAS; γ_k signifies the weight of the $k-$th evaluation indicator; u_{ik} is the value of the $k-$th evaluation indicator of town i.

Finally, the value of the accessibility of the study district equals:

$$\phi = \sum_{i \in T} M_i a_i, \tag{7}$$

2.2.4. Economy

The project cost is selected as the economic efficiency indicator of constructing ramps. Based on the density of ramps along the expressway, the project cost is calculated:

$$\Omega = l_c + \delta_c \rho^{-\beta_c}, \tag{8}$$

where Ω represents the project cost of constructing all ramps along the expressway section; l_c, δ_c, and β_c are parameters related to the project cost.

Table 2 describes the information of all the indicators employed in the evaluation indicator system.

Table 2. Evaluation indicator system of ramp spacing on expressway.

Indicator (Criterion)	Dimension	Indicator Source
Average speed ($\overline{v}$)	positive	VISSIM simulation
Average delay ($\overline{d}$)	negative	VISSIM simulation
Accident rate (ϵ)	negative	VISSIM simulation and calculation
Traffic accessibility (ϕ)	positive	Calculation based on data
Project cost (Ω)	negative	Evaluation based on ramp data and terrain conditions

3. Methodology

To evaluate the effect of different ramp spacing settings, we develop the entropy-weighted TOPSIS estimation method by employing the evaluation indicators in Section 2.2.

Based on the weight of each indicator determined by the entropy weight method, TOPSIS is used to compare different alternatives of ramp spacing settings.

3.1. Entropy Weight Method

The entropy weight method, which introduces the idea of information entropy, is an objective weighting method [37]. The method mainly uses the difference degree of indicators to estimate the effective information in the known data and calculate the entropy weight of each indicator. The entropy weight reflects the ability of each evaluation indicator to pass decision information [38,39]. The basic calculation steps of the entropy weight method are as follows:

Step 1: Initial data matrix normalization

It is set that the initial data matrix of the entropy weight evaluation system consists of m evaluation objects and n evaluation indicators,

$$X = \begin{bmatrix} x_{11} & x_{12} & \cdots & x_{1n} \\ x_{21} & x_{22} & \cdots & x_{2n} \\ \vdots & \vdots & \ddots & \vdots \\ x_{m1} & x_{m2} & \cdots & x_{mn} \end{bmatrix}, \tag{9}$$

where $x_{ij}(i = 1,2,\ldots,m; j = 1,2,\ldots,n)$ denotes the value of the $j-th$ indicator in the $i-th$ evaluation object.

In order to eliminate the influence of different indicator units on the evaluation results, each indicator is standardized. The step transformation method is a standardization method used commonly, and the equations are as follows:

$$x'_{ij} = \begin{cases} \frac{x_{ij}-x_{j\min}}{x_{j\max}-x_{j\min}}, x_{j\max} \neq x_{j\min} \\ 1 \quad , x_{j\max} = x_{j\min} \end{cases} \text{(applicable benefit indicators)}, \tag{10}$$

$$x'_{ij} = \begin{cases} \frac{x_{j\max}-x_{ij}}{x_{j\max}-x_{j\min}}, x_{j\max} \neq x_{j\min} \\ 1 \quad , x_{j\max} = x_{j\min} \end{cases} \text{(applicable cos t indicators)}. \tag{11}$$

Step 2: Estimating the proportion p_{ij} of x'_{ij} of the $i-th$ evaluation object for the $j-th$ indicator

$$p_{ij} = x'_{ij} / \sum_{i}^{m} x'_{ij} (j = 1,2,\ldots,n), \tag{12}$$

Step 3: Calculating the value of the information entropy e_j of the $j-th$ indicator

$$e_j = -k \sum_{i=1}^{m} p_{ij} \ln p_{ij} (j = 1,2,\ldots,n), \tag{13}$$

where $k = 1/\ln n$, is a non-negative constant related to the number of evaluation objects; when set $p_{ij} = 0$, $p_{ij} \ln p_{ij} = 0$.

Step 4: Calculating the weight of the indicators

The weight w_j of the $j-th$ indicator is calculated by:

$$w_j = \frac{1-e_j}{\sum_{j=1}^{n}(1-e_j)} (j = 1,2,\ldots,n) \tag{14}$$

3.2. TOPSIS Method

TOPSIS is a comprehensive distance-based evaluation method, first proposed by C. L. Hwang and K. Yoon in 1981 [40]. The method uses the proximity of the evaluation alternative to the idealized target to rank the merits of each evaluation alternative.

In this study, we design m spacing alternatives by varying the value of the ramp spacing and consider them as the evaluation objects of TOPSIS. The five evaluation indicators in Section 2.2 are introduced as criteria in TOPSIS. We express the set of alternatives as $A = \{A_1, A_2, \ldots, A_m\}$, and the set of criteria as $C = \{C_1, C_2, \ldots, C_n\}$, where $n = 5$. The construction process of the entropy-weighted TOPSIS estimation method is shown as follows.

Step 1: Establishing the decision matrix

According to the obtained performance value d_{ij}, we construct the decision matrix $D = [g_{ij}]_{m\times n}$ for evaluation:

$$D = \begin{array}{c} \\ A_1 \\ A_2 \\ \vdots \\ A_m \end{array} \begin{array}{c} \begin{array}{cccc} C_1 & C_2 & \cdots & C_n \end{array} \\ \begin{bmatrix} g_{11} & g_{12} & \cdots & g_{1n} \\ g_{21} & g_{22} & \cdots & g_{2n} \\ \vdots & \vdots & \ddots & \vdots \\ g_{m1} & g_{m2} & \cdots & g_{mn} \end{bmatrix} \end{array}, \tag{15}$$

where A represents the evaluated alternative and C represents the criterion.

Step 2: Determining the normalization decision matrix

Performance values are normalized to ensure that utility preferences have a consistent unit of measurement while avoiding extreme values affecting similarity distance measurement. A normalized performance value (z_{ij}) is calculated as below:

$$z_{ij} = \frac{g_{ij}}{\sqrt{\sum_{i=1}^{m} g_{ij}{}^2}} \forall i, j \tag{16}$$

The normalization decision matrix is expressed as:

$$Z = \begin{array}{c} \\ A_1 \\ A_2 \\ \vdots \\ A_m \end{array} \begin{array}{c} \begin{array}{cccc} C_1 & C_2 & \cdots & C_n \end{array} \\ \begin{bmatrix} z_{11} & z_{12} & \cdots & z_{1n} \\ z_{21} & z_{22} & \cdots & z_{2n} \\ \vdots & \vdots & \ddots & \vdots \\ z_{m1} & z_{m2} & \cdots & z_{mn} \end{bmatrix} \end{array} \tag{17}$$

Step 3: Determining the positive and negative ideal solutions

In the TOPSIS method, the evaluation criteria are divided into benefit criteria and cost criteria. Let the set of benefit criteria be expressed as B, and H denote a set of cost criteria. Z^+ represents the positive ideal solution (PIS) and Z^- denotes the negative ideal solution (NIS). According to the Equations (18) and (19), Z^+ and Z^- can be calculated:

$$Z^+ = \left(\left(\max_i z_{ij} \mid j \in B\right), \left(\min_i z_{ij} \mid j \in H\right)\right) = \left(\left(z_j^+ \mid j = 1, 2, \ldots, m\right)\right) \tag{18}$$

$$Z^- = \left(\left(\min_i z_{ij} \mid j \in B\right), \left(\max_i z_{ij} \mid j \in H\right)\right) = \left(\left(z_j^- \mid j = 1, 2, \ldots, m\right)\right) \tag{19}$$

Step 4: Computing the distance to the ideal solutions

The Euclidean distance from each alternative A_i to PIS or NIS can be calculated, respectively, by the following equations:

$$D_i^+ = \sqrt{\sum_{j=1}^{n} w_j \left(Z_j^+ - z_{ij}\right)^2}, i = 1, 2, \ldots, m \tag{20}$$

$$D_i^- = \sqrt{\sum_{j=1}^{n} w_j \left(Z_j^- - z_{ij}\right)^2}, i = 1,2,\ldots,m \quad (21)$$

where D_i^+ is the distance from the alternative A_i to PIS; D_i^- signifies the distance from the alternative A_i to NIS.

Step 5: Calculating the relative proximity to PIS

The relative proximity (S_i) of an alternative is expressed as follows:

$$S_i = \frac{D_i^-}{D_i^+ + D_i^-} \quad (22)$$

where $0 \leq S_i \leq 1$.

Step 6: Ranking the alternatives

Rank alternatives in descending order according to their proximity value (S_i). The alternative with larger relative proximity is closer to PIS. At last, the alternative with the highest relative proximity value is considered the most suitable setting of ramp spacing on expressway.

4. Case Study

In this section, we illustrate the applicability of the entropy-weighted TOPSIS estimation method in evaluating and comparing different ramp spacing settings along the expressway. The case study is conducted based upon the Beijing–Hong Kong–Macao Expressway within Henan Province, China.

4.1. Study Area

Henan Province is a provincial administrative region located in central China, which owns one of the most salient transportation hubs. Its population reached 98.83 million by 2021. By the end of 2020, the mileage of expressways in Henan Province reached 7100 km, with 17 national expressway arteries, such as the Beijing–Hong Kong–Macao Expressway, and more than 50 regional expressways.

In this study, we select the Beijing–Hong Kong–Macao Expressway as the object of case study (see Figure 1) as it is an important logistics channel in Henan Province. The total length of the Beijing–Hong Kong–Macao Expressway within Henan Province is 513 km. The overall environment is set based on the alignment of the Beijing–Hong Kong–Macao Expressway (Minggang Toll Station-Lingying Toll Station section in Henan Province), and an expressway with a total length of 175 km is constructed in ArcGIS for the study. The expressway studied subsumes eight lanes in both directions with a lane width of 3.75 m.

Figure 1. Schematic of the study scenario for the ramp spacing of expressway.

According to the highway design specification [41] conforming to the current status quo of road transportation in China: the auxiliary lanes are needed to be designed for too short ramp spacing, and U-turn lanes should be provided for excessively long ramp spacing. To ensure safety and reduce additional construction costs, we take the minimum spacing (4 km) stipulated in the specification as minimal ramp spacing and set maximal ramp spacing to 24 km. On this basis, a total of 21 alternatives of ramp spacing settings are generated, in which the ramp spacing ranges from 4 km to 24 km.

To obtain the data of the evaluation indicators of each alternative, we determine the expressway traffic demand and substitute it into VISSIM for simulation. The traffic demand of the expressway is divided into traffic demand along the route (TDAR) and transit traffic demand (TTD).

The town is set as the study unit, and "location potential" is used to characterize the locational advantage of a town over the standard town. The location potential of the town is calculated based on the OD data of the expressway and the point of interest (POI) data of towns. The steps for estimating the traffic demand of the expressway are shown as follows:

(i) Estimate the traffic demand of towns: using location potential [36], disperse the cross-sectional traffic volume of the toll stations along the expressway to each town;

(ii) Determine the TDAR: on the basis of distance decay theory [42,43], allocate the estimated traffic demand of the towns along the expressway to each ramp reset according to the spacing alternatives;

(iii) Calculate the TTD: the TTD can be obtained by setting the ratio of the TTD.

The Detailed Expressions for Estimating Traffic Demand are Given in Appendix A.

4.2. Results and Discussions

Based upon the entropy-weighted TOPSIS estimation method, the ramp spacing alternatives are evaluated and the ranking results are sequentially obtained. The weights of each indicator calculated using Equations (9)–(14) are shown in Table 3.

Table 3. The weight of the evaluation indicators of the spacing alternatives.

Indicator (Criterion)	Average Speed	Average Delay	Accident Rate	Traffic Accessibility	Project Cost
Weight (w_j)	0.20211	0.25285	0.16918	0.13340	0.24246

In order to discretize the indicator values of the normalization decision matrix in TOPSIS, we replace the normalization decision matrix with the matrix obtained through Equations (9)–(11). Table 4 shows the normalization decision matrix Z.

Since the elements of the obtained decision matrix Z are all positive values, the positive ideal solution Z^+ consist of the maximum value of each column element in Z, and the negative ideal solution Z^- denotes the opposite. They are shown as follows:

$$Z^+ = \left(C_1^+, C_2^+, \ldots, C_n^+\right) \tag{23}$$

$$Z^- = \left(C_1^-, C_2^-, \ldots, C_n^-\right) \tag{24}$$

Thus, PIS and NIS are:

$$Z^+ = \{1.00, 1.00, 1.00, 1.00, 1.00\} Z^- = \{0.00, 0.00, 0.00, 0.00, 0.00\}$$

Based on the normalized Euclidean distance, the distance from PIS or NIS to each alternative is measured. Next, each spacing alternative's comprehensive score (C_i) is obtained by calculating the relative proximity. Finally, the alternatives are ranked according to the descending order of C_i. Among all the alternatives, A_{11} (ramp spacing was 14 km) have the comprehensive score closest to 1 and the best overall effectiveness. Therefore, the optimal alternative of ramp spacing settings for this study is A_{11}.

Table 4. Normalization decision matrix for spacing alternatives evaluation.

Spacing Alternative	Decision Matrix				
	Average Speed	Average Delay	Accident Rate	Traffic Accessibility	Project Cost
A_1	0.47	0.66	0.54	1.00	0.00
A_2	0.00	0.29	0.66	0.98	0.05
A_3	0.65	0.72	0.92	0.99	0.10
A_4	1.00	0.80	0.67	0.96	0.15
A_5	0.43	0.60	0.87	0.94	0.20
A_6	0.81	0.74	0.80	0.92	0.25
A_7	0.55	0.70	1.00	0.91	0.30
A_8	0.96	1.00	0.41	0.86	0.35
A_9	0.58	0.92	0.67	0.83	0.40
A_{10}	0.44	0.46	0.29	0.85	0.45
A_{11}	0.89	0.68	0.64	0.71	0.50
A_{12}	0.04	0.49	0.00	0.71	0.55
A_{13}	0.70	0.64	0.42	0.60	0.60
A_{14}	0.19	0.08	0.19	0.56	0.65
A_{15}	0.79	0.35	0.59	0.55	0.70
A_{16}	0.65	0.43	0.21	0.51	0.75
A_{17}	0.35	0.00	0.42	0.44	0.80
A_{18}	0.52	0.36	0.32	0.37	0.85
A_{19}	0.13	0.12	0.24	0.24	0.90
A_{20}	0.28	0.05	0.32	0.20	0.95
A_{21}	0.45	0.07	0.34	0.00	1.00

4.2.1. Comparison of Ramp Spacing Alternatives

Figure 2 depicts the evaluation indicator values of several representative alternatives as compared to the optimal setting A_{11}. The ramp spacing in A_2 and A_{19} is close to the left endpoint and right endpoint of the spacing value interval, respectively. A_2 has a great performance in terms of traffic accessibility. However, the setting of the ramps too close leads to more weaving points along the expressway, which is not conducive to vehicle speed and traffic safety. Meanwhile, A_2 is not economical due to the need of building more ramps. These findings are consistent with those of previous studies [9,24]. In contrast, the ramp spacing of A_{19} is relatively long. This alternative reduces the project cost and it is not detrimental to traffic efficiency, accessibility, and safety. It can be observed that adjacent ramps are so far apart that traffic entering or leaving the expressway accumulated on the same ramp, causing traffic congestion in merging and diverging areas. Therefore, the two ramp spacing alternatives A_2 and A_{19} have relatively low comprehensive scores.

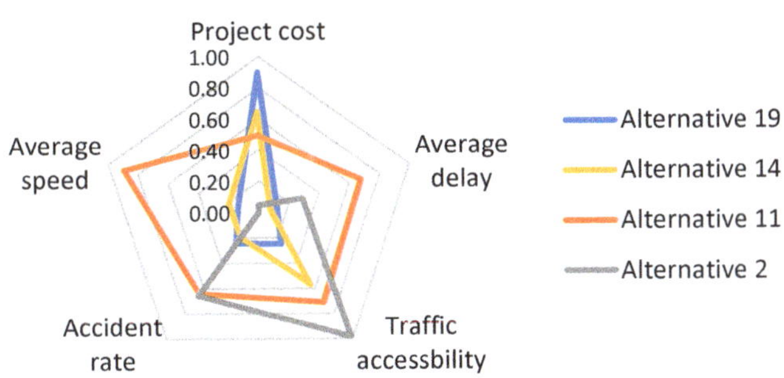

Figure 2. Comparison of evaluation indicators of several spacing alternatives.

Regarding alternative A_{14}, the ramp spacing value is 17 km, which is close to the ramp spacing of the optimal alternative A_{11}. However, the comprehensive score of the A_{14} (0.358) is apparently lower than that of the optimal alternative (0.663). As can be seen from Figure 2,

A_{14} is less effective in traffic efficiency and safety. The low comprehensive score of A_{14} may be due to the fact that the ramps it provides cannot effectively serve the demand-intensive area. As shown in Figure 3a,b, the density of demand units (towns) distributed along the two expressway sections are about 0.235 and 0.255, respectively, which are both greater than that along the whole studied section (0.177), where these two areas are both demand-intensive areas. In Figures 3a and 4a, the optimal setting A_{11} provides three ramps to serve the demand-intensive area I, while A_{14} sets only two ramps capable of carrying the traffic demand in the area. Additionally, as Figures 3b and 4b show, both A_{11} and A_{14} set two ramps to serve the demand-intensive area II. We further compare the distribution of ramps within the demand-intensive area II. On the one hand, ramps provided in A_{11} are located in the middle of this area and can effectively partake the traffic demand within the area. On the other hand, one of the ramps established in A_{14} is at the edge of the demand-intensive area II; according to the distance decay theory, more traffic demand is concentrated on the other ramp. In conclusion, the location of ramps in alternative A_{14} is not adapted to the practical circumstances of traffic demand distribution within the demand-intensive areas, resulting in potential traffic congestion in the merging and diverging areas; thus, ramp spacing alternative A_{14} yields relatively lower traffic efficiency and safety.

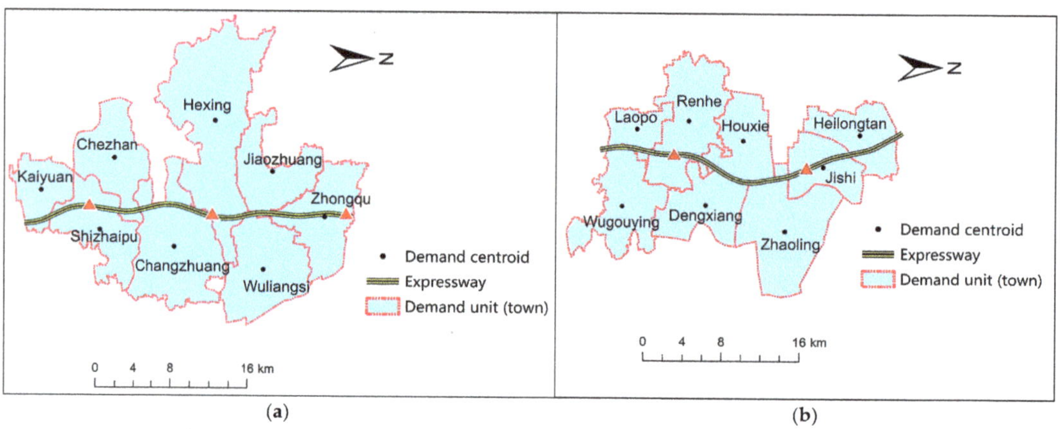

Figure 3. The distribution of the ramps set according to A_{11} within the demand-intensive area: (**a**) Demand-intensive area I; (**b**) Demand-intensive area II.

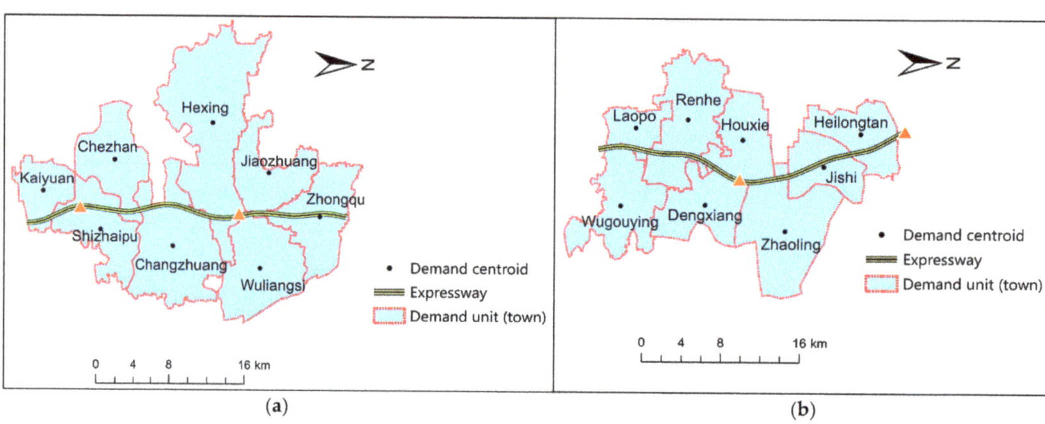

Figure 4. The distribution of the ramps set according to A_{14} within the demand-intensive area: (**a**) Demand-intensive area I; (**b**) Demand-intensive area II.

4.2.2. Sensitivity Analysis of TDARs

In this subsection, we conduct a sensitivity analysis on the effect of TDAR variation on the resulting optimal ramp spacing setting. First, assuming a constant value of TTD, the study scenarios with increased TDAR are simulated using VISSIM. Then, the simulation data of the spacing alternatives are evaluated using the method proposed in this research, and the variation of the optimal ramp spacing value with TDAR is analyzed. Figure 5 shows the heat map of the comprehensive score of spacing alternatives. The horizontal axis represents the average TDAR of the towns along the expressway; the vertical axis is the comprehensive score; the closer the color to blue in the figure indicates a lower score, and conversely, the closer the color to orange. As can be seen from Figure 5, the spacing values corresponding to the alternatives with high scores become shorter as TDAR increases. The reason may be that the traffic flow on the main line is mainly affected by the traffic flow in the weaving area when the ramp spacing is set larger than a specific value. Therefore, when the specific spacing condition is satisfied, adding the number of ramps to disperse the traffic flow entering or exiting the expressway is conductive to mitigating the disturbance of the more complex weaving behavior generated by increased TDAR and improving the overall performance of the expressway.

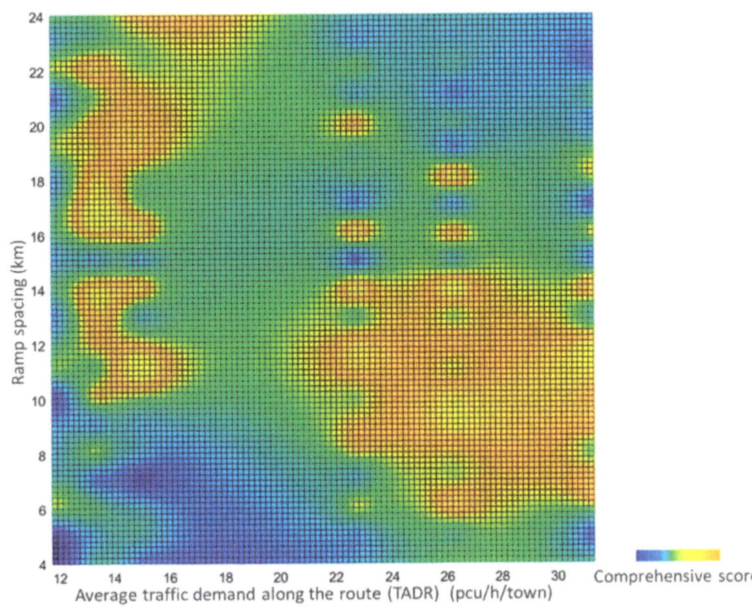

Figure 5. Evaluation of ramp spacing alternatives under different traffic demands along the expressway.

Overall, this study provides theoretical references and suggestions for the local authority to design appropriate ramp spacing along the expressway: (i) setting the ramp spacing of expressway should comprehensively consider the traffic efficiency, traffic accessibility, safety, and economy. It is not advisable to set the ramp spacing too short so as to improve accessibility, nor too long spacing in an effort to reduce construction costs, which can affect safety negatively. (ii) When setting the ramp spacing of the expressway from the regional level, the ramp spacing of the demand-intensive areas should be adjusted as per the actual demand distribution to improve the service quality of expressways. (iii) The practical implication of the proposed estimation method can indeed improve the practical operations of the expressway service via accommodating local considerations of ramp placement.

5. Conclusions

Expressways are the critical ingredients of the entire highway transportation system, which promotes economic growth and sustainable development. The spacing of ramps, which are the main channels connecting the general road network to the expressway, significantly affects the degree of function and overall efficiency of the expressway. Setting the ramp spacing that ensures the best overall benefits of the expressway is a complex issue that requires consideration of multiple influences.

This paper studies the problem of evaluating the ramp spacing of expressways in a specific district. The purpose is to propose a method that can comprehensively evaluate the ramp spacing of expressways from multiple aspects. The method presented in this study is the entropy-weighted TOPSIS estimation method, and the evaluation indicator system consists of traffic efficiency, safety, traffic accessibility, and economy. The settings of different ramp spacing are applied on the expressway through VISSIM.

The study case applying the entropy-weighted TOPSIS estimation method to the Beijing–Hong Kong–Macao Expressway in Henan Province aims to demonstrate the method's effectiveness and to explore potential principles for the rational setting of ramp spacing. Therefore, the evaluation results were analyzed after each spacing alternative was evaluated. The analysis was focused on the differences in spacing alternatives' performance: the influence of spacing values on each evaluation indicator, the significant differences in the comprehensive scores of the spacing values, and the requirements of TDAR for spacing value. The results of the analysis show: (1) spacing values that are too long or too short are detrimental to the overall benefit of the expressway; (2) spacing values that place the ramps at locations where they can effectively share TDAR are the foundation for making the best overall benefit of the expressway; (3) when TDAR increases, appropriately shortening the ramp spacing can keep the expressway operating well.

The method and setting principles proposed in this study aim to set a reasonable spacing that satisfies the regional traffic demand and improves the overall efficiency of the entire expressway system. This study overcomes the one-sidedness of the single-factor influence setting and the localization of adjacent ramps as the research object. The research results could provide theoretical references for the local authority to overall plan the ramp spacing and improve the practical operations of expressway. Since the transportation-related carbon emission and environmental costs are drawing considerable research attention [44], future work may consider the environmental costs of setting ramps.

Author Contributions: Conceptualization: D.C.; methodology: J.M. and Y.Z.; software: J.M.; investigation: D.C. and Y.Z.; writing—original draft preparation: J.M. and Y.Z.; visualization: Y.Z. All authors have read and agreed to the published version of the manuscript.

Funding: This research was funded by the Ministry of Education of the People's Republic of China Humanities and Social Sciences Youth Foundation, grant number 22YJCZH123, the Natural Science Foundation of Jiangsu Province, grant number BK20220846, the China Postdoctoral Science Foundation, grant numbers 2021M690614 and 2021T140112, and the Transportation Science and Technology Project of Henan Province, grant number 2022-2-2.

Data Availability Statement: Data sharing is not applicable to this article.

Acknowledgments: The authors would like to thank the students from the School of Transportation, Southeast University, for their help in collecting the data required for this study.

Conflicts of Interest: The authors declare no conflict of interest.

Appendix A

Table A1. List of notations.

Variable	Notation
i	index of towns
j	index of ramps
L_i	the comprehensive level of service of roads within the town i
l_{ij}	the distance from town i to ramp j
M_i	the comprehensive aggregation scale of town i
ε_i	the accessibility of town i when considering distance factor
β_{rj}	the weight of the ramp j attached to the toll station with location potential level r
λ_i	the accessibility of town i
λ_0	the accessibility of the standard town
Lp_i	the location potential of town i
Lp_0	the location potential of the standard town
ξ	the proportionality coefficient
χ	the elastic correction factor for the increase in location potential due to the traffic accessibility
φ	the elastic correction factor for the increase in location potential due to the comprehensive aggregation scale
$coef_i$	the location influence coefficient of town i
q_i	the traffic demand along the route allocated to town i
F_j	the cross-sectional flow of the toll station where ramp j is located
$z^t\left(l_{ij}\right)$	the cumulative probability of travel to ramp j from a town which is l_{ij} kilometers away from ramp j
ζ	parameter of the function related to distance decay theory
ψ	parameter of the function related to distance decay theory
V_t	the transit traffic demand of expressway
V_a	the traffic demand along the route of expressway
μ	the proportion of transit traffic demand
Sets	
T	the set of towns
R	the set of ramps set according to spacing alternative
R_i	the set of ramps that can serve the town i, $R_i \in R$
R_0	the set of the original ramps on the expressway
R_{0i}	the set of ramps that can serve the town i, $R_{0i} \in R_0$
T_j	the set of towns served by ramp j

(1) Estimate the traffic demand of the towns along the expressway using location potential

Estimating traffic demand using location potential should consider the accessibility of towns. The accessibility of towns is affected by the LOS of roads and the distance between towns and ramps. For the distance factor, it is necessary to consider the influence of the location potential of the toll station where the ramp is located. Therefore, we classify the level of the location potential of toll stations according to their cross-sectional traffic volume, and the accessibility of each town is calculated:

$$\varepsilon_i = \frac{\sum_{j \in R_{0i}} \beta_{rj} \frac{1}{l_{ij}}}{\sum_{j \in R_{0i}} \beta_{rj}} (i \in T), \quad \text{(A1)}$$

Then, combined with the description of the LOS of the road in Equation (4), the accessibility of town i can be estimated by:

$$\lambda_i = L_i \cdot \varepsilon_i (i \in T) \quad \text{(A2)}$$

Finally, the location potential model is described as follows:

$$\begin{cases} Lp_i = \xi \lambda_i^{\chi} M_i^{\varphi} \\ Lp_0 = \xi \lambda_0^{\chi} M_0^{\varphi} = \sum\limits_{i \in T} \overline{Lp_i} \\ coef_i = \frac{Lp_i}{Lp_0} \qquad (i \in T) \\ q_i = \sum\limits_{j \in R_{0i}} F_j \frac{coef_i}{\sum_{i \in T_j} coef_i} \end{cases} \tag{A3}$$

(2) Reset traffic demand along the route using distance decay theory

The traffic demand along the route is redistributed to each ramp based on distance decay theory. It can be expressed by:

$$z^t(l_{ij}) = \zeta \cdot l_{ij}^{-\psi} (i \in T, j \in R_i), \tag{A4}$$

where $z^t(l_{ij})$ is the cumulative probability of travel to ramp j from a town which is l_{ij} kilometers away from ramp j. Let $\zeta = 17.41$, $\psi = 1.022$.

(3) Determine the transit traffic demand by setting the proportion of transit traffic.

The relationship between the transit traffic demand and the traffic demand along the expressway is expressed as below:

$$V_t = \frac{\mu V_a}{1 - \mu} \tag{A5}$$

where the value of μ is taken as 15%.

References

1. National Academies of Sciences, Engineering, and Medicine. *An Expanded Functional Classification System for Highways and Streets;* The National Academies Press: Washington, DC, USA, 2018; p. 84.
2. Ma, J.; Li, D.; Tu, Q.; Du, M.; Jiang, J. Finding optimal reconstruction plans for separating trucks and passenger vehicles systems at urban intersections considering environmental impacts. *Sustain. Cities Soc.* **2021**, *70*, 102888. [CrossRef]
3. Ma, J.; Wu, X.; Jiang, J. Lane restriction system to reduce the environmental cost of urban roads. *Transp. Res. Part D: Transp. Environ.* **2023**, *115*, 103575. [CrossRef]
4. Liu, W.; Chen, X.; Hu, A. Study on Traffic Flow Parameters Model of Expressway Interchanges Opening Distance. *Appl. Mech. Mater.* **2012**, *178*, 2623–2628. [CrossRef]
5. Chen, S.-K.; Mao, B.-H.; Liu, S.; Sun, Q.-X.; Wei, W.; Zhan, L.-X. Computer-aided analysis and evaluation on ramp spacing along urban expressways. *Transp. Res. Part C Emerg. Technol.* **2013**, *36*, 381–393. [CrossRef]
6. Wang, R.; Hu, J.; Zhang, X. Analysis of the driver's behavior characteristics in low volume freeway interchange. *Math. Probl. Eng.* **2016**, *2016*, 2679516. [CrossRef]
7. Flintsch, A.M.; Guo, F.; Han, S.; Hancock, K.; Williams, B.; Li, Y.; Gibbons, R. *Impact of Access Spacing Standards on Crash Risk after Controlling for Access Volumes;* Virginia Transportation Research Council (VTRC): Charlottesville, VA, USA, 2020.
8. Pilko, P.; Bared, J.G.; Edara, P.K.; Kim, T. *Safety Assessment of Interchange Spacing on Urban Freeways: [Techbrief];* FHWA-HRT-07-031; United States. Federal Highway Administration. Office of Research, Development, and Technology: Washington, DC, USA, 2007.
9. Guo, Y.Q. Effects of Ramp Spacing on Freeway Mainline Crashes. *Appl. Mech. Mater.* **2011**, *97*, 95–99. [CrossRef]
10. Shea, M.S.; Le, T.Q.; Porter, R.J. Combined Crash Frequency–Crash Severity Evaluation of Geometric Design Decisions: Entrance–Exit Ramp Spacing and Auxiliary Lane Presence. *Transp. Res. Rec.* **2015**, *2521*, 54–63. [CrossRef]
11. Le, T.Q.; Porter, R.J. Safety evaluation of geometric design criteria for spacing of entrance–exit ramp sequence and use of auxiliary lanes. *Transp. Res. Rec.* **2012**, *2309*, 12–20. [CrossRef]
12. American Association of State Highway and Transportation Officials (AASHTO). *A Policy on Geometric Design of Highways and Streets, 2011;* AASHTO: Washington, DC, USA, 2011.
13. Ministry of Transport of the People's Republic of China. *Guidelines for Design of Highway Grade-Separated Intersections (JTG/T D21-2014);* People's Communications Publishing House: Beijing, China, 2014.
14. Board, T.R. *HCM2010: Highway Capacity Manual*, 5th ed.; Transportation Research Board: Washington, DC, USA, 2010.
15. Chen, H.; Lu, L.; Lu, J.; Zhu, S.; Wei, D. Development and applications of models of average speed in the combination area of urban expressway ramps. *Adv. Transp. Stud.* **2015**, *35*, 19–42.
16. Wang, F.; Li, Y.; Liu, J.; Li, Y.; Xia, L. Identification of extreme ramp spacing on Expressway Based on GPS Speed Data Driven Model. *J. Highw. Transp. Dev.* **2016**, *33*, 127–135.
17. Jang, M.-S. Methodology for evaluating freeway interchange spacing for high design speed based on traffic safety: Focused on analysis of acceleration noise using microscopic traffic simulations. *J. Korean Soc. Transp.* **2009**, *27*, 145–153.

18. Kim, H.R.; Kim, K.S.; Lee, G.H.; Shin, J.S.; Baek, J.G. Determining the Required Minimum Spacing between Freeway Interchange for High-speed Roadway. *Int. J. Highw. Eng.* **2017**, *19*, 45–55. [CrossRef]
19. Torbic, D.J.; Harwood, D.W.; Gilmore, D.K.; Richard, K.R.; Bared, J.G. Safety analysis of interchanges. *Transp. Res. Rec.* **2009**, *2092*, 39–47. [CrossRef]
20. Liu, Z.-P.; Zhou, X.-W. Research on Project Cost Calculation of Expressway. *Constr. Econ.* **2021**, *42*, 70–74. [CrossRef]
21. Molan, A.M.; Hummer, J.E.; Ksaibati, K. Introducing the super DDI as a promising alternative service interchange. *Transp. Res. Rec.* **2019**, *2673*, 586–597. [CrossRef]
22. Qiao, J.; Song, M. Study on Interchange Overpass Distance Based on Maximum Entropy Principle. In Proceedings of the ICTIS, Wuhan, China, 22–24 October 2011; pp. 481–487.
23. Winkler, M.; Fan, W. Evaluating impacts on freeway capacity using VISSIM: Accounting for truck lane restrictions, driver behavior, and interchange density. *Adv. Transp. Stud.* **2011**, *25*, 15–28.
24. Chen, D.; Mo, F.; Chen, Y.; Zhang, J.; You, X. Optimization of ramp locations along freeways: A dynamic programming approach. *Sustainability* **2022**, *14*, 9718. [CrossRef]
25. Broniewicz, E.; Ogrodnik, K. A comparative evaluation of multi-criteria analysis methods for sustainable transport. *Energies* **2021**, *14*, 5100. [CrossRef]
26. Roszkowska, E. Multi-criteria decision making models by applying the TOPSIS method to crisp and interval data. *Mult. Criteria Decis. Mak. /Univ. Econ. Katow.* **2011**, *6*, 200–230.
27. Zhang, M.; Sun, Q.; Yang, X. Research on the assessment of the capacity of urban distribution networks to accept electric vehicles based on the improved TOPSIS method. *IET Gener. Transm. Distrib.* **2021**, *15*, 2804–2818. [CrossRef]
28. Aljohani, K.; Thompson, R.G. A multi-criteria spatial evaluation framework to optimise the siting of freight consolidation facilities in inner-city areas. *Transp. Res. Part A: Policy Pract.* **2020**, *138*, 51–69. [CrossRef]
29. Hamurcu, M.; Eren, T. Strategic planning based on sustainability for urban transportation: An application to decision-making. *Sustainability* **2020**, *12*, 3589. [CrossRef]
30. Shishegaran, A.; Shishegaran, A.; Mazzulla, G.; Forciniti, C. A novel approach for a sustainability evaluation of developing system interchange: The case study of the Sheikhfazolah-Yadegar interchange, Tehran, Iran. *Int. J. Environ. Res. Public Health* **2020**, *17*, 435. [CrossRef] [PubMed]
31. Vavrek, R.; Bečica, J. Capital city as a factor of multi-criteria decision analysis—Application on transport companies in the Czech Republic. *Mathematics* **2020**, *8*, 1765. [CrossRef]
32. Li, Q.-l.; Wang, H.; Wu, Y.-l.; Wang, H.-t.; Zhu, S.-y. Expressway Interchange Minimum Clear Distance Model Based on Nonfree Lane Changing Behavior. *J. Highw. Transp. Dev.* **2022**, *39*, 165–173.
33. Subiantoro, W.; Mudiyono, R. Evaluation of minimum ramp distance in efforts to improve performance on Jakarta-Cikampek toll road. In Proceedings of the IOP Conference Series: Earth and Environmental Science, Semarang, Indonesia, 21–22 September 2021; p. 012015.
34. Pei, Y.-L.; Cheng, G.-Z. Research on the relationship between discrete character of speed and traffic accident and speed management of freeway. *Zhongguo Gonglu Xuebao* **2004**, *17*, 74.
35. Institute, B.G.M.E.D.R. *Code for Design of Urban Road Engineering (CJJ 37-2012)*; China Architecture & Building Press: Beijing, China, 2012.
36. Shuang, W.; Hao, J.; Ai, D.; Huang, X.; Zhang, L.; Meng, P.; Zhu, C. Zoning and Pattern of Rural Residential Land Consolidation Based on Locational Potential Theory. *Trans. Chin. Soc. Agric. Eng.* **2013**, *29*, 251–261+297.
37. Chen, C.-H. A novel multi-criteria decision-making model for building material supplier selection based on entropy-AHP weighted TOPSIS. *Entropy* **2020**, *22*, 259. [CrossRef]
38. Dong, X.; Lu, H.; Xia, Y.; Xiong, Z. Decision-making model under risk assessment based on entropy. *Entropy* **2016**, *18*, 404. [CrossRef]
39. Zhao, H.; Yao, L.; Mei, G.; Liu, T.; Ning, Y. A fuzzy comprehensive evaluation method based on AHP and entropy for a landslide susceptibility map. *Entropy* **2017**, *19*, 396. [CrossRef]
40. Tzeng, G.-H.; Huang, J.-J. *Multiple Attribute Decision Making: Methods and Applications*; CRC Press: Boca Raton, FL, USA, 2011.
41. Consultants, C.F.H. *Design Specification for Highway Alignment (JTG D20-2017)*; People's Communications Publishing House: Beijing, China, 2017.
42. Chen, Z.; Jin, F.; Yang, Y.; Wang, W. Distance-decay Pattern and Spatial Differentiation of Expressway Flow: An Empirical Study Using Data of Expressway Toll Station in Fujian Province. *Prog. Geogr.* **2018**, *37*, 1086–1095.
43. Gao, S.; Wang, Y.; Gao, Y.; Liu, Y. Understanding urban traffic-flow characteristics: A rethinking of betweenness centrality. *Environ. Plan. B Plan. Des.* **2013**, *40*, 135–153. [CrossRef]
44. Ma, J.; Xu, M.; Jiang, J. Mapping high-resolution urban road carbon and pollutant emissions using travel demand data. *Energy* **2023**, *263*, 126059. [CrossRef]

systems

MDPI

Article

High-Speed Rails and City Innovation System: Empirical Evidence from China

Jiafeng Gu

Institute of Social Science Survey, Peking University, Beijing 100871, China; isssgujf@pku.edu.cn

Abstract: The rapid development of high-speed rail has markedly shortened the travel time from one city to another. However, the impact of space–time compression brought about by high-speed rail on city innovation has not received sufficient attention. This paper examines the space–time compression phenomenon produced by high-speed railway networks and its impact on city innovation from 2000 to 2019 using a sample of 279 Chinese prefecture-level cities. The empirical results show that there was a strong space–time compression during this period. The development of high-speed rail can promote city innovation. However, the construction of high-speed rail also produces a siphon effect, which accelerates the convergence of innovative elements in cities with stronger innovation capabilities. Nevertheless, it has a negative spillover effect on cities with weaker innovation capabilities. Finally, policy recommendations for promoting the balanced development of city innovation and recommendations for future research are presented.

Keywords: high-speed rail; space–time compression; city innovation; patent; spatial difference-in-differences method

Citation: Gu, J. High-Speed Rails and City Innovation System: Empirical Evidence from China. *Systems* **2023**, *11*, 24. https://doi.org/10.3390/systems11010024

Academic Editors: Mahyar Amirgholy and Jidong J. Yang

Received: 23 November 2022
Revised: 26 December 2022
Accepted: 1 January 2023
Published: 4 January 2023

1. Introduction

The development of cities has become a major engine of regional and national development [1,2]. At present, 54% of the global population lives in cities, and the global urbanization process is accelerating [3]. According to a United Nations report, the proportion of the global urban population is expected to hit a record of 68% by 2050 [4]. Referred to as the soul of a city and its competitiveness, the capability for innovation is the source of a city's value creation process and the key to its comprehensive competitiveness [5–7]. With the rapid advancement of economic globalization and convenient transportation, as well as rapid advances in science and technology, time and space have been compressed and the world has shrunk into a "global village" [8,9]. In this context of space–time compression, cities have become increasingly important in the global system, acting as basic units of direct participation in international economic activities [10,11]. However, because innovation activities in cities are a complex systemic project, they involve time, space, and society. Therefore, studying the city innovation system in the context of space–time compression is of great significance for cultivating the innovation capability and competitiveness of cities.

According to the existing literature, researchers have reached a preliminary consensus on the basic components of the city innovation system [12–14]. However, some key factors, such as spatio-temporal context, transport infrastructure, and space–time compression, have not received much attention. The impact of space–time compression caused by transportation development on city innovation is still in a "black box" state. Most of the related studies on space–time compression are qualitative analyses [15–17], and they rarely measure space–time compression from a quantitative perspective. Quantitative research on space–time compression and its impact on city innovation is still in its infancy. With the large-scale development of high-speed rail (HSR), the increasing impact of HSR on regional economies has received widespread attention [18–20]. The introduction and rapid development of HSR overcomes geographical and spatial barriers and effectively improves city accessibility, resulting in space–time compression [21,22]. Most studies on the impact of HSR on innovation have been analyzed at the national, provincial, and company levels [23–25], lacking a level of analysis for cities. The impact of the introduction of HSR can be assessed at different levels (national, provincial, city, or company), but the focus of the assessment is often different at different levels. In fact, due to the heterogeneity of cities, the impact of HSR on city innovation can vary, a fact that has not received attention in the existing studies. To control for heterogeneity between cities, the term "cities" in this study refers to prefecture-level cities in China.

In this study, panel data for 279 Chinese prefecture-level cities from 2000 to 2019 were used to investigate the increasing impact of HSR construction on urban innovation from a space–time compression perspective. The upgrading of transportation infrastructure is an important driver of economic growth; specifically, the advent of HSR has far-reaching effects on traditional transportation patterns, economic development, and technological innovation, so it is of great practical significance to study the impact of high-speed rail on urban innovation. In addition, HSR, similarly to the Internet and airline facilities, has changed the spatiotemporal relationship between cities and regions, as well as people's travel concepts and lifestyles, bringing new development opportunities for urban innovation. With regard to the current uneven development of scientific research strength between cities and regions, how does the space–time compression brought about by HSR change the space–time pattern of innovation? This is a common concern among various countries.

The academic contributions of this paper are as follows. First, this paper plays an important role in enriching and expanding the existing literature on urban innovation systems. In the study of city innovation systems (CISs), the existing literature has focused on firms [26], industries [27], social innovation [28], dynamic capabilities [29], and networks [30]. However, these studies have ignored the intrinsic interconnectedness of innovation systems among neighboring cities. In this paper, we consider the innovation system among neighboring cities as a whole, i.e., a regional innovation system formed from the merging of intra-city innovation systems and cross-city innovation systems, and an important driving factor of this merger is the construction of HSR. Second, this study contributes to the theory of space–time compression by empirically measuring space–time compression and its impact on urban innovation. Although studies from Harvey [15] and Janelle [31] have elaborated upon the concept of space–time compression, measurements and empirical tests of space–time compression are lacking. The present study fills this knowledge gap. Finally, this paper enriches the study of the consequences of the introduction of HSR. The existing literature has examined the impact of HSR on regional economic development [32] and internal migration [33], but less attention has been paid to urban innovation systems. This paper complements the research on the economic consequences of HSR by linking the introduction of HSR to urban innovation based on an integrated local and neighborhood hierarchy from a holistic system perspective, which is an emerging field that has so far received relatively little attention from scholars.

The paper is structured as follows. The development of China's HSR and the concept of space–time compression are introduced in Section 2. The theoretical background, literature review, and development of the hypothesis are discussed in Section 3. Information about the data, variables, and methods is reported in Section 4. The empirical results are reported and some robustness checks are provided and discussed in Section 5. The final section concludes the paper.

2. High-Speed Rail in China and Space–Time Compression

HSR is a milestone in the integration of transportation technology and railway modernization. In 2003, the Qinhuangdao–Shenyang passenger-dedicated line was opened. It was China's first HSR in the true sense, marking the beginning of China's HSR era. The rapid development and construction of HSR in China has brought great convenience to people's travel. By the end of 2021, the total operating mileage of HSR in China reached 40,000 km, accounting for about 70% of the world's HSR network [34]. According to statistics from the

National Railway Administration of China, in 2019, electric multiple units (EMUs) carried 2.29 billion passengers, an increase of 14.1% compared to 2018, accounting for 64.1% of railway passenger traffic. In addition, the introduction of HSR has also greatly reduced the space–time distance, improved transport accessibility and economic connections, and accelerated the cross-regional flow of economic factors [35].

Cities with convenient transportation connections tend to be in close proximity to each other, and "space–time compression" occurs in geographic space along the direction of transportation connections. Empirical evidence from China and other countries shows that the development of HSR has facilitated the process of space–time compression [36]. In mainland China, the development of HSR has brought cities closer together, which, in turn, has contributed to space–time compression [18,19,37]. For example, in Taiwan, China, the construction and development of HSR has led to space–time compression with varying uniformity in the geographical areas along its route [38]. The introduction of HSR has reduced the space–time distances between cities in continental Europe [39]. European HSR has compressed time and space by compressing temporal distances and changing relative positions [40]. Similar empirical evidence has also been obtained from Japan [23]. However, there is still a lack of relevant research on the impact of this space–time compression caused by HSR on city innovation.

3. Theoretical Background and Hypothesis Development

With the development of modern transportation and communication technology, people generally have access to a completely different social space and time experience than before [17]. Scholars in different fields have begun to pay more and more attention to exploring the phenomenon of space–time compression caused by the development of transportation and communication technology [31,41]. The advancement of modern technology has led to a "global village" [15]. The theory of space–time compression refers to the shortening of the human travel time between two places due to the progress and development of transportation [42]. Space–time compression has led to the emergence of fluid space [16,43].

To assess the impact of space–time compression, it is necessary to delve into the measurement of the degree of space–time compression. Harvey [15] concept of space–time compression is only used as a framework for analyzing social, cultural, and institutional changes and lacks a measurable approach [44]. The approach based on the average rate of time–space convergence, proposed by Janelle [31], also has limitations. For example, the average rate of time–space convergence has the disadvantage of being unstable and vulnerable to distance. The use of an "isochrone map" is another method [45]. However, it does not fully show the overall pattern of regional multi-node space–time compression. Some scholars, such as Spiekermann and Wegener [39] and Vickerman, et al. [46], have tried to modify and improve the above methods from the perspective of traffic accessibility by plotting time–space maps but have failed to effectively overcome their limitations. A common weakness of these measures is their treatment of travel time as static. Consequently, these measures fail to capture the potential dynamics of social time and social space, which is the core of social-spatial theory [47,48]. Therefore, a breakthrough in the methods of measuring space–time compression is needed to advance the development of spatiotemporal social science.

The theory of "space–time compression" emphasizes the important influence of fast-developing transportation technologies and tools on people's communication in this modern and fast-paced society. Spatial distance is an important factor that affects interpersonal relationships between two places [17]. As Tobler's First Law of Geography says, everything is related to everything else, and near things are more related than distant things [49]. The construction of HSR has greatly optimized the original transportation network, shortened the space–time distance, and positively influenced regional economic growth [20,22]. The flow of innovation elements is limited by spatial distance [50]. With the compression of spatial distances, the effectiveness of tacit knowledge dissemination and face-to-face communication gradually increases and the frequency of information exchange increases as well [51]. Innovative entities can acquire more tacit knowledge in face-to-face communication, resulting in better performance in cooperative research and development [52,53] and purchases of technology [54]. These activities have been shown to be significantly and positively correlated with patent citations [55].

Hypothesis 1 (H1). *The introduction of HSR promotes city innovation.*

The introduction of HSR leads to time and space compression between cities along the route, which intensifies the innovation competition among these cities. The rapid development of HSR intensifies competition between cities and regions, thereby changing the spatial patterns of cities [37]. The introduction of HSR can facilitate the flow of innovation-related information and innovative talent between cities along the route [19,56]. Qin [57] found that HSR helps to boost economic activities between large cities due to significantly shorter travel times, and it may actually hurt smaller counties along accelerated railway lines. The introduction of HSR exacerbates the imbalance of urban centrality, forming a hierarchical spatial structure with big cities as innovation hubs, but it also impedes local economic growth and innovation in peripheral areas [18]. HSR connections and networks significantly increase the cost of debt by 2.2% for firms that are located in those non-node cities [58]. Innovation and economic growth are inevitably boosted in those regional centers but suppressed in those adjacent and peripheral regions due to the agglomeration effect of transport infrastructure [59–61]. With the introduction of HSR, innovation elements, such as talent, information, and funding, are more likely to flow to cities with stronger innovation capabilities, which will have a negative impact on the innovation capabilities of cities with relatively weaker innovation capabilities. Therefore, the introduction of HSR will intensify inter-city competition and lead to negative spillover effects in terms of innovation.

Hypothesis 2 (H2). *There is a negative spillover effect of inter-city innovation.*

The analytical framework for understanding the impact mechanism of HSR on city innovation is illustrated in Figure 1. The test of the so-called spatial spillover effect proposed in the abovementioned Hypothesis 2 was realized empirically by means of a spatial lag term [62–64]. This analysis framework was constructed using the spatial difference-in-differences method, and its purpose was to handle three different treatment effects that existed at the same time in the process of natural experiments: the treatment effect due to the introduction of HSR, the spillover effect within the treated cities, and the spillover effect on the control cities. In other words, in terms of innovation, there is a spatial dependence between cities along the HSR route [21]. These cities interact strategically with each other, rather than independently of each other [62,63]. In evaluating the impact and effect of HSR on urban innovation, it is necessary to include and examine the potential impact of inter-city spatial dependence.

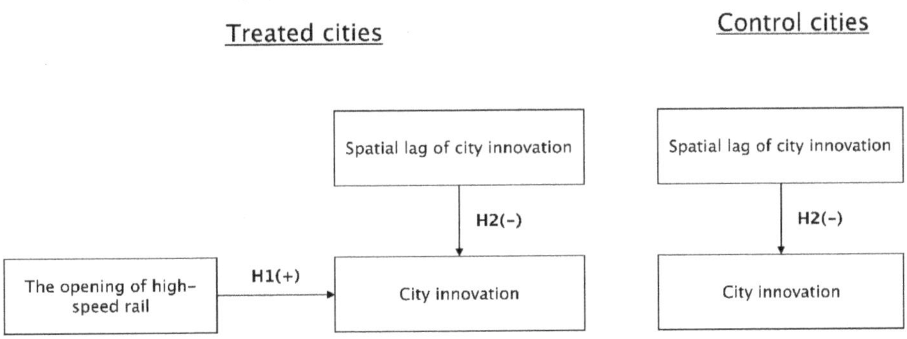

Figure 1. Conceptual framework for the effect of HSR on city innovation.

4. Data, Variables, and Empirical Strategies

4.1. Data

The research object of this study was China's prefecture-level cities, with a total sample consisting of 279 prefecture-level cities. The study period spanned two decades from 2000 to 2019. Therefore, the total number of observations was 5580. Data were obtained from the statistical yearbooks of Chinese cities over these years. The key data in this paper were the data concerning the introduction of HSR, which was manually compiled from the official website of the Chinese National Railway Administration. Panel data for 279 prefecture-level cities across the country from 2000 to 2019 were used in this study as balanced data.

4.2. Variables and Measurement

4.2.1. Dependent Variables

Generally, inventions refer to new technical solutions, breakthrough ideas, or new processes proposed for products, methods, or improvements [65]. Inventions, together with utility models and designs, constitute the object of protection under China's patent law [66]. Due to their high technological value, invention patents can be used more objectively to measure a city's innovation capability and to evaluate the technological competitiveness of a city more accurately [67]. Therefore, in this study we used the number of invention patents granted (NIP) as the dependent variable.

4.2.2. Independent Variables

The location of the HSR network was determined based on information from the National Development and Reform Commission and the railway company, according to a comprehensive plan. The local government has little influence on its location. Therefore, the introduction of HSR can be used as a treatment in a quasi-natural experiment. To capture the impact of the introduction of HSR on the changes in city innovation output, cities that introduced HSR between 2000 and 2019 were selected as the treatment group in this paper, and cities that did not introduce HSR during the same period were selected as the control group. The treatment group included a total of 190 cities, all of which introduced HSR before 2020. The remaining 89 cities belong to the control group. In this study, in order to test the impact of the introduction of HSR on the city's innovation capability, we introduced an independent variable, DID. Here, DID is a combined and newly created dummy variable, and it is the product of the dummy variable of the group and the dummy variable of the introduction of HSR is calculated as follows: $DID = \left(DID^{(1)} - \overline{DID}^{(1)}\right) \times \left(DID^{(2)} - \overline{DID}^{(2)}\right)$, where $DID^{(1)}$ is a dummy variable that is used to distinguish the experimental group from the control group: $DID^{(1)} = 1$ if they are cities in the treatment group (cities where HSR was introduced), and $DID^{(1)} = 0$ if they are not cities in the treatment group; $DID^{(2)}$ is a dummy variable used to reflect the local patent policy: $DID^{(2)} = 1$ if they are cities in the year when HSR was opened and every year thereafter, and $DID^{(2)} = 0$ in the previous year; $\overline{DID}^{(1)}$ and $\overline{DID}^{(2)}$ are the mean values of those two dummy variables.

In addition to the abovementioned core independent variable, several other independent variables were included in this study. The number of employees in each city in the comprehensive scientific and research technical services (NES) industry was used to measure the development of urban technical services, expressed per 10,000 people. Scientific expenditure (SE) was used to measure the city's investment in science and technology, expressed in CNY 100 million. The number of college students per 10,000 people (NCS) was used to measure the city's stock of human resources. The number of ordinary colleges and universities (NCU) was used to measure the degree of development of higher education in the city. The number of R&D personnel (NRD) was used to measure the city's R&D human resources, expressed in number of people. The export value of goods (EVG) was used to measure the city's foreign trade export capacity, expressed in CNY 100 million. The number of foreign-invested enterprises (NFI) was used to measure the city's ability to attract foreign investment. The total industrial output value of domestic-funded enterprises (TVD) was used to measure the city's own industrial strength, expressed in CNY 100 million. Per capita GDP (PGDP) was used as a measure of a city's overall economic strength, expressed in CNY. The sample size and descriptive statistics of these variables are reported in Table 1 below.

Table 1. Sample size and descriptive statistics of variables.

Variable	Obs	Mean	S.D.	Min	Max
NIP	5580	940.118	2438.795	1.000	33,202.000
DID	5580	0.090	0.147	−0.090	0.229
NES	5580	0.426	0.863	0.010	7.360
SE	5580	4.285	17.186	0.000	554.982
NCS	5580	182.200	217.131	1.900	1293.700
NCU	5580	6.483	11.114	1.000	84.000
NRD	5580	16,758.450	29,455.000	14.000	281,369.000
EVG	5580	494.182	1558.180	0.003	16,708.950
NFI	5580	80.720	231.414	0.000	3818.000
TVD	5580	1522.297	2114.499	0.601	16,046.310
PGDP	5580	32,501.820	30,261.460	99.000	467,749.000

4.3. Methods

4.3.1. Standard Deviation Ellipse Method

D. Welty Lefever's famous standard deviation ellipse (SDE) method was used to identify the spatial center of gravity of each city's innovation [68]. This method has been widely used in innovation research [69,70]. This method analyzes the spatial distribution of the target variable by calculating the standard deviation ellipse of the target variable and its center of gravity [71,72].

4.3.2. Difference-in-Differences Method

The so-called difference-in-differences model is widely used in quasi-experimental studies [73,74]. However, the traditional difference-in-differences model involves the premise that the observed samples are independent of each other [75]. In this study, we used the difference-in-differences model. Since the introduction of HSR occurred in the form of sequential events, in the experimental group we could distinguish between pre- and post-events, whereas in the control group we could not. In addition, in this study we assumed that the studied cities were not independent of each other but performed strategic interactions that led to the spillover effects of innovation [76]. The spatial difference-in-differences (SDiD) model has been proven in previous studies to deal effectively with the abovementioned problems [77,78]. Therefore, in this study, we drew on the work of Heckert [75] and Gu [79] and modified the difference-in-differences equation as follows:

The model can be expressed as follows:

$$NIP_{it} = C + \rho W \times NIP_{it} + \gamma DID_{it} + \beta_1 NEC_{it} + \beta_2 SE_{it} + \beta_3 NCS_{it} + \beta_4 NCU_{it} + \beta_5 NRD_{it} + \beta_6 EVG_{it} + \beta_7 NFI_{it} + \beta_8 TVD_{it} + \beta_9 PGDP_{it} + \varepsilon_{it}, \varepsilon_{it} \sim N(0, \sigma^2 I_n) i = 1, \ldots, 31 \quad (1)$$

where W is the constructed spatial weight matrix. Traditionally, the spatial weight matrix is defined by t and the inverse distance is constructed as follows: $w_{ij} = 1/d_{ij}$, where d_{ij} is the spatial distance from observation point i to observation point j. What is used here is not the geographical distance between cities but the innovation distance between cities. In calculating the innovation distance, the difference in the number of patents granted between cities is used. The calculation formula is as follows: $d_{ij} = |d_j - d_i|$. Here, d_i and d_j represent the average annual number of patents granted from 2010 to 2019 for city i and city j, respectively.

In this study, our model was focused on the regression coefficient γ and the regression coefficient ρ. The former reflects the changes in the innovation ability of cities before and after the introduction of HSR in cities with HSR and compares the differences in the innovation capabilities of cities without HSR. The latter reflects the spatial spillover effect of inter-city innovation. According to Hypothesis 1, γ in Formula (1) should be significant and positive. According to spatial econometrics, ρ is usually used to identify and test the spatial spillover effect [80,81]. If ρ is statistically significant and negative, a negative spatial spillover effect exists [69,70]. According to Hypothesis 2, ρ in Formula (1) should be significant and negative.

5. Results

5.1. Space–Time Compression Process Analysis

Space–time compression is a very abstract concept. Here, the SDE method was used to intuitively reflect the process of space–time compression. Figure 2 shows the ellipses created by SDE and the centers of gravity, calculated based on the number of invention patents of 279 prefecture-level cities in 2010 and 2019. As shown in Figure 2 below, from 2010 to 2019, the spatial center of gravity of city innovation moved to the southwest. The ellipse also showed a similar trajectory of movement. It can be seen that the introduction of HSR changed the spatial pattern of urban innovation, causing the center of gravity to move towards the southwest.

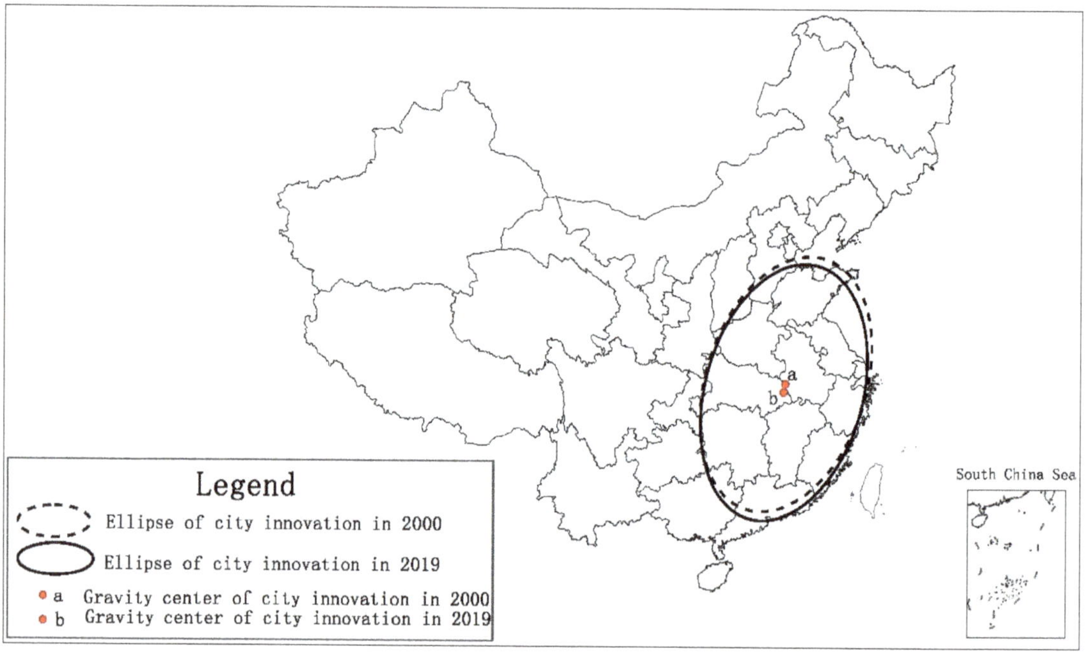

Figure 2. Center of gravity and ellipse of city innovation in 2010 and 2019.

However, according to Figure 2, the process of space–time compression cannot be tested directly. The traditional measurement of space–time compression is also static and partial [31,40,46]. It is a challenging task to comprehensively measure the compression process of the social space–time pattern. To test the social space–time compression process, the following steps are required: first, calculate the center of gravity of the city's innovation for each year; second, calculate the geographical distance from the spatial center of gravity to each city in each year; third, calculate the average distance from the spatial center of gravity to each city in each year. Finally, based on the variation in this average distance over time in a year, the presence of space–time compression is judged and tested. This principle is similar to compressing a plastic cup filled with water by hand. The greater the compression, the more water overflows from the plastic cup. Therefore, the degree of compression of a plastic cup can be measured by measuring the amount of overflowing water, which is the reverse side of plastic cup compression. This is the method of reverse measurement and testing. This analogy of the plastic cup is actually consistent with Newton and Kant's argument that space is a container [82]. Following the same principle here, we can measure the degree of space–time compression by measuring the change in the average distance from the spatial center of gravity of city innovation to each city. This average distance actually measures the spillover of knowledge from the core to the periphery of city innovation, which is the reverse side of the physical space–time compression due to the availability of convenient transportation. This average distance becomes increasingly larger over time, indicating increasingly stronger space–time compression. The average distance from the spatial center of city innovation to each city is summarized in Figure 3.

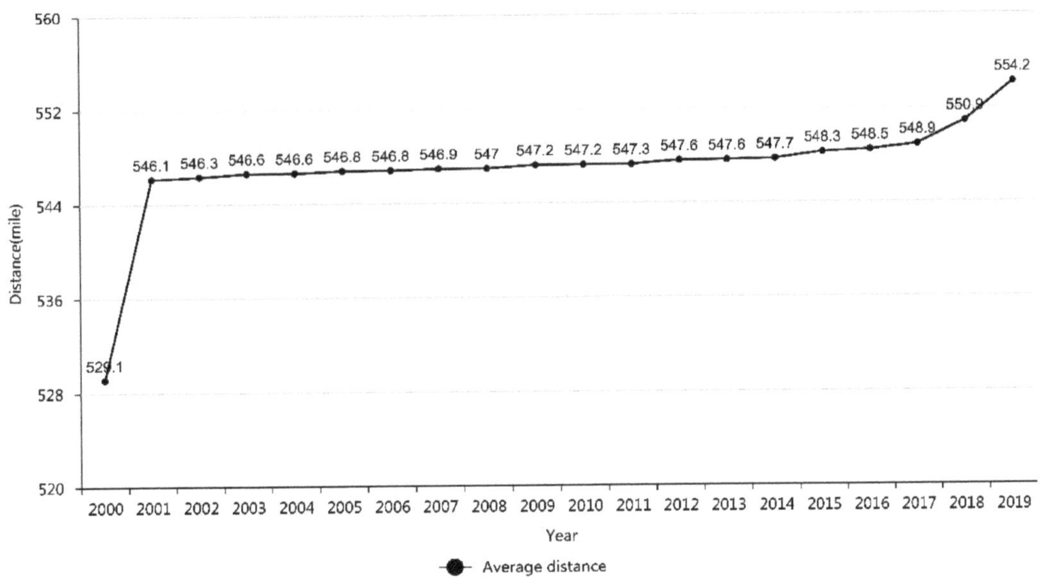

Figure 3. Average distance from the spatial center to each city in different years.

As shown in Figure 3, the average distance showed a continuous upward trend since 2000. This indicates that with the successive introduction of HSR and the increase in the mileage of HSR, the compressibility of time and space was strengthened. After 2001, the average distance showed a steady and continuous upward trend, whereas, after 2016, the growth trend of the average distance was markedly accelerated. This shows that after 2016, there was a certain accumulation of HSR mileage and a leap from quantitative to qualitative change, which led to the acceleration and intensification of space–time compression. The impact of space–time compression on regional urban innovation systems is also demonstrated graphically in Figure 4.

In Figure 4, the small circles represent the innovation elements of the city and the ellipses represent the innovation system within the city. The red-dashed ellipse represents a larger picture of the overall innovation system, including the innovation system of the city under study and the neighboring cities. With the advancement of space–time compression brought about by the construction of HSR, the innovation elements of the city are rapidly fissioned and continue to spill outward, thus influencing the innovation of other neighboring cities. Space–time compression leads to closer spatial and temporal distances between neighboring cities, reducing the cost of collaborative innovation between cities and creating a clustering effect of innovative cities. It is evident that space–time compression affects not only intra-city innovation systems but also cross-city innovation systems. Therefore, it is necessary to have a larger systematic view of urban innovation systems that can include both local and nearby urban innovation systems. This is a holistic, big-picture systematic view.

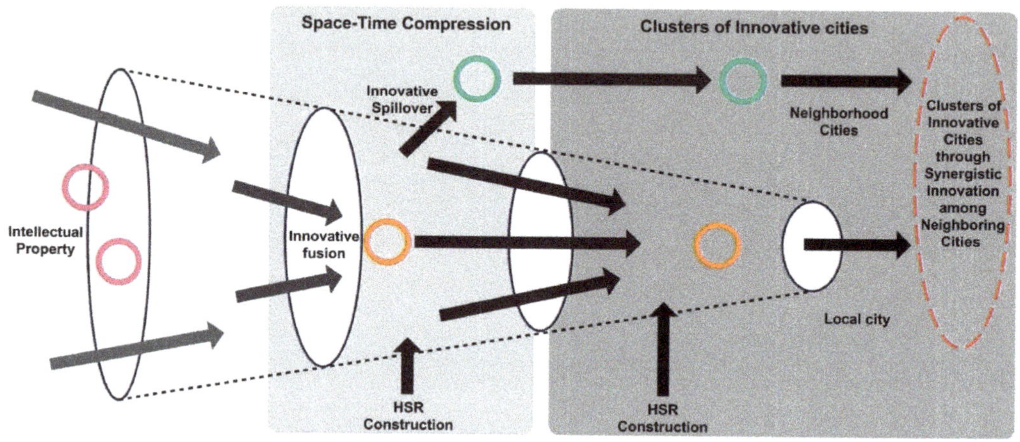

Figure 4. Mechanisms of space–time compression's effect on urban innovation.

Furthermore, it must be noted that this is still a dynamic system. The space–time compression brought about by the introduction and construction of high-speed rail leads to a fusion of innovation systems within cities, allowing urban innovation to focus more on technological areas related to core competencies. At the same time, space–time compression also leads to the fission of urban innovation systems, and some innovation elements spill over to neighboring cities, leading to the formation of regional innovation synergy. As demonstrated in Figure 4, this dynamic mechanism is an iterative process, leading to the continuous fusion and fission of urban innovation systems and eventually to the organic integration of intra-city and cross-city innovation systems and the emergence of innovative urban agglomerations.

5.2. Empirical Results

Figure 3 intuitively illustrates the space–time compression process that occurs in city innovation. However, the question of whether this space–time compression is driven by the introduction and development of HSR requires further research and testing. Here, this was tested empirically by means of the SDiD method. The empirical results are reported in Table 2. Model 1 in Table 2 is the model with fixed effects, whereas Models 2, 3, and 4 in Table 2 are models with random effects. In Model 2, only the year was fixed. In Model 3, the year and city were fixed. In Model 4, the year, city, and province were fixed. The Hausman test was insignificant, indicating that the model with random effects here was much better than the model with fixed effects. The empirical results are reported and summarized in Table 2.

As shown in Table 2, the coefficients of the key variable (DID) were significant in all four models. Thus, Hypothesis 1 was confirmed. The number of patented inventions in cities increased with the introduction of HSR. That is to say, the introduction of HSR enhanced the innovation capability of cities. This finding is obviously consistent with the research results of Agrawal, et al. [83] regarding highway construction, which showed that transport infrastructure may lead to an increase in regional patenting. The introduction of HSR facilitated the flow of talent between cities along the route, thereby increasing the likelihood that innovators in different cities would engage in collaborative innovation to improve their innovation capabilities [53]. With the introduction of HSR, researchers often use HSR instead of flying for intercity transportation, which reduces the cost of transportation between cities, and this decrease in transportation cost has a pivotal role in promoting the production and reconstruction of scientific knowledge [84]. HSR certainly facilitates face-to-face communication among scientists in different cities, which is considered to be one of the most important and lasting drivers of knowledge diffusion and innovation [85]. These are driving forces of urban innovation due to the space–time compression brought about by the introduction of HSR.

Table 2. Empirical results of empirical estimations.

	NIP		NIP	
	Model 1 Fixed-Effect	**Model 2 Random-Effect**	**Model 3 Random-Effect**	**Model 4 Random-Effect**
DID	225.816 **(2.02)	208.793 *(1.91)	231.129 **(2.11)	229.308 **(2.1)
NES	211.458 ***(3.02)	400.495 ***(6.47)	287.366 ***(4.49)	247.927 ***(3.85)
SE	5.06 ***(5.94)	4.759 ***(5.69)	4.897 ***(5.86)	4.735 ***(5.66)
NCS	0.262 **(2.35)	0.358 **(3.32)	0.274 **(2.52)	0.281 **(2.59)
NCU	−37.222 ***(−10.62)	−28.482 ***(−8.41)	−32.117 ***(−9.39)	−32.319 ***(−9.44)
NRD	0.009 **(2.16)	0.022 ***(6.22)	0.018 ***(4.91)	0.018 ***(5.04)
EVG	0.713 ***(2.79)	0.923 ***(12.18)	0.968 ***(13.11)	1.078 ***(13.75)
NFI	0.004(0.03)	−0.121(−1.01)	−0.123(−1.03)	−0.111(−0.93)
TVD	−0.013(−1.15)	−0.014(−1.28)	−0.01(−0.95)	−0.008(−0.65)
PGDP	−0.001(−0.79)	−0.001(−1.65)	−0.001(−1.63)	−0.001(−1.59)
_cons		7139.197 ***(13.98)	6711.118(13.87)	7058.596(11.42)
ρ	−1.007 ***(−10.58)	−3.326 ***(−14.46)	−3.145 ***(−14.36)	−3.414 ***(−14.54)
Sigma_u		1285.512	1231.867	1168.933
Sigma_e	720.213	713.062	712.189	711.507
Province FE	YES	NO	NO	YES
City FE	YES	NO	YES	YES
Year FE	YES	YES	YES	YES
Hausman Test	−18.00			
Log likelihood	−42,420	−45,150	−45,130	−45,110
Wald χ^2	315.49 ***	1138.92 ***	1222.72 ***	1334.9 ***
Pseudo R^2	0.66	0.588	0.61	0.659
Wald test of spatial terms	111.96 ***	209.05 ***	206.21 ***	211.52 ***

Note: *, **, and *** in Table 2 indicate statistical significance at 0.1, 0.05, and 0.01.

Distance is a natural measure of information asymmetry and a natural factor that hinders the diffusion of innovation [50,86]. Although the geographical location cannot be changed, the development of HSR has greatly reduced city accessibility times through space–time compression and has helped to promote city innovation. This is predicted by the theory of space–time compression [17,31,87]. In this regard, Drolc, et al. [87] once urged researchers to take the effects of time and space seriously. However, space–time compression may also be a double-edged sword. Although space–time compression can have a positive effect on innovation, it can also introduce some negative effects. Harvey [15], a proponent of the concept of space–time compression, had a clear understanding of this concept and criticized the many drawbacks brought about by space–time compression for society [44]. Therefore, the compression of time and space needs to be viewed dialectically.

The possible impact of space–time compression caused by HSR on city innovation also needs to be viewed dialectically. As can be seen in Table 2, the regression coefficients of the spatial lag term in all four models were significant and negative. Thus, Hypothesis 2 was confirmed. In other words, there was a negative spillover effect of inter-city innovation. The introduction of HSR has promoted the process of regional integration [37]. The innovation gradient formed by the difference in innovation levels between regional central cities and peripheral cities tends to cause the transfer of innovative elements, such as human capital, transportation conditions, funds, and information from peripheral cities to central cities [18,19,56]. As a result, the innovative development of central cities inhibits the innovative development of peripheral cities and ultimately enhances the polarization of regional innovation [59–61]. In other words, this has a siphon effect on surrounding cities or cities with gaps in their innovation capabilities, thus inhibiting the improvement of innovation levels in other cities [58]. The space–time compression caused by the introduction of HSR is likely to lead to a polarization of "the strongest and the weakest" in urban innovation. The so-called Matthew effect in city innovation is not necessarily good for regional innovation and development [88].

In addition, in this study we found that the number of employees in scientific technological research and the integrated technical services industry, scientific expenditure, the number of college students per 10,000 people, the number of R&D personnel, and the export value of goods contributed to promoting city innovation. However, the number of ordinary colleges and universities had a negative and significant impact on city innovation. The number of foreign-invested enterprises, the gross industrial output value of domestic-funded enterprises, and the impact of GDP per capita on city innovation were not significant.

5.3. Placebo Test

To test the validity of the difference-in-differences model, placebo tests are often conducted, using false treatment times rather than real treatment times. Here, two false HSR introduction times were constructed, one with the introduction of HSR one year in advance and the other with the introduction of HSR one year later. According to the two false HSR introduction times, the regression and simulation of the difference-in-differences models were carried out according to Formula (1). In the case of both false HSR introduction time models, the regression coefficients of DID were significant, which showed that the introduction of HSR was not a factor leading to increased urban innovation but merely a proxy variable for other factors reflecting inter-city gaps. In the case of both false HSR introduction time models, one of the regression coefficients of DID was not significant, which showed that the introduction of HSR was a factor contributing to the increase in city innovation. In other words, the improvement in city innovation capabilities could be attributed to the introduction of HSR. These empirical results are reported in Table 3.

Table 3. Empirical results of placebo tests.

	Placebo Test: One Year after		**Placebo Test: One Year before**	
	Model 5 Fixed-Effects	**Model 6 Random-Effects**	**Model 7 Fixed-Effects**	**Model 8 Random-Effects**
DID	88.556(0.72)	141.953 (1.18)	238.841 **(2.11)	234.39 **(2.12)
NES	208.645 ***(2.98)	244.87 ***(3.8)	210.906 ***(3.02)	247.483 ***(3.84)
SE	5.028 ***(5.9)	4.709 ***(5.63)	5.047 ***(5.93)	4.721 ***(5.64)
NCS	0.264 **(2.37)	0.281 **(2.59)	0.263 **(2.36)	0.282 **(2.6)
NCU	−36.193 ***(−10.43)	−31.276 ***(−9.22)	−37.269 ***(−10.63)	−32.332 ***(−9.44)
NRD	0.009 **(2.21)	0.018 ***(5.08)	0.009 **(2.15)	0.018 ***(5.03)
EVG	0.71 ***(2.78)	1.074 ***(13.91)	0.713 ***(2.8)	1.078 ***(13.95)
NFI	0.015(0.12)	−0.107(−0.89)	0.003(0.02)	−0.111(−0.93)
TVD	−0.012(−1.05)	−0.006(−0.56)	−0.013(−1.18)	−0.007(−0.67)
PGDP	−0.001(−0.8)	−0.001(−1.61)	−0.001(−0.77)	−0.001(−1.57)
_cons		7044.466 ***(11.4)	6711.118(13.87)	7053.312 ***(11.41)
ρ	−1.007 ***(−10.57)	−3.412 ***(−11.4)	−1.007 ***(−10.57)	−3.412 ***(−14.54)
Sigma_u		1167.455		1168.783
Sigma_e	720.457	711.752	720.189	711.503
Province FE	YES	YES	YES	YES
City FE	YES	YES	YES	YES
Year FE	YES	YES	YES	YES
Hausman Test	−15.85		−18.01	

Table 3. *Cont.*

	Placebo Test: One Year after		Placebo Test: One Year before	
	Model 5 Fixed-Effects	**Model 6 Random-Effects**	**Model 7 Fixed-Effects**	**Model 8 Random-Effects**
Log likelihood	−42,420	−45,110	−42,420	−45,110
Wald χ^2	311.58 ***	1333.82 ***	315.75 ***	1335.43 ***
Pseudo R^2	0.662	0.56	0.66	0.659
Wald test of spatial terms	111.73 ***	211.05 ***	111.75 ***	211.45 ***

Note: **, and *** in Table 3 indicate statistical significance at 0.05, and 0.01.

According to Models 5 and 6 in Table 3, the regression coefficient of DID was not significant if the introduction time of HSR was delayed by one year. This indicates that the artificial delay in the introduction of the HSR by one year did not affect city innovation, and that the variable of HSR introduction time was not a proxy variable for other factors. In other words, it is reasonable to attribute the improvement of city innovation capabilities to the introduction of HSR.

5.4. Robustness Test: Spatial Error

Robustness testing of the spatial difference-in-differences model needed to be carried out in relation to multiple aspects. One of the most important aspects is to see if the empirical results would change if the settings of the spatial model were to change. The empirical results in Table 2 were obtained based on the spatial lag model. Unlike the previous method, the spatial error model was used here to test whether these results were still valid. In particular, this was achieved by removing the spatial lag term $W \times NIP_{it}$ from Formula (1) and then assuming that the error term had a spatial effect: $\varepsilon_{it} = C + \delta W \times \varepsilon_{it} + \mu_{it}, \mu_{it} \sim N(0, \sigma^2 I_n)$. The empirical results are summarized in Table 4, where Model 11 and Model 12 are placebo tests.

Table 4. Robustness test results for spatial errors.

	NIP		Placebo Test: One Year after	
	Model 9 Fixed-Effects	**Model 10 Random-Effects**	**Model 11 Fixed-Effects**	**Model 12 Random-Effects**
DID	234.708 **(2.1)	191.143 *(1.73)	157.217(1.3)	122.776(1.02)
NES	187.335 ***(2.71)	332.589 ***(5.3)	183.694 ***(2.66)	330.092 ***(5.26)
SE	0.731(0.554)	2.995 **(2.49)	0.713(0.58)	2.991 **(2.48)
NCS	0.28 **(2.5)	0.345 ***(3.13)	0.28 **(2.5)	0.345 ***(3.13)
NCU	−37.923 ***(−10.4)	−29.117 ***(−8.04)	−37.071 ***(−10.22)	−28.407 ***(−7.89)
NRD	−0.004(−0.88)	0.022 ***(6.18)	−0.004(−0.85)	0.022 ***(6.22)
EVG	0.51 **(1.97)	0.729 ***(10.68)	0.519 **(1.98)	0.728 ***(10.66)
NFI	−0.153(−1.06)	−0.062(−0.45)	−0.171(−1.17)	−0.073(−0.53)
TVD	0.002(0.17)	−0.003(−0.22)	0.002(0.19)	−0.002(−0.29)
PGDP	−0.001(−1.49)	−0.001(−1.16)	−0.001(−1.52)	−0.001(−1.19)
_cons		−163.003(−0.56)		−170.351(−0.58)
δ	−1.172 ***(−23.72)	−1.905 ***(−27.73)	−1.173 ***(−23.81)	−1.905 ***(−27.73)
Sigma_u		949.316		948.241
Sigma_e	708.411	711.129	708.586	711.299
Province FE	YES	YES	YES	YES
City FE	YES	YES	YES	YES
Year FE	YES	YES	YES	YES
Hausman Test	402.03 ***		391.33 ***	

Table 4. *Cont.*

	NIP		Placebo Test: One Year after	
	Model 9 Fixed-Effects	**Model 10 Random-Effects**	**Model 11 Fixed-Effects**	**Model 12 Random-Effects**
Log likelihood	−42340	−45080	−42340	−45080
Wald χ^2	880.89 ***	2769.5 ***	878.32 ***	2769.48 ***
Pseudo R^2	0.23	0.759	0.237	0.759
Wald test of spatial terms	562.52 ***	769.06 ***	566.8 ***	769.13 ***

Note: *, **, and *** in Table 4 indicate statistical significance at 0.1, 0.05, and 0.01.

Comparing the empirical results of Model 9 and Model 10 in Table 4 with those of Models 1 and 4 in Table 2, they are essentially the same. In these four models, it is clear that the coefficients of DID are all significantly positive. In addition, the ρ values in Table 2 are all significantly negative. Similarly, the δ values in Table 4 are consistently significantly negative. This consistency shows that the negative spatial spillover effect of urban innovation was universal. Based on the placebo test, the empirical results of Model 11 and Model 12 in Table 4 were consistent with those of Models 5 and 6 in Table 3, and this indicates that the findings of this study were robust.

5.5. Robustness Tests: Different Spatial Weight Matrices

The test of the robustness of the SDiD model also includes the investigation of whether the empirical results would change as a result of changes in the spatial weight matrix. The spatial distance, in the previous part of this paper, was defined as the gap in the average annual number of city patents granted. Next, the spatial distance was defined as the gap in the average annual number of city patents filed. The empirical results are reported in Table 5, and the detailed empirical results are summarized in Appendix A.

Table 5. Robustness test for another measure of innovation distance.

	NIP		Placebo Test: One Year after	
	Model 13 Fixed-Effects	**Model 14 Random-Effects**	**Model 15 Fixed-Effects**	**Model 16 Random-Effects**
DID	229.241 **(2.07)	212.685 *(1.93)	104.648(0.86)	107.879(0.89)
ρ	−4.717 ***(−16.08)	−2.726 ***(−13.74)	−4.715 ***(−16.08)	−2.722 ***(−13.74)
Province FE	YES	YES	YES	YES
City FE	YES	YES	YES	YES
Year FE	YES	YES	YES	YES
Hausman Test	356.45 ***		−522.67 ***	
Log likelihood	−42,350	−45,160	−42,350	−45,160
Wald χ^2	464.86 ***	1365.58 ***	461.09 ***	1365.02 ***
Pseudo R^2	0.659	0.702	0.665	0.7
Wald test of spatial terms	258.48 ***	188.89 ***	258.43 ***	188.8 ***

Note: *, **, and *** in Table 5 indicate statistical significance at 0.1, 0.05, and 0.01.

The results of Models 13 and 14 in Table 5 are consistent with those of Models 1 and 4 in Table 2. Based on the placebo test, the results of Models 15 and 16 in Table 5 are consistent with those of Models 5 and 6 in Table 3. The empirical results were similar in that spatial distance could also be defined in terms of the gap in the city's annual average GDP per capita. The empirical results are summarized in Table 6, and the detailed empirical results are reported in Appendix B.

Table 6. Robustness test of economic distance.

	NIP		Placebo Test: One Year after	
	Model 17 Fixed-Effects	**Model 18 Random-Effects**	**Model 19 Fixed-Effects**	**MODEL 20 Random-Effects**
DID	217.778 *(1.93)	185.077 *(1.66)	98.356(0.73)	78.848(0.65)
ρ	−2.867 ***(−18.44)	−0.985 ***(−6.68)	−2.867 ***(−18.42)	−0.985 ***(−6.68)
Province FE	YES	YES	YES	YES
City FE	YES	YES	YES	YES
Year FE	YES	YES	YES	YES
Hausman Test	379.35 ***		358.68 ***	
Log likelihood	−42,420	−45,160	−42,420	−45,160
Wald χ^2	940.66 ***	1744.32 ***	937.02 ***	1745.74 ***
Pseudo R^2	0.501	0.765	0.506	0.765
Wald test of spatial terms	339.87 ***	44.61 ***	339.28 ***	44.64 ***

Note: * and *** in Table 6 indicate statistical significance at 0.1, and 0.01.

It can be seen that despite the change in the definition of spatial distance, the results of this study did not change significantly. In other words, the adjustment and changing of the spatial weight matrix did not affect the results of this study, which further illustrates the robustness of the results of this study. Regarding endogeneity problems, as shown in the tests presented above, various types of control variables and fixed-effects models were adopted to address the potential problem of omitted variables. These treatments have controlled the endogeneity problem to some extent.

6. Conclusions

In this paper, the introduction of HSR in China was used as a treatment in a quasi-natural experiment and panel data from 279 prefecture-level cities from 2000 to 2019 was investigated to empirically test the city innovation hypothesis in relation to transportation improvement from the perspective of space–time compression. In this study, we adopted an inverse measurement method to measure physical space–time compression by calculating the average spatial distance from each city's innovation spatial center to each city. The average distance was found to increase year by year, which implies that the production of HSR networks led to physical space–time compression between cities and also led to an overflow of city innovation. This study also showed that the introduction of HSR was beneficial to face-to-face communication between people, which promoted the dissemination of "soft information", accelerated knowledge spillover, and was able to significantly improve the level of city innovation. However, the introduction of HSR could also lead to the transfer of innovation factors from cities with weaker innovation capabilities to cities with stronger innovation capabilities, resulting in a negative spillover effect. As a policy implication, the results of this paper suggest that the construction of HSR networks within city agglomerations should be strengthened to form an innovative development space, with city agglomerations as the mainstay of this approach. Central cities have already been connected to the national high-speed railway network. The government should make full use of this advantage to further promote the ability of central cities to radiate and facilitate the development of small- and medium-sized cities with weak innovation capabilities.

However, this study also has some limitations. For example, the samples studied were all prefecture-level cities, excluding smaller county-level cities; the measurement indicators of city innovation were relatively simple and singular; and international comparative studies were lacking. These are some possible directions for future research. Before concluding this paper, it is necessary to emphasize the difference between space–time compression and spacetime compression. Spacetime is a concept related to time and space in quantum physics [89]. However, this concept was gradually introduced into the social sciences [90]. Currently, no research has been published on how to distinguish the difference between space–time compression and spacetime compression and how to evaluate the impact of spacetime compression. This is also an important direction for future research.

Funding: This research was funded by the Social Science Foundation of China grant number 17BSH122 And The APC was funded by the Social Science Foundation of China.

Data Availability Statement: The data presented in this study are available on request from the corresponding author.

Acknowledgments: The authors gratefully acknowledge the Social Science Foundation of China (17BSH122) for the support provided to this research.

Conflicts of Interest: The authors declare no conflict of interest.

Appendix A

Table A1. Empirical results of robustness tests of another measure of innovation distance.

	NIP		Placebo Test: One Year after	
	Model 13 Fixed-Effects	Model 14 Random-Effects	Model 15 Fixed-Effects	Model 16 Random-Effects
DID	229.241 **(2.07)	212.685 *(1.93)	104.648(0.86)	107.879(0.89)
NES	193.138 ***(2.78)	298.423 ***(4.62)	190.205 ***(2.74)	295.661 ***(4.57)
SE	5.168 ***(6.11)	5.292 ***(6.28)	5.135 ***(6.07)	5.266 ***(6.24)
NCS	0.253 **(2.29)	0.274 ***(2.5)	0.255 **(2.3)	0.274 **(2.5)
NCU	−37.009 ***(−10.63)	−32.779 ***(−9.45)	−35.962 ***(−10.43)	−31.814 ***(−9.26)
NRD	0.004(0.87)	0.02 ***(5.61)	0.004(0.92)	0.02 ***(5.65)
EVG	0.47 *(1.85)	0.958 ***(12.92)	0.468 *(1.84)	0.954 ***(12.88)
NFI	−0.084(−0.69)	−0.079(−0.66)	−0.075(−0.6)	−0.072(−0.6)
TVD	−0.004(−0.35)	−0.014(−1.25)	−0.003(−0.25)	−0.013(−1.16)
PGDP	−0.001(−1.42)	−0.001(−1.4)	−0.001(−1.43)	−0.001(−1.42)
_cons		5622.767 ***(10.2)		5607.706 ***(10.18)
ρ	−4.717 ***(−16.08)	−2.726 ***(−13.74)	−4.715 ***(−16.08)	−2.722 ***(−13.74)
Sigma_u		1127.713		1126.325
Sigma_e	715.213	718.57	715.451	718.808
Province FE	YES	YES	YES	YES
City FE	YES	YES	YES	YES
Year FE	YES	YES	YES	YES
Hausman Test	356.45 ***		−522.67 ***	
Log likelihood	−42,350	−45,160	−42,350	−45,160
Wald χ^2	464.86 ***	1365.58 ***	461.09 ***	1365.02 ***
Pseudo R^2	0.659	0.702	0.665	0.7
Wald test of spatial terms	258.48 ***	188.89 ***	258.43 ***	188.8 ***

Note: *, **, and *** indicate statistical significance at 0.1, 0.05, and 0.01.

Appendix B

Table A2. Empirical results of robustness tests of economic distance.

	NIP		Placebo Test: One Year after	
	Model 17 Fixed-Effects	Model 18 Random-Effects	Model 19 Fixed-Effects	Model 20 Random-Effects
DID	217.778 *(1.93)	185.077 *(1.66)	98.356(0.73)	78.848(0.65)
NES	183.04 ***(2.6)	331.962 ***(5.24)	180.211 ***(2.56)	329.595 ***(5.43)
SE	5.12 ***(5.64)	4.877 ***(5.46)	5.102 ***(5.61)	4.857 **(2.48)
NCS	0.27 **(2.37)	0.327 ***(2.94)	0.272 **(2.36)	0.328 ***(2.94)

Table A2. *Cont.*

	NIP		Placebo Test: One Year after	
	Model 17 Fixed-Effects	**Model 18 Random-Effects**	**Model 19 Fixed-Effects**	**Model 20 Random-Effects**
NCU	−38.039 ***(−10.63)	−30.517 ***(−8.61)	−37.114 ***(−10.46)	−29.711 ***(−8.45)
NRD	0.008 **(1.97)	0.027 ***(7.89)	0.009 **(2.01)	0.027 ***(7.93)
EVG	0.664 **(2.57)	0.654 ***(10.03)	0.663 **(2.56)	0.651 ***(9.99)
NFI	−0.068(−0.52)	−0.041(−0.32)	−0.058(−0.44)	−0.035(−0.27)
TVD	−0.013(−1.17)	−0.016(−1.48)	−0.012(−1.1)	−0.016(−1.41)
PGDP	−0.001(−0.95)	−0.001(−0.69)	−0.001(−1.01)	−0.001(−0.74)
_cons		−225.333(−0.78)		−232.287(−0.81)
ρ	−2.867 ***(−18.44)	−0.985 ***(−6.68)	−2.867 ***(−18.42)	−0.985 ***(−6.68)
Sigma_u		930.796		929.433
Sigma_e	720.044	724.012	720.261	724.226
Province FE	YES	YES	YES	YES
City FE	YES	YES	YES	YES
Year FE	YES	YES	YES	YES
Hausman Test	379.35 ***		358.68 ***	
Log likelihood	−42,420	−45,160	−42,420	−45,160
Wald χ^2	940.66 ***	1744.32 ***	937.02 ***	1745.74 ***
Pseudo R^2	0.501	0.765	0.506	0.765
Wald test of spatial terms	339.87 ***	44.61 ***	339.28 ***	44.64 ***

Note: *, **, and *** indicate statistical significance at 0.1, 0.05, and 0.01.

References

1. Liu, Y.; Zhang, X.; Pan, X.; Ma, X.; Tang, M. The spatial integration and coordinated industrial development of urban agglomerations in the Yangtze River Economic Belt, China. *Cities* **2020**, *104*, 102801. [CrossRef]
2. Comunian, R. Rethinking the Creative City: The Role of Complexity, Networks and Interactions in the Urban Creative Economy. *Urban Stud.* **2015**, *48*, 1157–1179. [CrossRef]
3. Ni, P.; Kamiya, M.; Shen, L.; Gong, W.; Xu, H. Reviews of Global Urban Competitiveness 2017–2018 Driving Force, Agglomeration, Connectivity and the New Global City. In *House Prices: Changing the City World*; Ni, P., Kamiya, M., Wang, H., Eds.; Springer: Clam, Switzerland, 2019; pp. 49–131.
4. UN. *The 2018 Revision of the World Urbanization Prospects*; The Population Division of the United Nations Department of Economic and Social Affairs (UN DESA): New York, NY, USA, 2018.
5. Feldman, M.P.; Audretsch, D.B. Innovation in cities: Science-based diversity, specialization and localized competition. *Eur. Econ. Rev.* **1999**, *43*, 409–429. [CrossRef]
6. Acs, Z.J. Innovation and the Growth of Cities. *Contrib. Econ. Anal.* **2010**, *266*, 635–658.
7. Visvizi, A.; Lytras, M.; Damiani, E.; Mathkour, H. Policy making for smart cities: Innovation and social inclusive economic growth for sustainability. *J. Sci. Technol. Policy Manag.* **2018**, *9*, 126–133. [CrossRef]
8. Forman, C.; Goldfarb, A.; Greenstein, S. How did location affect adoption of the commercial Internet? Global village vs. urban leadership—ScienceDirect. *J. Urban Econ.* **2005**, *58*, 389–420. [CrossRef]
9. Bedir, M.; Hilgefort, J. The global village. *Archit. Rev.* **2019**, *244*, 24–27.
10. Acuto, M.; Leffel, B. Understanding the global ecosystem of city networks. *Urban Stud.* **2021**, *58*, 1758–1774. [CrossRef]
11. Jenkins, P.; Wilkinson, P. Assessing the Growing Impact of the Global Economy on Urban Development in Southern African Cities: Case Studies in Maputo and Cape Town. *Cities* **2002**, *19*, 33–47. [CrossRef]
12. Dvir, R.; Pasher, E. Innovation engines for knowledge cities: An innovation ecology perspective. *J. Knowl. Manag.* **2004**, *8*, 16–27. [CrossRef]
13. Yao, L.; Li, J.; Li, J. Urban innovation and intercity patent collaboration: A network analysis of China's national innovation system. *Technol. Forecast. Soc. Chang.* **2020**, *160*, 120185. [CrossRef]
14. Boikova, M.; Ilyina, I.; Salazkin, M. Urban Futures: Cities as Agents of Globalization and Innovation. *Foresight STI Gov.* **2011**, *5*, 32–48. [CrossRef]
15. Harvey, D. *The Condition of Postmodernity an Enquiry into the Origins of Cultural Change*; Wiley-Blackwell: Oxford, UK, 1989.
16. Jin, B.; Yang, W.; Li, X.; Sha, J.; Wang, X. A literature review on the space of flows. *Arab. J. Geosci.* **2021**, *14*, 1–24. [CrossRef]
17. Janelle, D.G. Time, Space, and the Human Geographies of Opportunity. In *Time, Space, and the Human Geographies of Opportunity*; Wuppuluri, S., Ghirardi, G., Eds.; Springer: Cham, Switzerland, 2017; pp. 487–501.

18. An, Y.; Wei, Y.D.; Yuan, F.; Chen, W. Impacts of high-speed rails on urban networks and regional development: A study of the Yangtze River Delta, China. *Int. J. Sustain. Transp.* **2021**, *16*, 483–495. [CrossRef]
19. Zhang, X.; Wu, W.; Zhou, Z.; Yuan, L. Geographic proximity, information flows and corporate innovation: Evidence from the high-speed rail construction in China. *Pac.-Basin Financ. J.* **2020**, *61*, 101342. [CrossRef]
20. Coto-Millán, P.; Inglada, V.; Rey, B. Effects of network economies in high-speed rail: The Spanish case. *Ann. Reg. Sci.* **2007**, *41*, 911–925. [CrossRef]
21. Shi, W.; Lin, K.-C.; McLaughlin, H.; Qi, G.; Jin, M. Spatial distribution of job opportunities in China: Evidence from the introduction of the high-speed rail. *Transp. Res. Part A Policy Pract.* **2020**, *133*, 138–147. [CrossRef]
22. Ahlfeldt, G.M.; Feddersen, A. From periphery to core: Measuring agglomeration effects using high-speed rail. *J. Econ. Geogr.* **2018**, *18*, 355–390. [CrossRef]
23. Komikado, H.; Morikawa, S.; Bhatt, A.; Kato, H. High-speed rail, inter-regional accessibility, and regional innovation: Evidence from Japan. *Technol. Forecast. Soc. Chang.* **2021**, *167*, 120697. [CrossRef]
24. Dobruszkes, F.; Dehon, C.; Givoni, M. Does European high-speed rail affect the current level of air services? An EU-wide analysis. *Transp. Res. Part A Policy Pract.* **2014**, *69*, 461–475. [CrossRef]
25. Gao, Y.; Zheng, J. The impact of high-speed rail on innovation: An empirical test of the companion innovation hypothesis of transportation improvement with China's manufacturing firms. *World Dev.* **2020**, *127*, 104838. [CrossRef]
26. Burström, T.; Peltonen, J. The role of knowledge-intense high-impact firms in city innovation systems. *Innovation* **2018**, *20*, 377–392. [CrossRef]
27. Maye, D. 'Smart food city': Conceptual relations between smart city planning, urban food system. *City Cult. Soc.* **2019**, *16*, 18–24. [CrossRef]
28. Andion, C.; Alperstedt, G.D.; Graeff, J.F.; Ronconi, L. Social innovation ecosystems and sustainability in cities: A study in Florianópolis, Brazil. *Environ. Dev. Sustain.* **2022**, *24*, 1259–1281. [CrossRef]
29. Linde, L.; Sjödin, D.; Parida, V.; Wincent, J. Dynamic capabilities for ecosystem orchestration A capability-based framework for smart city innovation initiatives. *Technol. Forecast. Soc. Chang.* **2021**, *166*, 120614. [CrossRef]
30. Garcia, B.C.; Chavez, D. Network-based innovation systems: A capital base for the Monterrey city-region, Mexico. *Expert Syst. Appl.* **2014**, *41*, 5636–5646. [CrossRef]
31. Janelle, D.G. Central place development in a time–space framework. *Prof. Geogr.* **1968**, *20*, 5–10. [CrossRef]
32. Liang, Y.; Zhou, K.; Li, X.; Zhou, Z.; Sun, W.; Zeng, J. Effectiveness of high-speed railway on regional economic growth for less developed areas. *J. Transp. Geogr.* **2020**, *82*, 102621. [CrossRef]
33. Xu, Z.; Sun, T. The Siphon effects of transportation infrastructure on internal migration: Evidence from China's HSR network. *Appl. Econ. Lett.* **2021**, *28*, 1066–1070. [CrossRef]
34. Lu, Y.; Cheng, S. China's total high-speed railway mileage to reach 39,000 kilometers by year end. *People's Daily*, 26 May 2020; p. 4.
35. Jia, S.; Zhou, C.; Qin, C. No difference in effect of high-speed rail on regional economic growth based on match effect perspective? *Transp. Res. Part A Policy Pract.* **2017**, *106*, 144–157. [CrossRef]
36. Zhang, Y.; Liu, J.; Wang, B. The impact of High-Speed Rails on urban expansion: An investigation using an SDID with dynamic effects method. *Socio-Econ. Plan. Sci.* **2022**, *82*, 101294. [CrossRef]
37. Jiao, J.; Wang, J.; Jin, F. Impacts of high-speed rail lines on the city network in China. *J. Transp. Geogr.* **2017**, *60*, 257–266. [CrossRef]
38. Chou, J.-S.; Chien, Y.-L.; Nguyen, N.-M.; Truong, D.-N. Pricing policy of floating ticket fare for riding high speed rail based on time-space compression. *Transp. Policy* **2018**, *69*, 179–192. [CrossRef]
39. Spiekermann, K.; Wegener, M. The shrinking continent: New time—space maps of Europe. *Environ. Plan. B Plan. Des.* **1994**, *21*, 653–673. [CrossRef]
40. Gutiérrez, J.; González, R.; Gómez, G. The European high-speed train network: Predicted effects on accessibility patterns. *J. Transp. Geogr.* **1996**, *4*, 227–238. [CrossRef]
41. Janelle, D.G. Time–space Convergence. In *Handbook of Research Methods and Applications in Spatially Integrated Social Science*; Stimson, R., Ed.; Edward Elgar Publishing, Inc.: Northampton, UK, 2014; pp. 43–60.
42. McKenzie, R.D. *The Metropolitan Community*; McGraw-Hill Book: New York, NY, USA, 1933.
43. Castells, M. *The Informational City: Economic Restructuring and Urban Development*; Wiley-Blackwell: Oxford, UK, 1989.
44. Sheppard, E. David Harvey and Dialectical Space-Time. In *David Harvey: A Critical Reader*; Castree, N., Gregory, D., Eds.; Blackwell Publishing Ltd.: Oxford, UK, 2006.
45. Hyman, G.; Mayhew, L. Advances in travel geometry and urban modelling. *GeoJournal* **2004**, *59*, 191–207. [CrossRef]
46. Vickerman, R.; Spiekermann, K.; Wegener, M. Accessibility and Economic Development in Europe. *Reg. Stud.* **1999**, *33*, 1–15. [CrossRef]
47. Lefebvre, H. *The Production of Space*; Wiley-Blackwell: Hoboken, NJ, USA, 1992.
48. Bosch, S.; Schmidt, M. Wonderland of technology? How energy landscapes reveal inequalities and injustices of the German Energiewende. *Energy Res. Soc. Sci.* **2020**, *70*, 101733. [CrossRef]
49. Westlund, H. A brief history of time, space, and growth: Waldo Tobler's first law of geography revisited. *Ann. Reg. Sci.* **2013**, *51*, 917–924. [CrossRef]
50. Graevenitz, G.; Graham, S.J.H.; Myers, A.F. Distance (still) hampers diffusion of innovations. *Reg. Stud.* **2022**, *56*, 227–241. [CrossRef]

51. Breschi, S.; Lissoni, F. Mobility of skilled workers and co-invention networks: An anatomy of localized knowledge flows. *J. Econ. Geogr.* **2009**, *9*, 439–468. [CrossRef]
52. Wei, H.; Su, Y.S. The effect of institutional proximity in non-local university–industry collaborations: An analysis based on Chinese patent data. *Res. Policy* **2013**, *42*, 454–464.
53. Dong, X.; Zheng, S.; Kahn, M.E. The role of transportation speed in facilitating high skilled teamwork across cities. *J. Urban Econ.* **2020**, *115*, 103212. [CrossRef]
54. Cai, Y.; Tian, X.; Xia, H. Location, Proximity, and M&A Transactions. *J. Econ. Manag. Strategy* **2016**, *25*, 688–719.
55. Duguet, E.; Macgarvie, M. How well do patent citations measure flows of technology? Evidence from French innovation surveys. *Econ. Innov. New Technol.* **2005**, *14*, 375–393. [CrossRef]
56. Francis, B.B.; Hasan, I.; John, K.; Waisman, M. Urban agglomeration and CEO compensation. *J. Financ. Quant. Anal.* **2016**, *51*, 1925–1953. [CrossRef]
57. Qin, Y. 'No county left behind?' The distributional impact of high-speed rail upgrades in China. *J. Econ. Geogr.* **2017**, *17*, 489–520. [CrossRef]
58. Wang, Y.; Liang, S.; Kong, D.; Wang, Q. High-speed rail, small city, and cost of debt: Firm-level evidence. *Pac.-Basin Financ. J.* **2019**, *57*, 101194. [CrossRef]
59. Thompson, A.C. Does public infrastructure affect economic activity?: Evidence from the rural interstate highway system. *Reg. Sci. Urban Econ.* **2000**, *30*, 457–490.
60. Faber, B. Trade Integration, Market Size, and Industrialization: Evidence from China's National Trunk Highway System. *Rev. Econ. Stud.* **2014**, *81*, 1046–1070. [CrossRef]
61. Hodgson, C. The effect of transport infrastructure on the location of economic activity: Railroads and post offices in the American West. *J. Urban Econ.* **2018**, *104*, 59–76. [CrossRef]
62. Gu, J. Spatiotemporal context and firm performance: The mediating effect of strategic interaction. *Growth Chang.* **2021**, *52*, 371–391. [CrossRef]
63. Gu, J. Spatial Dynamics between Firm Sales and Environmental Responsibility: The Mediating Role of Corporate Innovation. *Sustainability* **2021**, *13*, 1648. [CrossRef]
64. Anselin, L.; Griffith, D.A. Do Spatial Effects Really Matter in Regression Analysis? *Pap. Reg. Sci.* **1988**, *65*, 11–34. [CrossRef]
65. Zhang, G.; Chen, X. The value of invention patents in China: Country origin and technology field differences. *China Econ. Rev.* **2012**, *23*, 357–370.
66. Parker, R.A.; Ridge, C.; Cong, C.; Appelbaum, R. China's Nanotechnology Patent Landscape: An Analysis of Invention Patents Filed with the State Intellectual Property Office. *Nanotechnol. Law Bus.* **2009**, *6*, 524–539.
67. Castaldi, C.; Los, B. Geographical patterns in US inventive activity 1977–1998: The "regional inversion" was underestimated. *Res. Policy* **2017**, *46*, 1187–1197. [CrossRef]
68. Gong, J. Clarifying the Standard Deviational Ellipse. *Geogr. Anal.* **2002**, *34*, 155–167. [CrossRef]
69. Gu, J. Spatiotemporal dynamics of the patent race: Empirical evidence from listed companies in China. *Asian J. Technol. Innov.* **2020**, *30*, 106–133. [CrossRef]
70. Gu, J. Determinants of biopharmaceutical R&D expenditures in China: The impact of spatiotemporal context. *Scientometrics* **2021**, *6*, 1–21.
71. Wang, B.; Shi, W.; Miao, Z. Confidence Analysis of Standard Deviational Ellipse and Its Extension into Higher Dimensional Euclidean Space. *PLoS ONE* **2015**, *10*, e0118537. [CrossRef] [PubMed]
72. Gu, J. Spatial Interactions and the Commercialisation of Academic Patents: The Chinese Experience. *Sci. Technol. Soc.* **2022**, *27*, 543–562. [CrossRef]
73. O'Neill, S.; Kreif, N.; Grieve, R.; Sutton, M.; Sekhon, J.S. Estimating causal effects: Considering three alternatives to difference-in-differences estimation. *Health Serv. Outcomes Res. Methodol.* **2016**, *16*, 1–21. [CrossRef] [PubMed]
74. Bonhomme, S.; Sauder, U. Recovering Distributions in Difference-in-Differences Models: A Comparison of Selective and Comprehensive Schooling. *Rev. Econ. Stat.* **2011**, *93*, 479–494. [CrossRef]
75. Heckert, M. A Spatial Difference-in-Differences Approach to Studying the Effect of Greening Vacant Land on Property Values. *Cityscape* **2015**, *17*, 51–60.
76. Gu, J. Spatial recruiting competition in Chinese higher education system. *High. Educ.* **2012**, *63*, 165–185. [CrossRef]
77. Kosfeld, R.; Mitze, T.; Rode, J.; Wlde, K. The Covid-19 containment effects of public health measures A spatial difference-in-differences approach. *J. Reg. Sci.* **2021**, *61*, 799–825. [CrossRef]
78. Jia, R.; Shao, S.; Yang, L. High-speed rail and CO_2 emissions in urban China: A spatial difference-in-differences approach. *Energy Econ.* **2021**, *99*, 1873–6181. [CrossRef]
79. Gu, J. Effects of Patent Policy on Outputs and Commercialization of Academic Patents in China: A Spatial Difference-in-Differences Analysis. *Sustainability* **2021**, *13*, 13459. [CrossRef]
80. Anselin, L.; Florax, R.J.G.M.; Rey, S.J. *Advances in Spatial Econometrics: Methodology, Tools and Applications*; Springer: Berlin/Heidelberg, Germany, 2004.
81. Sage, J.L.; Pace, R.K. *Introduction to Spatial Econometrics*; Chapman and Hall, CRC: New York, NY, USA, 2009.
82. Friedman, M. Newton and Kant on Absolute Space: From Theology to Transcendental Philosophy. In *Constituting Objectivity: Transcendental Perspectives on Modern Physics*; Bitbol, M., Kerszberg, P., Petitot, J., Eds.; Springer: Clam, Switzerland, 2009; pp. 35–50.

83. Agrawal, A.; Galasso, A.; Oettl, A. Roads and Innovation. *Rev. Econ. Stat.* **2017**, *99*, 417–434. [CrossRef]
84. Catalini, C.; Fons-Rosen, C.; Gaull, P. Did Cheaper Flights Change the Direction of Science? *SSRN Electron. J.* **2016**, *3*, 1–27. [CrossRef]
85. Antonelli, C. Collective knowledge communication and innovation: The evidence of technological districts. *Reg. Stud.* **2000**, *34*, 535–547. [CrossRef]
86. Holm, E.J.; Jung, H.; Welch, E.W. The impacts of foreignness and cultural distance on commercialization of patents. *J. Technol. Transf.* **2021**, *46*, 29–61. [CrossRef]
87. Drolc, C.A.; Gandrud, C.; Williams, L.K. Taking Time (and Space) Seriously: How Scholars Falsely Infer Policy Diffusion from Model Misspecification. *Policy Stud. J.* **2021**, *49*, 484–515. [CrossRef]
88. Bonitz, M.; Bruckner, E.; Scharnhorst, A. Characteristics and impact of the Matthew Effect for Countries. *Scientometrics* **1997**, *40*, 407–422. [CrossRef]
89. Amati, D.; Ciafaloni, M.; Veneziano, G. Can spacetime be probed below the string size? *Phys. Lett. B* **1989**, *216*, 41–47. [CrossRef]
90. Janelle, K.-H. Constructing Carbon Market Spacetime: Climate Change and the Onset of Neo-Modernity. *Ann. Assoc. Am. Geogr.* **2010**, *100*, 953–962.

MDPI AG
Grosspeteranlage 5
4052 Basel
Switzerland
Tel.: +41 61 683 77 34

Systems Editorial Office
E-mail: systems@mdpi.com
www.mdpi.com/journal/systems

www.ingramcontent.com/pod-product-compliance
Lightning Source LLC
Chambersburg PA
CBHW061503210926
51850CB00008B/1804
* 9 7 8 3 7 2 5 8 1 9 8 0 5 *